Climate and Fisheries

Climate and Fisheries

D.H. CUSHING

ACADEMIC PRESS 1982
A Subsidiary of Harcourt Brace Jovanovich, Publishers
London New York
Paris San Diego San Francisco
São Paulo Sydney Tokyo Toronto

ACADEMIC PRESS INC. (LONDON) LTD
24/28 Oval Road, London NW1 7DX

United States Edition published by
ACADEMIC PRESS INC.
111 Fifth Avenue, New York, New York, 10003

British Library Cataloguing in Publication Data

Cushing, D.H.
Climate and fisheries.
1. Fishing resources—Climate factors
I. Title
639'.22 SH327.5

ISBN 0-12-199720-0

Printed in Great Britain by
St Edmundsbury Press,
Bury St Edmunds, Suffolk

Introduction

The industrialization of fisheries began towards the end of the nineteenth century, and as a consequence stocks were reduced. Since then considerable efforts have been made both nationally and internationally to manage such stocks properly, although success in this direction remains limited. As the data sets lengthen, fisheries biologists have become interested in the variation of the vital parameters of the stocks with time, and so there is potential interest in any connexions with climatic change. Stocks have appeared and disappeared in the past, and fishermen have credited or blamed the weather.

The climate changes on a longer scale than the weather, months to millennia, or longer, back to the Ice Ages. The study of climate is a fluid one: part is a branch of physical science, and part is a historical study. The historical part is the assessment of documents—the catalogue of data which overwhelms us within recent decades but which attenuates as the observer's eye reaches back across the centuries. The scientific study of climate today is one gorged on data constrained by the general circulation models of the atmosphere. The scientific study of fisheries is based on annual increments or decrements of biomass, which means that the data base is relatively meagre. Hence the events to be described are restricted to those which are prominent.

There is a mass of material published on the effects of environmental factors upon the lives of fishes which is not covered in this book. Obviously the effects of climatic change upon the fish stocks must be in some sense indistinguishable from those of the environment more extensively. However, the phenomena discussed are very pronounced and selected for that reason, and so many of the minor signals are necessarily excluded.

I have been helped considerably by my former colleagues at the Fisheries Laboratory in Lowestoft. Dr R.R. Dickson and Mr T. Wyatt read certain chapters; Dr John Shepherd advised me on certain aspects of periodic analysis and upon some questions of oceanic circulation; Mr P. Scholes provided me with material on dead cod from the Southern North Sea; Mrs Irene Gooch drew many of the figures.

198 Yarmouth Road
Lowestoft
Suffolk

D.H. Cushing
September 1982

Contents

Acknowledgements

The following figures are reproduced with permission:

Figure 1 from Fig. 1 in Cushing, D. H. (1975) "Fisheries Resources of the Sea and their Management" Oxford University Press, Oxford, 87 pp.

Figure 2 from Fig. 6 in Cushing, D. H. (1981) "Fisheries Biology" 2nd edn. University of Wisconsin Press, Madison, 295 pp.

Figure 4 from Fig. 141 in Hardy, A. C. and E. R. Gunther (1935) "The Plankton of the South Georgia Whaling grounds and adjacent areas" *Discovery Rep.* **XI,** 1–453.

Figure 5 from Fig. 9 in Cushing, D. H. (1952) "Echo surveys of fish" *J. Cons. int. Explor. Mer* **18,** 45–60.

Figure 7 from Fig. 9 in Laurs, R. M. and R. J. Lynn (1977) "Seasonal migration of albacore, *Thunnus alalunga,* into North American coastal waters, their distribution, relative abundance and association with Transition Zone waters" *Fish. Bull. Nat. Mar. Fish. Serv.* **75,** 795–822.

Figure 8 from Fig. 3 in Ahlstrom, E. H. (1967) "Co-occurrences of sardine and anchovy larvae in the California current region off California and Baja California" *Calcofi. Rep.* **XI,** 117–135.

Figure 9 from Charts I–III in Townsend, C. H. (1935) "The distribution of certain whales as shown by log book records of American whaleships" *Zoologica* **19,** 1–50.

Figure 11 from Fig. 1 in Koslow, J. A. (1976) "Alaska pollock—already overfished?" *Sea Frontiers* **22,** 2, 98–105.

Figure 12 from Fig. 1 in Cushing, D. H. (1980) "The biology of a fish population" Marine Sciences and Ocean Policy Symposium pp. 51–69. University of California, Santa Barbara.

Figure 13 from Fig. 2 in Cushing, D. H. (1975) "The natural mortality of plaice" *J. Cons. int. Explor. Mer* **36**(2), 150–7.

Figure 14 from Fig. 3 in Cushing, D. H. (1980) "The biology of a fish population" Marine Sciences and Ocean Policy Symposium pp. 51–69. University of California, Santa Barbara.

Figure 19 from Fig. 11 in Cushing, D. H. (1981) "Fisheries Biology" 2nd edn. University of Wisconsin Press, Madison 295 pp.

Figure 22 from Fig. 18 in Cushing, D. H. (1981) "Fisheries Biology" 2nd edn. University of Wisconsin Press, Madison 295 pp.

Figure 23 from Fig. 9 in Jones, F. R. Harden *et al.* (1978) "Tactics of fish movement in relation to migration and water circulation" pp. 185–207 *In* Proc. Joint Oceanogr. Assembly (ed. Deacon and Charnock) Plenum Press, New York.

Figure 24a from Fig. 3 in Shepherd, J. G. and D. H. Cushing (1980) "A mechanism for density dependent survival of larval fish as the basis of a stock-recruitment relationship" *J. Cons. int. Explor. Mer* **39**. (2) 160–7.

Figure 24b from Fig. 4 in Shepherd, J. G. and D. H. Cushing (1980) "A mechanism for density dependent survival of larval fish as the basis of a stock-recruitment relationship" *J. Cons. int. Explor. Mer* **39**. (2) 160–7.

Figure 26 from Fig. 1 in Devold, F. (1963) "The life history of the Atlanto-Scandian herring." *Rapp. Procès-Verb. Cons. int. Explor. Mer* **154,** 98–108.

Figure 39 from Fig. 3 in Mason, B. J. (1976) "Towards the understanding and prediction of climatic variation" *Quart. J. Roy. Meteorol. Soc.* **102,** 473–498.

Figure 40a from Fig. 4 in Dansgaard, W. (1981) "Palaeoclimatic studies on ice cores" pp. 193–206 *In* "Climatic Variations and Variability: Facts and Theories" (ed. A. Berger) Reidel, Dordrecht.

Figure 40b from Fig. 8 in Dansgaard, W. (1981). "Palaeoclimatic studies on ice cores" pp. 193–206 *In* "Climatic Variations and Variability: Facts and Theories" (ed. A. Berger) Reidel, Dordrecht.

Figure 41b from Fig. 3a in Kukla, G. J. and D. A. Robinson (1981) "Temperature changes in the last hundred years" pp. 287–302 *In* "Climatic Variations and Variability: Facts and Theories," (ed. A. Berger) Reidel, Dordrecht.

Figure 47a from Chart 1 in Schmidt, J. (1909) "The distribution of the pelagic fry and the spawning regions of the gadoids in the North Atlantic from Iceland to Spain" *Rapp. Procès-Verb. Cons. int. Explor. Mer* **10,** 4.

Figures 48, 49 from Figs 7, 8 in Cushing, D. H. (1982) "A simulacrum of the Iceland cod stock" *J. Cons. int. Explor. Mer* **40,** 1, 27–36.

Figure 50a from Fig. 1 in Russell, F. S. (1973) "A summary of the observations of the occurrence of planktonic stages of fish off Plymouth, 1924–72" *J. Mar. Biol. Assn UK* NS **53,** 347–355.

Figures 50b, c, d from Figs 1a, b, c in Southward, A. J. (1980) "The Western English Channel—an inconstant ecosystem" *Nature* **285,** (5784), 361–6.

Figures 51a, b from Figs 6, 7 in Cushing D. H. (1961) "The failure of the Plymouth herring fishery" *J. Mar. Biol. Assn UK* NS **41,** 3, 799–816.

Figure 52 from Fig. 1 in Ottestad, P. (1969) "Forecasting the annual yield in sea fisheries." *Nature* **185,** (4707), 183.

Figure 57 from Fig. 118 in Garrod, D. J. and J. M. Colebrook (1978) "Biological effects of variability in the North Atlantic Ocean" *Rapp. Procès-Verb. Cons. int. Explor. Mer* **173,** 128–144.

Figure 61b from Fig. 11 in Glover, R. S. (1957) "An ecological survey of the drift net herring survey off the north east coast of Scotland II; the planktonic environment of the herring." *Bull. Mar. Ecol.* **5**, 1–43.

Figure 62 from Fig. 150 in Anon. (1973) "Continuous plankton records; a plankton atlas of the North Sea and the North Atlantic" *Bull. Mar. Ecol.* **7,** 1–174.

Figure 63b from Fig. 2 in Glover, R. S. (1979) "Natural fluctuations of populations" *Ecotoxicology and Environmental Safety* **3,** 190–203.

Figures 67a, b from Figs 5, 4 in Glover, R. S. (1979) "Natural fluctuations of populations" *Ecotoxicology and Environmental Safety* **3,** 190-203.

Figure 70 from Fig. 4 in Anon. (1965) "The North Sea Herring" *Coop. Res. Rep. Int. Council. Explor. Sea* **4,** 57.

Figure 72 from Fig. 140 in Burd, A. C. (1978) "Long term changes in herring stocks" *Rapp. Cons. int. Explor. Mer* **172,** 137–53.

Figure 73 from Fig. 2 in Beverton, R. J. H. (1962) "Long term dynamics of certain North Sea fish populations" pp. 242–62 *In* "The Exploitation of Natural Animal Populations" (ed. E. D. Le Cren and M. W. Holdgate) 398 pp. Blackwell, Oxford.

Figures 75a, b from Figs 2, 4 in Cushing, D. H. (1980) "The decline of the herring stocks and the gadoid outburst." *J. Cons. int. Explor. Mer* **39,** 1. 70–81.

Figure 80 from Fig. 2 in Lamb, H. H. (1965) "Frequency of weather types" *Weather* **20,** 1, 9–12.

Figure 84a, b from Fig. 7 in Woodhead, P. M. J. (1964) "The death of North Sea fish during the winter of 1962–3, particularly with reference to the sole, *Solea vulgaris*." *Helgol. Wiss. Meeresunters* **10,** 283–300.

Figure 85 from Fig. 2 in Fonselius, S. H. CM, 1971, C8 6pp. Mimeo

Figure 86 from Fig. 1 in Lindquist, A. (1960) "Salthatlen och djurvälden i Baltiska havet" *Fisk. Tidskr. Finl.* NS **3,** 90–94.

Figures 90, 91, 93 from Figs 3, 4, 50 in Fonselius, S. H. (1969) "Hydrography of the Baltic Deep Basins III" *Fish. Bd. Swed. Hydrogr.* **23,** 97.

Figure 99a from Fig. 6 in Wooster, W. S. and O. Guillen (1974) "Characteristics of El Niño in 1972" *J. Mar. Res.* **32,** 3, 387–404.

Figure 101 from Fig. 1 in Quinn, W. H. (1974) "Monitoring and predicting El Niño invasions" *J. Appl. Meteorol.* 13, 825–904.

Figure 102 from Fig. 11 in Wyrtki, K. (1975) "Fluctuations of the dynamic topography in the Pacific Ocean" *J. Phys. Oceanogr.* **5**, 450–459.

Figure 103 from Fig. 8 in Hickey, Barbara (1975) "The relationship between fluctuations in sea level, wind stress and sea surface temperatures in the Equatorial Pacific" *J. Phys. Oceanogr.* **4,** 460–475.

Figure 104 from Fig. 7 in Wyrtki, K. and G. Meyers (1976) "The trade wind field over the Pacific Ocean" *J. Applied Meteorol.* **15,** 698–704.

Figure 105 from Fig. 5 in Wyrtki, K. (1977) "Sea level during the 1972 El Niño." *J. Phys. Oceanogr.* **7,** 779–87.

Figure 113 from Fig. 4 in Cushing, D. H. (1979) "The monitoring of biological effects; the separation of natural changes from those induced by pollution" *Phil. Trans. Roy. Soc. B* **286,** 597–607.

Figure 118 from Fig. 3 in Dickson, R. R. and H. H. Lamb (1972) "A review of recent hydrometeorological events in the North Atlantic Sector" pp. 35–62 *In Spec. Publ. 8 Int. Commn N.W. Atlant. Fisheries.*

Chapter One

The Nature of Fisheries

Wherever fishes can be found, they represent food of the right size which can be taken. Certain gears could become useful almost without development, although even a hook, once successful, would have the potential for centuries of modification and adaption. The simple gears, such as hooks and spears, must have first appeared after Stone Age man had looked at the range of sizes of convenient packets of food idly stemming the clear streams below, as does any boy today in limestone country. In the same way, identical methods must have arisen on the edge of the sea where the water was clear, much as shoals of tuna have been visible from Mediterranean cliffs ever since history has been written (Radcliffe, 1921).

From such primitive and early starting points, fisheries developed on parallel lines in various parts of the world as part of the early recorded culture of mankind. As ships became large and reliable enough to cruise and fight offshore, fishing vessels also began to work far from land: indeed decked boats from Dunwich, in Suffolk, went line fishing off Iceland in summer time in the sixteenth century (Houldsworth, 1874). Conflicts arose between nations, and from time to time wars broke out at sea about the presumed right to fish.

Industrialization came late to the fishing industry towards the end of the nineteenth century and the beginning of the twentieth. Today the annual catches amount to about 60 000 000 tonnes and the potential annual yield from the conventional gears described below is not likely to be much greater than about 100 000 000 tonnes (Gulland, 1971), although there are also additional reserves of protein, such as krill in the Antarctic. Although such quantities represent only a small part of the annual world food requirement, it is a special one: part of it supplies the essential amino acids, lysine and methionine, in the form of fish meal. A large proportion of the stocks of fish throughout the world ocean are overexploited and remain in need of good management. Fishermen have always known that fish stocks appear and subsequently vanish, as did the Loch Fyne herring. In the last decade large shoals of mackerel have appeared off the coasts of Cornwall, where they had not been recorded before (Lockwood, 1978). Much of fisheries research is concerned with management, but it is concerned also with the more subtle changes associated with long term climatic changes.

EARLY HISTORY

The early and primitive gears used by the first known fishermen are preserved in the recent sediments, and those in Europe and North America were reported by Rau (1884). In neolithic deposits, hooks, harpoons, sinkers and anchor stones have been found. The hooks from the lake dwellings are barbed in various ways, indicating that more than one species were caught. Bone harpoons with a serration of barbs suggest the need to retain the fish, perhaps on the end of a line. Sinkers are grooved or holed like "lucky stones", which means that ropes were used on cast nets to weight them to the lake bed or that sinkers were used on baited lines or on drift or gill nets. Nets and bark floats found in the softer remains of lake dwellings suggest that perhaps a drift net was used.

Radcliffe (1921) gives an account entitled *Fishing from the Earliest Times* and is based on the classical literature. Spears, nets, handlines, and rods with hooks are all recorded in Homer, and Herodotus noted the cast net, a circle of netting weighted at the edges which is flung skilfully over a shoal of fish in shallow water, thus entrapping it. Aristophanes refers to the sighting of tuna from a cliff top. From Theocritus we learn that lines were made of horse hair, rods were made of reed and that lobster pots were woven with rushes. He recorded the existence of seines, and there is an Egyptian drawing of a large seine being hand hauled by a number of men from a boat. Oppian's *Halieutica* describes reed fences, traps, tridents and of nets made of twisted hair or flax and even of a horn lantern for tunny fishing at night. According to Radcliffe, the most esteemed fishes in Roman times were red mullet, wrasse, sturgeon, turbot, bass, Moray eel, and sole; he also refers to *Asellus,* which is unidentified but may have been a cod-like species.

The classical record recalls many features of modern fisheries, and we can imagine that some of the choice fish came from the northern Adriatic, where turbot and sole are found today. However, probably all fisheries were in sight of the shore, because vessels were relatively small and perhaps not very manoeuvrable. Thus the early history represents a development from capture at the edge of the lake, river or sea to a genuine inshore fishery where the fish were perhaps no longer visible but were merely expected.

THE FOUR GEARS

The major fisheries in the ocean are exploited by four gears: the hook and line, the drift net, the trawl, and the purse seine. There are other engines of capture but they are usually found in minor or specialized fisheries, for instance the various traps—lobster pots and eel traps, the fixed engines that trap salmon in English and Scottish estuaries, and the great madragues for tuna found at various coastal points in the Mediterranean (for example, near Ceuta, the Spanish enclave in Morocco, and near Trapani in Sicily). Again, tridents are still used by fishermen in Tokyo Bay and the fishermen of the Sea of Galilee are using cast nets as they did two thousand years ago.

The hook and line is still used in its most primitive way by children all over the world. The long line is a development used in many inshore fisheries; the line is baited with mussels on hooks every fathom or so, laid on the sea bed, and left for a tide while demersal or bottom living fishes take the baits. It is hauled by a mechanical line hauler and is shot in the same way, but the flying hooks are baited by hand. In deeper water off the Faroe Islands, Iceland, Greenland and northern Norway, the great long lines were used in the same way. They were two or three miles in length, and many crew were needed to handle them; a line fisherman could be recognized by his hook-scarred hands. The Pacific halibut is caught today with such a gear, the hooks set in a group of six, or a skate, which is the unit of effort in that fishery (Fig. 1a).

The hook and line is also used for pelagic fishes, principally tuna. Hooks are trailed, glittering near the surface from high rods for albacore in the Bay of Biscay and for yellowfin tuna in the eastern tropical Pacific; such trollers are seen in many parts of the world, for example in the Strait of Georgia, where they catch the sockeye salmon on their way back to the

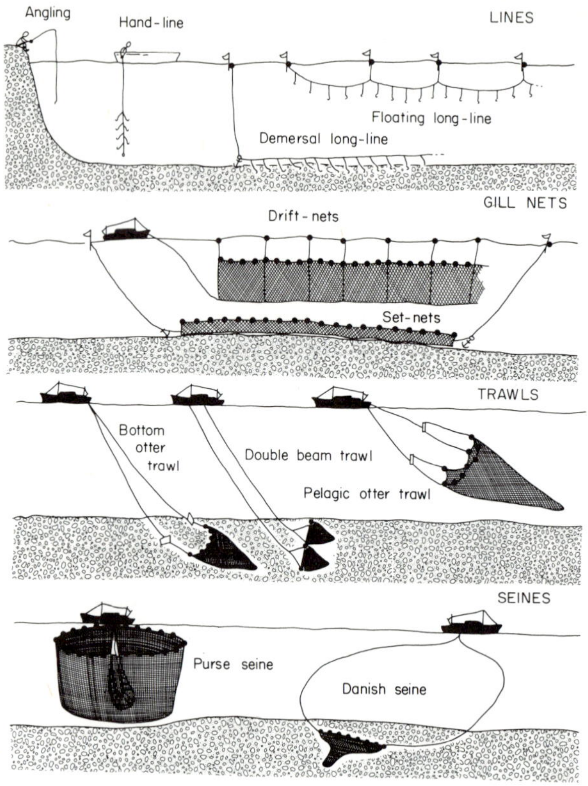

Fig. 1 Gears used by fishermen:
(*a*) the demersal long line with baited hooks and the tuna long line of unbaited hooks;
(*b*) the drift net which used to be used in the northeast Atlantic to catch herring and the bottom set net;
(*c*) the beam trawl and the otter trawl which have been used to catch demersal fish throughout the world and the pelagic trawl;
(*d*) the purse seine, the most powerful and most unselective gear used to catch fish and the Danish seine.

Fraser river. Across the subtropical ocean the Japanese and Koreans fish for all tuna species with pelagic long lines. Hooks of different shapes and sizes are hung at variable depths from a main line of 80 km in length. It takes all day to shoot such a long line; when shot, the liner steams back to the other end and it takes all night to haul it. Such vessels work in every Marsden square of the subtropical ocean, but they need large crews to handle the fish and discard the sharks.

Another labour-intensive gear is the drift net or gill net. A drifter shoots a fleet of seventy nets each floated up by buoys from a heavy messenger rope which remains attached to the vessel. The whole system, boat and fleet of nets like a long curtain hung from the floats at the surface, drifts or drives with the tide. Because the system drags a little in the sea, the fish migrating with the tide swim into it. In the North Sea, herring were caught by drift nets until the drifters were displaced by trawlers or purse seiners; none now survives. Spawning herring used to be caught by gill nets, which were drift nets anchored to the sea bed and which worked in an essentially different way as they did not drift with the tide. Drift nets still survive, as small fleets for salmon shot from a net drum in the Straits of Georgia and the Strait of San Juan de Fuca, between Vancouver Island and the mainland of British Columbia and the United States. Japanese salmon fishermen work shallow nets of up to 80 km in length at the surface in the western Pacific and off the Aleutian Islands. Such equipments survive particularly in inshore fisheries, but in the large fisheries of the open sea they tend to decline because so many crew members are needed to haul the long curtains of net by hand and because the nets themselves are now very expensive (Fig. 1b).

The trawl is today used widely throughout the world. It was recorded in the Middle Ages as the "wondrychoun", but the beam trawl was that commonly dragged by smacks across the North Sea in the nineteenth century. The beam was fastened to the tops of the triangular trawl heads which skidded along the sea bed as the equipment was towed on a single warp. The trawl itself was a long conical bag spread back from the beam to a long cod-end where the fish were gathered as the vessel progresses on its long haul. Such a trawl was used by the sailing smacks of all European nationalities as they explored the North Sea during the nineteenth century. In the 1890s the steam trawler tended to replace the sailing smack, although smacks sailed from Lowestoft as late as 1939. With the steam trawler came the otter trawl, spread by otter boards on separate warps reeled on a double-barrelled winch; the essential step was the use of the steam engine to drive that winch as well as the propeller. Then the trawl was shot on twin warps over the side. This rather complex step was made in one, in the steam trawlers built in British, German, Dutch and French yards in the 1880s and 90s.

The result of this development was a trawl made at Granton near Edinburgh which was used throughout the world as a standard gear, modified in detail on different grounds for a variety of fishes. During the 1930s, trawlermen from Germany experimented with headlines kited high to catch herring on their spawning grounds; towed at relatively high speeds they caught the spawning herring in large quantities. After the Second World War, a variety of trawling gears was developed for many purposes, box-like pelagic trawls for herring, mackerel, and sprats, high headline trawls for inshore vessels and heavy versions of the old Granton trawl to crush the rough sponges of arctic seas. The trawl was one of the essential gears of industrialization, adapted to mechanization and able to be handled by a smaller crew (Fig. 1c). This account is taken from Cushing (1966).

The last major gear is the ring net, purse seine or lampara, which is a curtain of net shot in a ring from the fishing vessel till the circle at the surface is complete and the hemisphere is closed below or pursed, and the shoal of pelagic fish is contained to be brailed aboard the seiner. A *senne tournante* was used on the French Atlantic coast for sardines a century ago, and mackerel were taken by developed purse seines in the fisheries off the eastern seaboard of the United States in the 1870s (Goode *et al.*, 1895). Such nets were used until the mid-1950s in many fisheries for herring-like and sardine-like fish. They were shot and hauled by hand, and many crew were used for this purpose. In Seattle in the late 1950s, the power block, by which the huge net was shot and hauled hydraulically, was invented. Not only were the numbers of the crew reduced, but purse seiners could put to sea and work in bad weather. Norwegian and Icelandic purse seiners extinguished the Norwegian herring fishery by fishing in the East Icelandic Current in winter, one of the roughest seas in the world ocean. Today the purse seine is the most powerful fishing instrument, and the managers of fish stocks should treat it with care and caution (Fig. 1d) because small and adult fish are caught indiscriminately.

The echosounder with the trawl and the sonar with the purse seine have made enormous differences to fishing techniques. Although echosounders were used in the 1930s by certain pioneering fishermen, it was the development of the recording echosounder during and after the Second World War that led to its use by post-war fishermen. In the North Sea large distant-water trawlers from Germany, Holland and France discovered the spawning and assembly grounds of herring with the echosounder before they caught them. At the same time driftermen searched for herring with echosounders in the North East and the North West Atlantic and off the coasts of British Columbia. Later, trawlermen throughout the world used echosounders to detect fish ahead of the trawl on the sea bed.

Sonar has been used since the Second World War to find submarines. Although sonar was used by fisheries biologists in the late 1940s and early 1950s, it was not until the power block released purse seiners on to the open ocean that fishermen started to use sonar in earnest. Today purse seiners find the pelagic fish with sonar, and the skippers make use of special skills to shoot their large seines about the shoals recorded by it. For the herring-like and sardine-like fishes the combination of purse seine and sonar is the most efficient catching equipment yet devised. An account of echo-sounding and sonar in fisheries is given in Cushing (1974).

Although the hook and line and the drift net survive in inshore fisheries and in the oceanic fisheries for tuna, they demand more crew than is currently economically feasible. Trawlers and purse seiners may use large crews, but with new machinery they can be used efficiently with relatively small crews, and so many fisheries survive today with trawlermen and purse seiners.

THE FOUR TYPES OF FISHERY

There are many ways in which fisheries may be classified, for example either as inshore, shelf and oceanic, or as demersal and pelagic. Biological categories will be used here, however, because fisheries depend upon the migration of the stocks and the stocks themselves are contained within the dynamic structure of the oceans. The classification used comprises four groups: (*a*) on spawning grounds, (*b*) on feeding grounds, (*c*) at hydrodynamic boundaries, and (*d*) in upwelling areas.

Fisheries on spawning grounds

The spawning season of fishes lasts for a relatively long time—two, three or four months for herring, plaice and cod in temperate waters and up to nine months for albacore in subtropical seas. An individual may spawn in two or three weeks in high and middle latitudes but may spread such activity over months in low latitudes. Hence fishermen have plenty of opportunity to find fish on the spawning grounds where they are abundant. Spawning grounds are sometimes small where the density of fish must be high, but they are also sometimes extensive where fishes are spread and less dense.

The Downs stock of herring (*Clupea harengus harengus* L.) spawns on very restricted grounds, one of which is 300 m across the tide and 3600 m along it (Bolster and Bridger, 1957). In the eastern English Channel and the southern North Sea, there are about half-a-dozen such grounds on strips of gravel on which the herring lay their sticky eggs. The herring migrate southwards towards such grounds in autumn and were exploited by drifters from Lowestoft, Great Yarmouth and IJmuiden and other ports in the open waters of the Southern Bight of the North Sea. Most of the spawning grounds lie between shallow banks along the tidal streams and as the fish approached them they were exploited by pair trawlers with midwater trawls from Holland and Belgium (Fig. 2). On the spawning grounds themselves, large distant-water vessels towed bottom trawls in line abreast; on leaving Dover on a clear November morning, I saw 150 large trawlers 24 km away on the Sandettié grounds and the scene was like an old naval battle. After spawning the fish drifted north-easterly and were caught by pair trawlers working between the banks. Such events took place between November

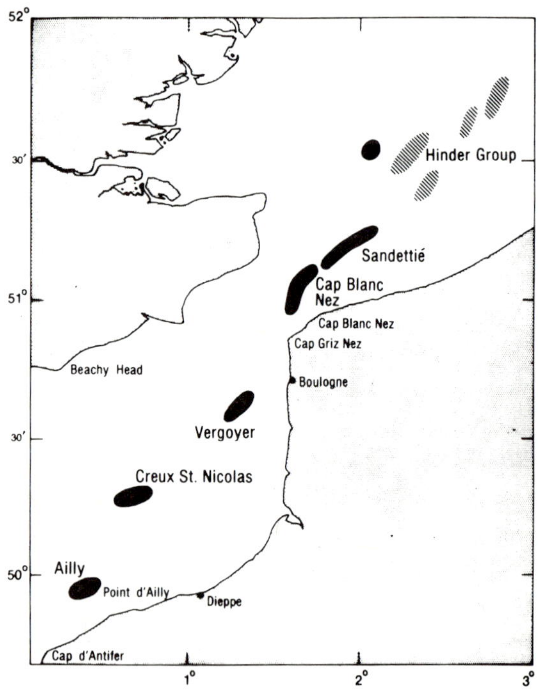

Fig. 2 Spawning grounds of the herring in the Southern North Sea (black areas) (Cushing, 1981); hatched areas are prespawning aggregations.

and the beginning of January every year between 1951 and 1963 after which the large vessels started to find the fishery unprofitable (Cushing and Bridger, 1966).

The Arctonorwegian cod (*Gadus morhua* L.) stock spawns predominantly in the Vestfjord in northern Norway where Georg Ossian Sars first discovered the transparent floating eggs of fishes (Sars, 1869). They were cod eggs laid by their grunting parents in a transition layer at 140 m and the spawning cod lie there in a patch about 48 km along on the northern edge of the fjord not far from the fishing port of Svolvaer (Fig. 3). Here live the line fishermen who have sailed for centuries into the fjord to catch the spawning cod from their small boats. In Svolvaer the cod are split and are dried in the sun and wind before export to West Africa as *bacalão* (the old Basque word for dried cod). The fish have also been caught by purse seines and by drift nets and gill nets and on the offshore banks by trawlers. The fishery lasts between late January and early April, after which the surviving cod spread out from the fjord into the Atlantic stream which carries them towards the Barents Sea. A fuller account is given in Cushing (1966).

The Hokkaido herring (*Clupea harengus pallasi* Valenciennes) appears in February and March to the north of the island of Hokkaido and enters coastal waters to spawn on seaweeds very close to the shoreline. The herring is of the *pallasi* group, and one of its distinguishing characters is that it spawns on the shoreline, as does the British Columbian herring. The fish were caught with trap nets or set nets near the tide lines. At the height of the fishery, in the 1890s, set nets were spread along the Western coast of Hokkaido from the Tsuguru Straits to the Soya straits and nearly 1 000 000 tonnes of herring were caught each year at the turn of the century (Motoda and Hirano, 1963).

The broadest spawning fishery is that for albacore (*Thunnus alalunga* Bonnaterre) spawning in the North Equatorial Current in June and July between Hawaii and the Philippines. As there might well be a stock in each subtropical gyre, it is interesting to find that spawning albacore are caught in the South Equatorial Current in October to December. Matsumoto (1966) found that tuna larvae are spread across most of the Pacific Ocean for nine months of the year. Hence the spawning season may last for two-thirds of the year as the adults drift through a spawning ground thousands of miles in length and hundreds of miles in breadth. The Japanese and South Korean long liners have worked on this enormous spawning ground for two or three months in each gyral, presumably when the density of fish is most intense.

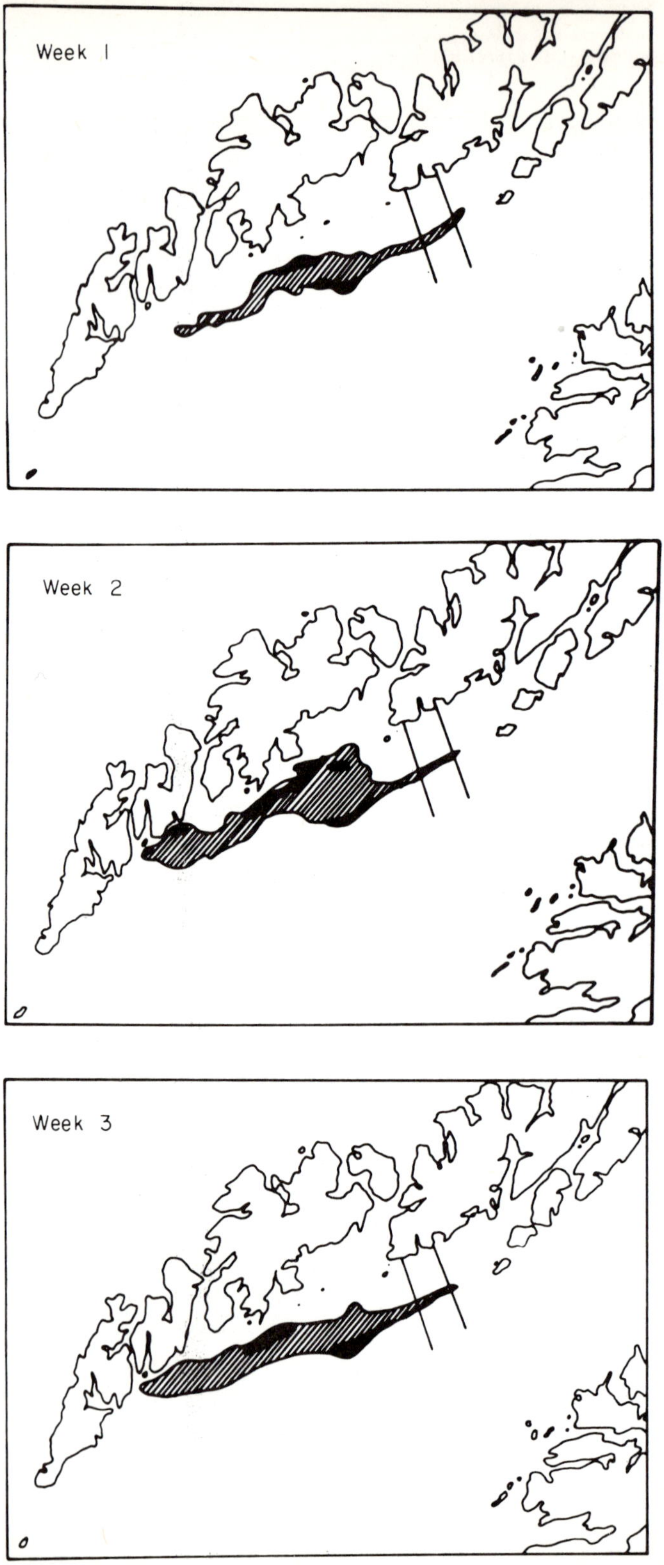
Week 1
Week 2
Week 3

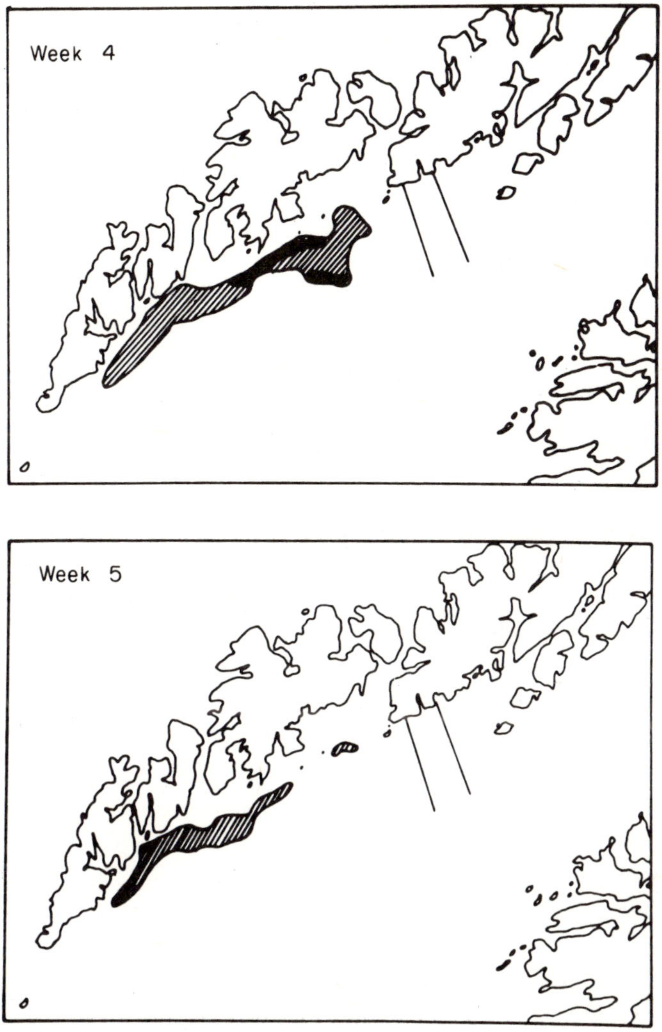

Fig. 3 Patches of spawning cod in successive weeks which lie on the northern edge of the Vestfjord not far from the port of Svolvaer (Bostrøm, 1955); the patches were recorded by echo sounder.

A spawning fishery occurs at a recurrent position each year where fish gather at the highest density, and therefore such a fishery is potentially the most profitable. In temperate and high latitude waters such fisheries last for two or three months, but in upwelling areas and in other parts of the subtropical seas, they are not identified so readily. This may be a consequence of the extended spawning seasons where spawning may evince a serial character. There is also a difficulty in definition; the Peruvian anchoveta fishery takes place mainly during the spawning season, which is also a feeding season, that of the major upwelling, and to describe it as a spawning fishery may obscure part of its real nature.

Fisheries on feeding grounds

If fish are to be caught outside their spawning season in middle and high latitude they must be taken on their feeding grounds, where they tend to be more dispersed. Hence most effort is exerted in fact on the extensive seas where fishes feed, plaice in the Central North Sea, Atlantoscandian herring in the Norwegian Sea, Atlantic salmon off West Greenland, Arcto-norwegian cod in the Barents Sea. In subtropical seas and in upwelling areas, fish are serial spawners and feed between egg batches, and the distinction between feeding grounds and spawning grounds is blurred, as noted above.

Fish gather on to patches of their food, and if they are dense enough and the fish sufficiently numerous, local and transient fisheries may appear and after a relatively short time vanish. The patch of food appeared in the first place merely because the fish were absent and fish gather there because they slow down to feed. As they feed, the food density is reduced until a level is reached at which the fish no longer slow down to feed and the shoals disengage. The dynamics of the processes have only been described sketchily (Cushing, 1955), but there are numerous charts that illustrate single snapshots of the kinematic sequence.

The phenomenon was first observed in the northern whale fisheries: Hjort and Ruud (1929) followed the temporal sequence in the catches of sei whales (*Balaenopter borealis* Lesson) feeding on "red water" (*Calanus finmarchicus* Gunner), the predominant copepod of northern seas, off West Greenland. Similar sequences were observed on the larger whales feeding on euphausids. Figure 4 shows the catches of blue whales off South Georgia and the distribution of krill (*Euphausia superba* Dana) in the same region (Hardy and Gunther, 1935). The fishery lasted for two months and the quantities of krill were dense for most of that period. The food demand of the baleen whales is enormous and was sustained by the krill.

The same processes were observed in the North Sea herring fishery. Skipper Ronnie Balls described how he steamed for a great distance shooting and hauling the little plankton net called a plankton indicator and how

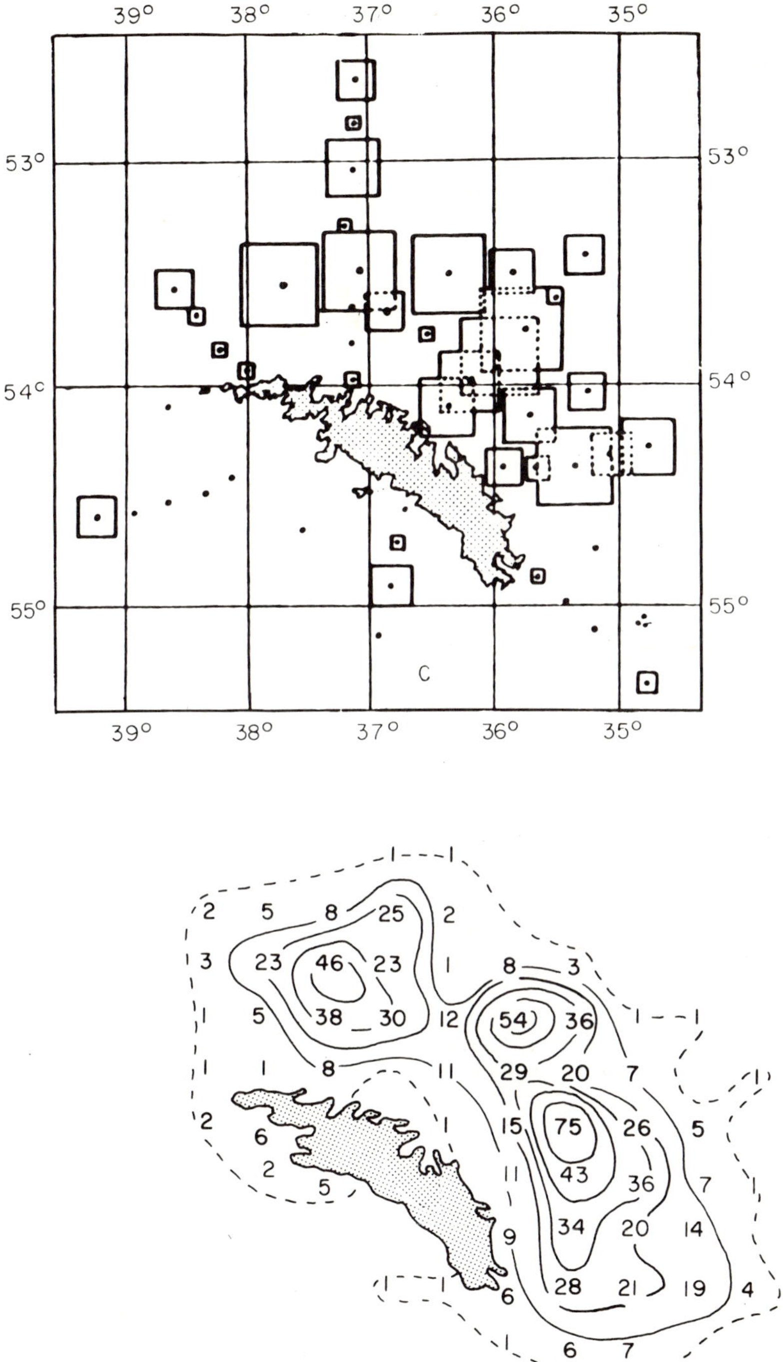

Fig. 4 Distributions of krill (*Euphausia superba* Dana) (Dec.–Jan. 1926–7) (*above*) and (*below*) blue whales off South Georgia (Hardy and Gunther, 1935).

he found nothing but the stinking green of dense diatom patches. In the late afternoon he found a small dense patch of *Calanus,* the preferred food of the herring; he steamed beyond the patch and again found the stinking green, so he turned and shot his drift nets and caught thirty tonnes of herring (Balls, 1951). The *Calanus* had eaten the algae and the herring had found the *Calanus.*

Figure 5 shows the distribution of pelagic fish, probably herring, as recorded by an echosounder, and the patches of *Calanus* on which they were

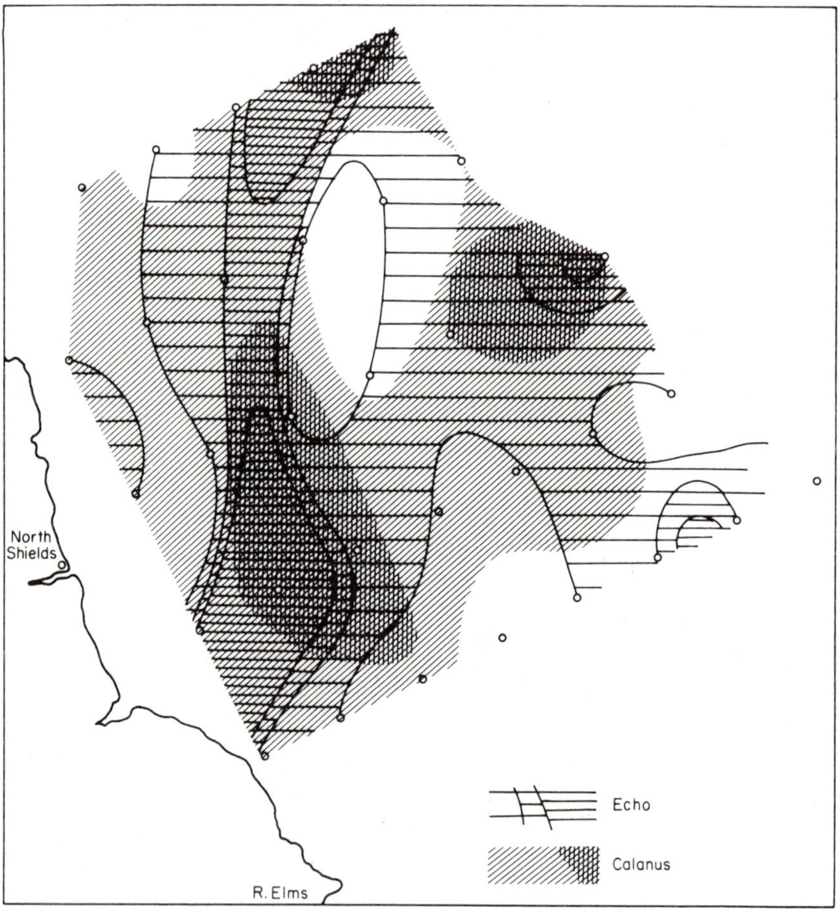

Fig. 5 Distribution of pelagic fish (perhaps herring) as recorded by an echosounder and patches of *Calanus* on which they were feeding in May 1949 (Cushing, 1952).

feeding in May 1949 (Cushing, 1952). The figure represents an instantaneous single shot of the aggregative processes of a feeding fishery which lasted for about three weeks or so. Jakobsson (1963) described a similar relationship with sonar in the herring fishery that used to take place north of Iceland in summer time as the Atlanto-Scandian herring assembled near the polar front. Because the quantities of *Calanus* are much denser than in the North Sea, the fishery lasted for longer.

The processes illustrated in the pelagic fisheries are the more intense because fishes in midwater are active and migrate across considerable distances. But the same principles must apply to other fishes. Cod in the Barents Sea shoal near the bottom as detected by echosounders. Differences are found from trawl haul to trawl haul, from drift net to its neighbour in a long fleet, from hook to hook on a pelagic long line or on a great long line set along the sea bed. Such observations are often classed in a study of patchiness in the sea, which implies an innate variance (which it is), but there may well be a dynamic base to it.

Fisheries at oceanic boundaries

Many oceanic boundaries are invisible and can only be detected with instruments, but others can be seen quite prominently. The Florida Current is detected by a change in colour and a difference in drift. At the western edge of the Kuroshio off Japan there is a pronounced dynamic boundary seen at the surface as a long rip so intense that it makes a grinding noise as one great water mass is thrust against another. It is used by fishermen who find fish gathered near the boundary which the Japanese call a *siome* (Uda, 1959). However, this is an extreme example and most boundaries outside the western boundary currents and the major upwelling areas are not very obvious.

Sharp temperature boundaries are found in high latitudes where warm and cool currents meet or at a front near an ice field. As the ice melts, a shallow stable layer is formed and production starts early and as the ice recedes a productive belt is formed in the stable layer (Marshall, 1957). At such a polar front off Jan Mayen during the fifties a large fleet of Russian drifters used to catch herring of the Atlantoscandian stock. A different sort of boundary is found off Bear Island where the warm West Spitsbergen current rises and presses against cool water on top of the Svalbard shelf between Bear Island and Spitsbergen. Trawlermen and scientists found that cod were found close to the boundary between the two water masses on

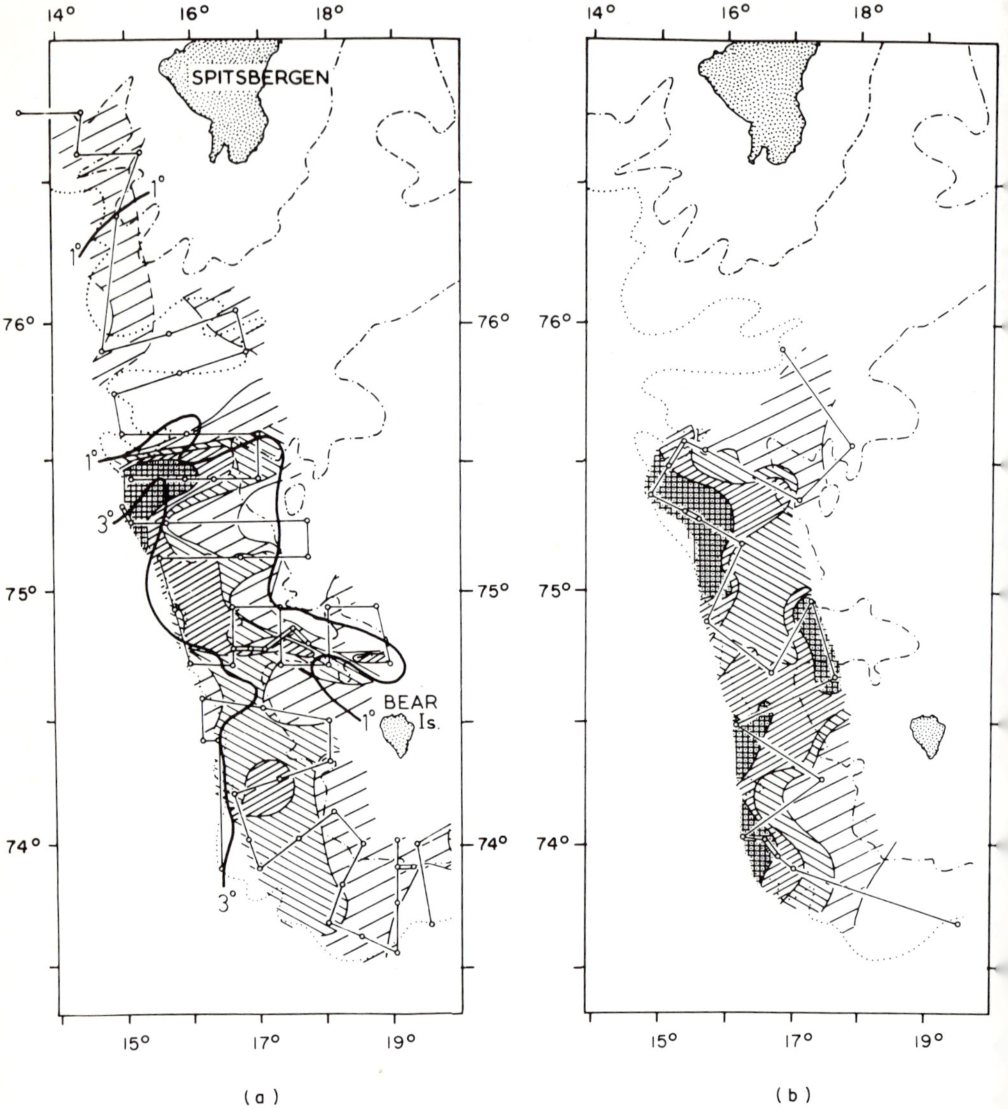

the bottom (Lee, 1952). Three echo surveys in June 1956 (Fig. 6) show how the fish recorded on the seabed aggregated against the 2°C isotherm on the bottom and also how the mass presses eastward during the short period of the surveys, about three weeks (Richardson *et al.,* 1959). Similar relationships were established by Templeman and Hodder (1965) for the haddock fishery (*Melanogrammus aeglefinus* (Linnaeus)) fishery on the Grand

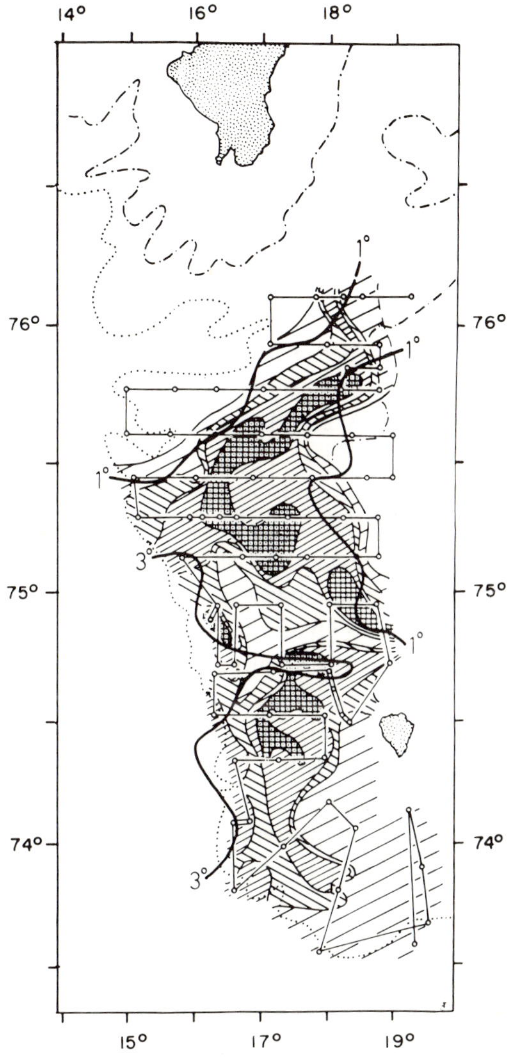

(c)

Fig. 6 Three echo surveys in the Barents Sea in June 1956 of fish near the seabed, showing how fish pack against the 2°C isotherm on the bottom and how the echopatches shift eastwards across the Svalbard Shelf (Richardson *et al.*, 1959).

Banks off Newfoundland. Such aggregations are probably partly mechanical in origin in the sense that water masses are pressed against each other, but there must be also a physiological component because the fish do not penetrate the cold water very much. Lee (1952) pointed out that the boundary was sharp (1½°C per km) which could be detected by a cod in about thirty seconds swimming on a straight course normally to the boundary.

A less obvious boundary is that between the Baltic outflow and the North Sea over the Norwegian deep water in the North Eastern North Sea. Because the Baltic outflow forms a stable layer, production starts earlier there than in the neighbouring North Sea waters. Autumn spawning herring over-winter in this region and gather at the edge of the Baltic outflow to feed on euphausids which feed on the higher production at night near the surface (Steele, 1961). Another boundary lies south of the island of Kyushu in Japan on the edge of the Kuroshio current as it flows north easterly, east of Japan. The bluefin tuna (*Thynnus thynnus* Linnaeus) ride in the current and are caught relatively close to the shore when the boundary is sharp; if it is less sharp the catches are lower. Laurs and Lynn (1977) have shown that albacore catches in June off California lie between fronts delineating transition zones between cool low salinity sub-arctic water in the eastern boundary current and the warm highly saline water of the central North Pacific; an example is given in Fig. 7. Analogous phenomena have recently been discovered around the British Isles where fronts are formed between stratified waters in summer and those dominated by the tides (Pingree *et al.*, 1975); almost certainly the distribution of pelagic fish such as herring must be bounded by such fronts.

The hydrodynamics of the shelf waters and those of the shallow seas are often less well described than those of the ocean proper. Such are the areas where fish grow, reproduce and place their nurseries, and in the future we might expect more detailed studies to reveal new fronts and boundaries of which fish may take advantage as they make their intricate migrations (see Chapter 3).

Fisheries in upwelling areas

The four major upwelling areas lie in the eastern boundary currents of the four anticyclones, i.e. off California, Peru, North-West Africa and off South-West Africa. Each moves poleward as spring gives way to summer and each is two or three hundred kilometres broad in biological terms, even

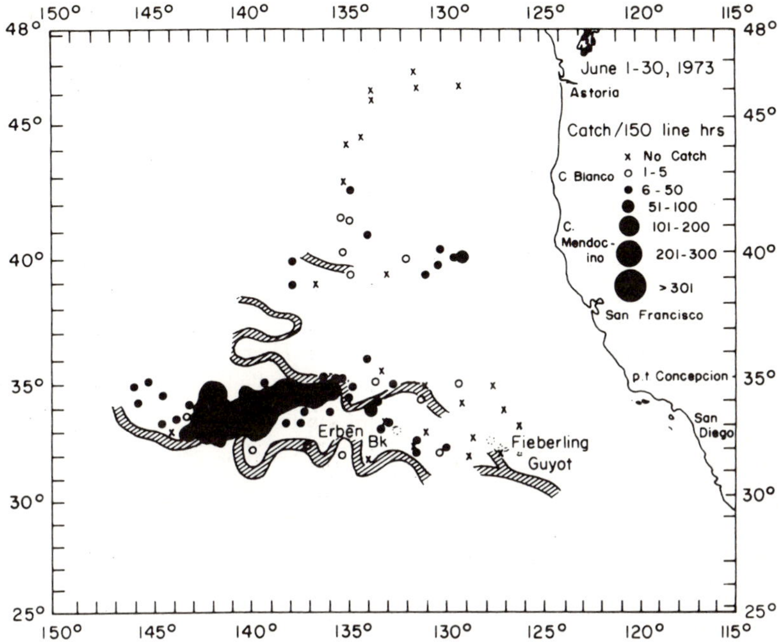

Fig. 7 Distribution of albacore catches in June 1973 off California which lie between the eastern boundary current and the warmer water mass of the Central North Pacific (Laurs and Lynn, 1977).

if the prominent physical processes are confined to a band within about 50 km of the coast. Bernal and McGowan (1981) show that differences in the zooplankton each year up to three hundred miles off California depend, not on indices of upwelling, but on indices of southward transport; the latter may be linked to the offshore divergences. The wind blows towards the equator alongshore (or at a slight angle to the coast) and by Coriolis force, water is drifted offshore and is replaced by cooler water from below. The cold water at the surface often generates fog, and the land is dry because the spring and summer winds are longshore. Because the cold water is dragged from below the euphotic layer, the production cycle starts in the rising water and continues as the water is drifted offshore. Fish gather and spawn on the patches of algae and both primary production and fish production are high.

The physical processes generate a dynamic boundary about 50 km offshore where a set of fronts may occur at convergences and divergences. The boundary occurs where the upwelled water meets the eastern boundary current. As a consequence of the physical processes a countercurrent flows poleward near the edge of the continental shelf. A swift inshore countercurrent is found in the Californian system and perhaps in other major areas. The important point about such a hydrodynamic structure is that the upwelling area provides a set of currents within which fish populations can exploit the quantities of food to their best advantage.

There are other upwelling areas particularly in the Indian Ocean and Indonesian area. There are minor areas such as the Gulf of Tehuantepec in Mexico where the *tehuantepequero* blows straight offshore, but in general they are of little importance to the fisheries. There are seasonal upwellings in the Indian Ocean off Somalia, off Southern Arabia and off the Malabar coast of India during the south-west monsoon between April and September, off Burma, the Andaman Islands and off the Orissa coast of India during the north-east monsoon. They yield less fish than the major areas, but remain of considerable importance.

The most studied upwelling area is that off the west coast of North America. The Californian sardine fishery flourished from the first two decades of the century until it collapsed in 1949/50. It was worked by lamparas and later by purse seiners, and the sardines were canned as described in Steinbeck's novel *Cannery Row.* The sardines (*Sardinops caerulea* Girard) spawn in spring in the broad upwelling region between Point Concepción and Baja California. The collapse of the fishery was probably due to recruitment overfishing (Murphy, 1966), although the replacement of the sardine stock by the anchovy (*Engraulis mordax* Girard) (Ahlstrom, 1966) may have started in the 1950s (Soutar and Isaacs, 1974) as shown by the distribution of scales in the anoxic sediments off Santa Barbara.

Jack mackerel (*Trachurus symmetricus* Ayres) and hake spawn at the same season offshore of the spawning grounds of the sardine. The hake migrate south to spawn during the winter (Chapter 2) and far offshore the albacore are caught in the divergences at the offshore edge of the eastern boundary current (see Fig. 7). Figure 8 shows the egg distributions of sardine and anchovy in 1954 and in 1962, showing the replacement. A pattern is shown in any of the major upwelling areas, clupeids inshore feeding on algae just before they spawn (Cushing, 1978), hake and mackerel feeding on zooplankton probably outside the dynamic boundary. Far to seaward are worked the tuna-like fisheries.

Between 1729 and 1919, sperm whales (*Physeter catodon* L.) were caught all over the world's subtropical ocean by whalers from Nantucket. Figure

9, which shows the positions of the catches of 36 908 sperm whales (Townsend, 1935), remains one of the best representations of the world upwelling systems—so long as it is recalled that the whales feed in the offshore divergences. One of the remarkable points about this distribution is that the great Line fishery at the equator was confined to the divergences of the South Equatorial Current both in the northern summer and in the northern winter. The Line is also the site of a fishery for yellowfin tuna (*Thunnus albacares* Bonnaterre) taken by Japanese fishermen working eastwards along the Line from the East Indies. They are caught in the divergences on the northern edge of the South Equatorial current and in the convergences between them and the Equatorial Countercurrent (King and Hida, 1954). It is possible that the divergences at the equator itself owe much of their productivity to the equatorial undercurrent which flows eastward in a depth of 150 m and from which water upwells to the surface.

The upwelling areas in the broadest sense are where fisheries are found in the subtropical ocean, in the four main areas in the eastern boundary currents, in the equatorial complexes and in the Indian Ocean. Off the Malabar coast of India, the oil sardine is caught as it gathers to spawn in a brief period toward the end of the southwest monsoon. This fishery is of considerable importance, but the upwelling system is a limited one (Darbyshire, 1967) and as our knowledge of hydrodynamic events increases we would expect the minor events to evince important aspects in the fisheries much as fronts will become of importance on the temperate shelves. Much of this account is taken from Cushing (1971).

THE INDUSTRIALIZATION OF FISHERIES

The industrial revolution permeated all sections of society in Europe and North America. As steam threshing machines and traction engines appeared on the fields of late Victorian England, steam vessels might have been expected fishing offshore. In fact it took many decades to have steam engines installed in fishing vessels—coal costs money and the wind and tide were free. Today all fishing operations on the high seas are highly industrialized and indeed some vessels use energy in very large quantities. We shall examine the first steps in industrialization and then the development of the great fisheries for fish meal and the equally important fisheries for processed and frozen fish.

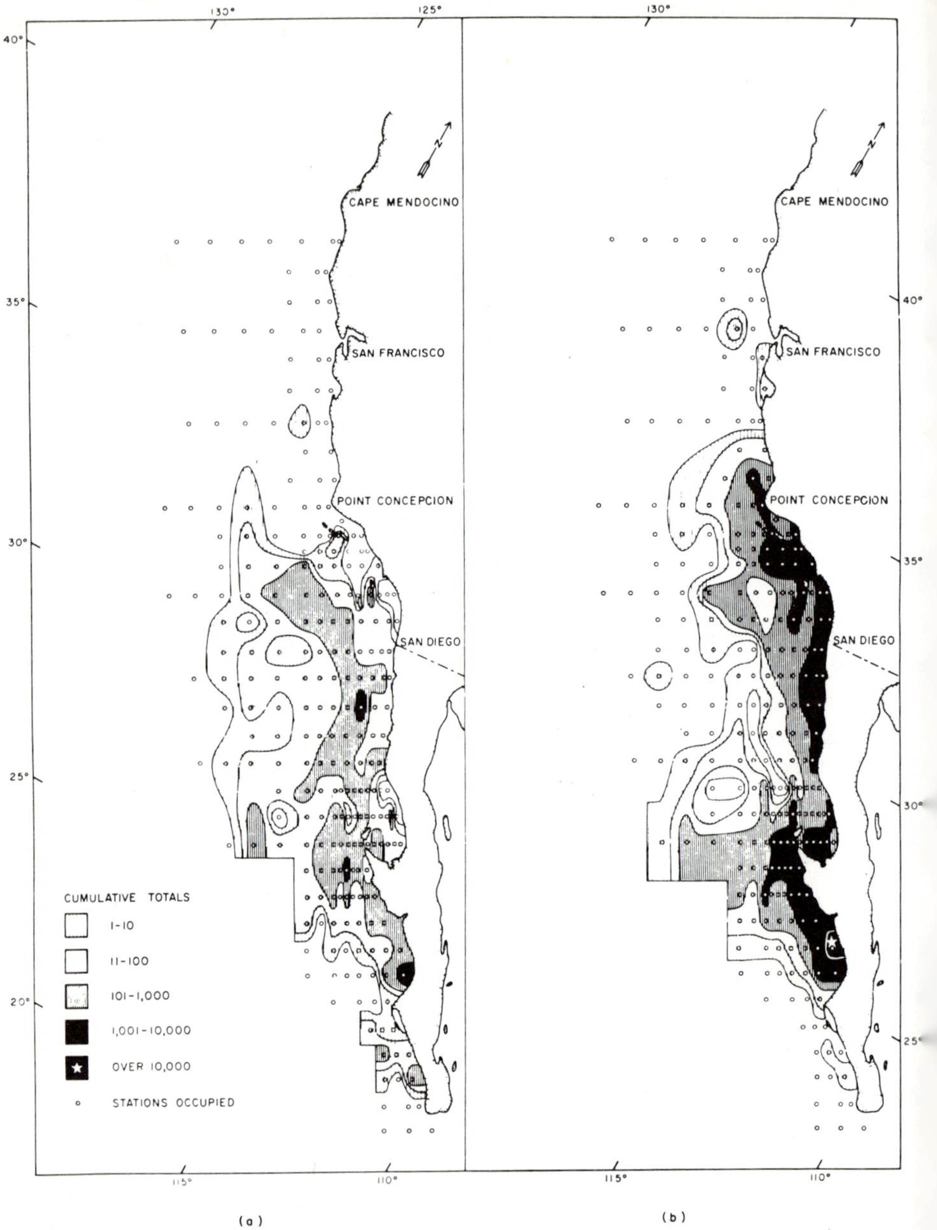

130°
125°
40°
35°
30°
25°
20°
CAPE MENDOCINO
SAN FRANCISCO
POINT CONCEPCION
SAN DIEGO
CUMULATIVE TOTALS
1-10
11-100
101-1,000
1,001-10,000
OVER 10,000
STATIONS OCCUPIED
115°
110°
(a)
(b)

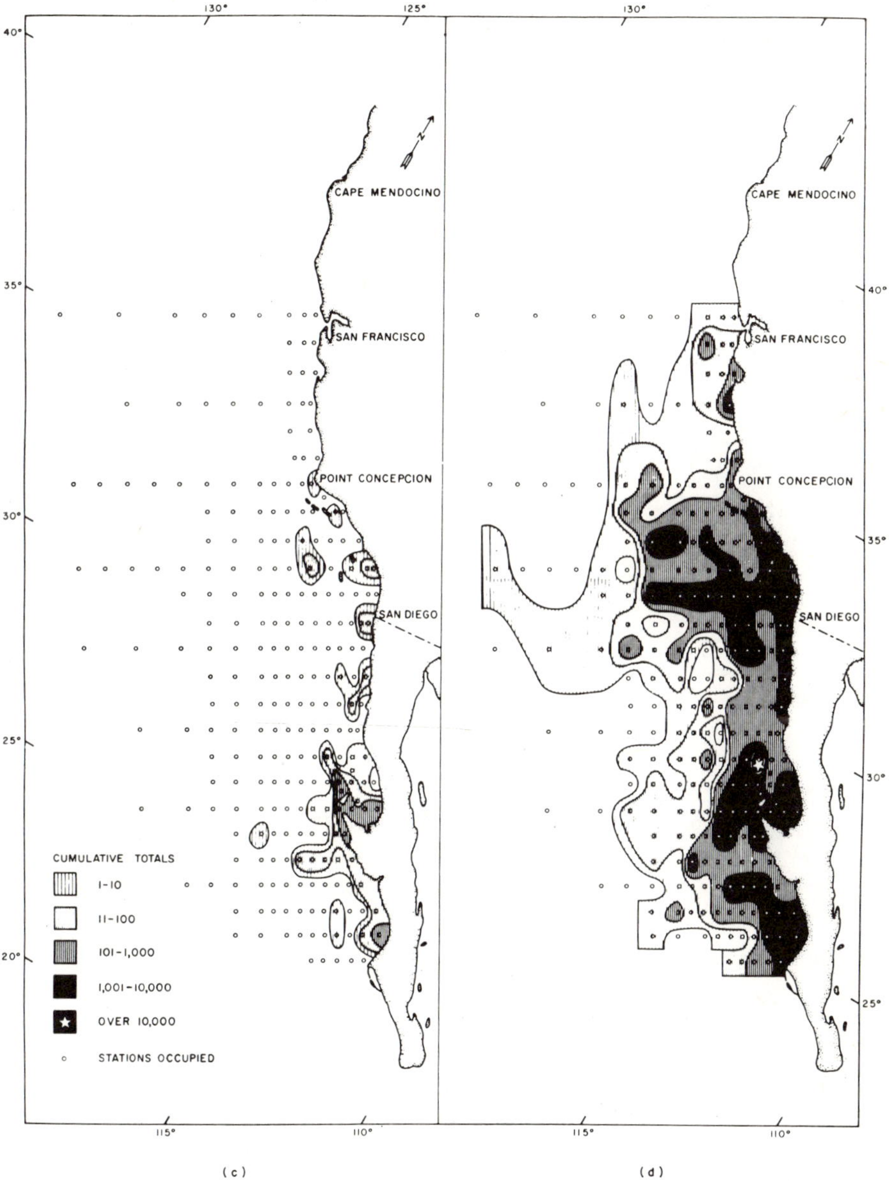

Fig. 8 Egg distributions of sardine and anchovy off California in 1954 and in 1962 (Ahlstrom, 1967).

(*a*) sardine 1954 (*b*) anchovy 1954

(*c*) sardine 1962 (*d*) sardine 1962

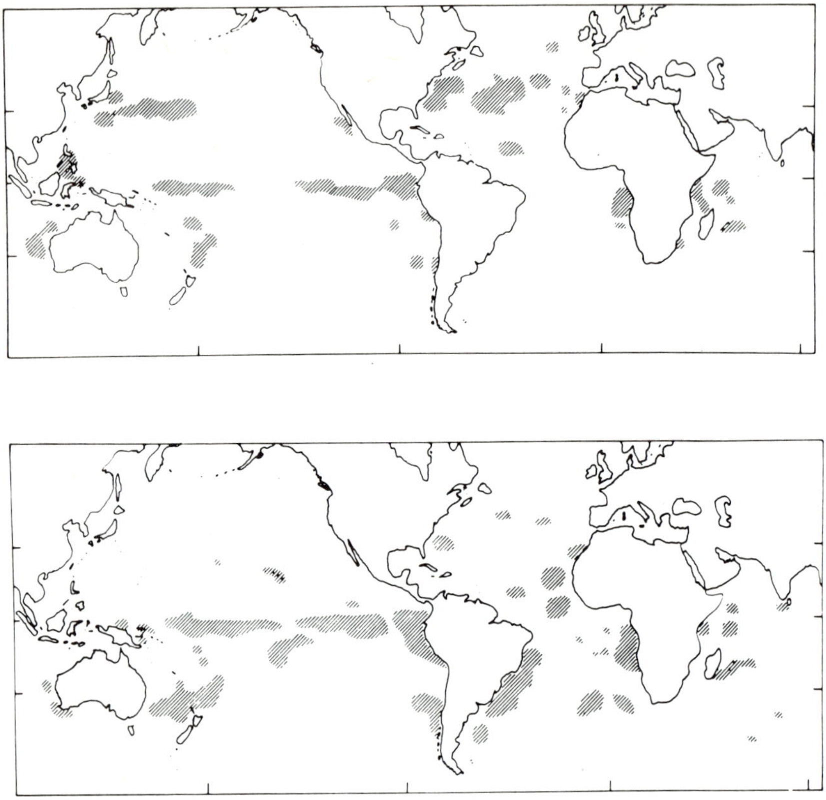

Fig. 9 Positions of catches of sperm whales between 1729 and 1919 by whalers from Nantucket (Townsend, 1935); (*above*) the southern summer and (*below*) the southern winter.

Early steps in industrialization

The first industrial fishery was that for menhaden (*Brevoortia tyrannus* Latrobe) in the 1870s. Before 1860 the menhaden was not of great importance, being caught mainly in inshore fisheries for bait used in the Canadian and Newfoundland cod fisheries. There is some dispute about who discovered that menhaden oil was useful, but many factories were erected in Maine between 1867 and 1874. The menhaden were caught by purse seines worked by special boats developed from whalers. But the important point was that there was a long-standing mackerel fishery off the eastern seaboard of the United States worked by schooners using purse seines. It is likely that the seines, much like those used today, were developed from shore seines in the early decades of the nineteenth century. The most important development in the menhaden fishery was the use of steamers to carry the nets and whalers to sea and to bring the fish back to the factories (Fig. 10).

The fish appeared off Cape Hatteras and Chesapeake Bay in late March and early April and left in December. They arrived in the Boothbay region of Maine in late May and early June and left in late October and early November. The fishery progressed up the coast, mainly within sight of the shore; the factories were distributed between Chesapeake Bay and Penobscot Bay in Maine, with the majority in Maine. In Maine between 1873 and 1878 1 to 2 000 000 gallons of oil were processed each year and about 20 000 tonnes of guano were landed. The oil was used primarily for currying leather, which is part of the softening process, and probably replaced whale oil to some degree; another use was as lamp oil in the coal mines. The menhaden fishery in the 1870s and 1880s was the first industrial fishery. It arose from a demand for animal oil and may have played a part in the opening of the West. This account is taken from Goode (1877).

One form of the early industrialization of fisheries in British waters was the use of fleets. In 1864 Hewett's home fleet of 120 sailing smacks worked in spring off the Dutch coast, in summer off Holland and Germany and in autumn around the Dogger and the Silver Pits. They shot their beam trawls three times in 24 hours and towed for 5 or 6 hours; it took 2–3

Fig. 10 Early menhaden fishery off the eastern seaboard of the United States was prosecuted by the first steamers to work in the fisheries (Goode, 1887).

hours to haul the trawl on the man-powered capstan. The fish was taken to Billingsgate in London every day in carriers, but it was first transported from smack to carrier by a small boat manned by boys, many of whom drowned. Ice was imported from Norway, much from Lake Oppegaard. Much of this development depended on the spread of the railway, and the fisheries supplied cheap protein to the growing industrial towns in midland England.

The first steam trawlers were the *Zodiac* and *Aries,* and because they had capstans below decks run by the steam engine the time spent hauling was reduced from two or three hours to twenty minutes. They caught four times as much as a smack. Between 1881 and 1902, 1573 steam trawlers were built in England and Wales and by 1895 the otter trawl had replaced the beam trawl. The otter trawl was spread by two otter boards and hauled by two warps, which entailed the invention of the double-barrelled steam winch. The standard otter trawl was the Granton trawl, modifications of which trawl are today used throughout the world in all the major trawl fisheries (see Fig. 1d). This account was taken from Cushing (1966).

Steam trawlers were used to explore the northern North Sea and more distant waters. They remained the backbone of European and North American fleets until oil replaced coal as a primary fuel in the late 1920s and early 1930s, and they developed the great trawl fisheries as we know them today. An offshoot of the development of the steam trawler was the steam drifter. Herring had been caught by drift nets for centuries, and the sailing drifters returned to port each morning. Steam drifters reached market more quickly and between 1890 and 1910 hundreds were built, particularly in England and Scotland; they vanished with the collapse of the North Sea herring fisheries.

The industrialization of fisheries started as part of the more general process. The revolution in fisheries was equally profound and led to the incorporation of bigger engines, more complex trawls made of synthetic twines, radio and acoustic aids, and eventually factory processing at sea.

The development of the fishmeal industry

Fish have been used for animal food for a long time and in times of glut, herring and sprats were put directly on the land. The British Columbian herring were used for fish meal during the 1930s, but in general the fishmeal industry developed in the years after the Second World War. One of the earliest of such industries was the Danish which exploited young herring for fishmeal. During the 1950s in European waters, herring were caught by large trawlers on or near their spawning grounds. Indeed the greatest quantities of North Sea herring during the 1950s were converted to fishmeal. In general, bottom trawls were used by the Danes both on the nursery grounds in the eastern North Sea and by the Germans, Dutch and French on the Fladen and on the Dogger and Southern North Sea spawning grounds (Cushing and Bridger, 1966).

In other parts of the world ocean, menhaden were caught by purse seine, once again off the eastern seaboard of the United States, pilchards were taken by purse seine off South Africa partly for fish meal and partly for canning, Norwegian purse seiners took herring in the Norwegian Sea and capelin in the Barents Sea, the Japanese fished for Alaska pollock with trawls in the Bering Sea, and the Peruvians caught anchoveta with purse seines in the Peruvian upwelling system. Up to 15–20 000 000 tonnes of fish were converted to fishmeal to be converted again to animal flesh. The conversion is costly because there is a loss of water as the fish is dried and a loss of protein in metabolic conversion, a factor of 13–15 depending on the methods of estimation. But the crews are not very numerous although they were on the large herring trawlers in the North Sea in the 1950s. Further the use of fishmeal throughout the world is really part of the industrialization of agriculture which started in the 1930s in the prairie dustbowl in the United States, continued in Britain during the Second World War, and is being sustained in Europe today.

The most remarkable industrial fishery in the world is that practised by the Danes in the North Sea. With light high-headline trawls originally used for catching herring in the eastern North Sea they explored the Northern North Sea for a small cod-like fish called Norway pout (*Gadus esmarki* Nilsson) and by the early 1970s were landing between 500 000 and 700 000 tonnes into the ports of Esbjerg, Friderikshaven and Hirtshals. Unfortunately large quantities of immature haddock and whiting (*Odontogadus merlangus* L.) were caught at the same time. Another form of trawl with a lengthened cod end was used on the Norfolk Banks and on the banks of the Eastern North Sea close to the Danish coast and in Northern North Sea close to the Norwegian deep water, to catch sandeels and up to 800 000 tonnes were caught in the early 1970s. Fortunately the by-catches of protected species are low. Light high-headline trawls were also used to catch sprats in the eastern North Sea in winter time, again up to 500 000 tonnes each year in the early 1970s. Young herring are caught as a by-catch in this fishery and such catches may delay the recovery of the herring in the

North Sea. The total catches for fish meal from the North Sea have reached nearly 2 000 000 tonnes and they cause some problems in the management of the protected species (Madsen *et al.*, 1978).

A number of fish stocks exploited by industrial fisheries have collapsed, the Japanese sardine, the British Columbian herring, the Atlantoscandian herring, the North Sea herring, the Barents Sea capelin, the South African pilchard and the Peruvian anchoveta. Of course it was not the use for fish meal that caused the collapse but the vulnerability of pelagic stocks like herring or sardines to excessive fishing pressure particularly with purse seines. Purse seines are the most efficient gears used for catching fish and the small pelagic fish comprise stocks that are vulnerable to recruitment overfishing, the condition under which the incoming year classes are reduced under the pressure of fishing.

THE DISTANT-WATER FISHERIES

The Portuguese white fleet fished for cod from dories on the Grand Banks off Newfoundland from the sixteenth century onwards. This fishery started when the early explorers discovered the cod; indeed by 1506 several hundred men from Brittany and the Basque country were working each summer off Cape Breton and Prince Edward Island. By 1527 fishermen from Devon had to employ merchant ships to carry the sun-dried cod back to Bristol (Lyons, 1964), but later distant-water fisheries started when fishermen had exhausted grounds nearer home. British fishermen had fished off Iceland in summertime for centuries, but when the steam trawlers had reduced the catch rates in the North Sea in the 1890s (Garstang, 1901–03), they sailed for Icelandic waters in greater numbers at all seasons; any trawlerman who works in the stormy seas east of Iceland in winter earns his money. In the 1920s the big distant-water trawlers rediscovered the rich cod grounds in the Barents Sea. They were the traditional "freshers", who packed their catches on ice for up to three weeks before they landed them.

In the 1950s a new trawler was invented. Instead of dragging the trawl laboriously over the rail along the side of the vessel, it was hauled directly and much more easily up a ramp in the stern. Almost as a consequence it became possible to rig a factory deck below the working deck, where fish were frozen, immediately they were caught. Such freezers, largely developed in the British Isles, could stay at sea for up to two months and bring ashore hundreds of tonnes of frozen fillets. They worked across the

North Atlantic from the Grand Banks and West Greenland to Iceland and the Barents Sea.

However, the greatest development in the distant-water fisheries took place in the fleets from Russia and Japan. In the 1920s both countries decided to satisfy their protein demand from the sea. During the 1920s and 30s the Japanese made exploratory voyages for salmon and tuna across all the North Pacific. By the 1960s the Japanese and South Koreans were catching tuna in all the subtropical oceans and salmon from the western north Pacific. The Russians made exploratory voyages on the continental shelves of the Atlantic and Pacific; they worked for hake off Oregon, silver hake off New England, hake off South Africa and South America, and mackerel and horse mackerel west of the British Isles. The Japanese caught Alaska pollock off Alaska and the Aleutian Islands, squid on the Grand Banks and off New Zealand, and hake off South Africa and South America. Lastly the Californian tuna boats from San Diego in California steamed to Angola, Namibia and New Zealand.

Between them the Russians and Japanese in the early 1970s were landing nearly 15 000 000 tonnes each year. In 1977 the U.N. Law of the Sea became the tacit rule if not the international law and distant-water vessels are by now, to some degree, excluded from the 200-mile limits of those countries which they had earlier exploited. Although in the long term this must mean the transfer of protein to countries where it is needed, it also implies the disappearance of the large stern-trawler freezer. The freezing at sea may even become reduced in scope.

SOME EXAMPLES OF FISHERIES

So far fisheries have been classified in somewhat arbitrary groupings for didactic purposes, but to describe the fisheries as they are, we need to examine a few in greater detail.

The Pacific salmon

Five species of Pacific salmon—sockeye *Oncorhynchus nerka* (Walbaum); pink *O. gorbuscha* (Walbaum); chinook, *O. tschawytscha* (Walbaum); chum, *O. keta* (Walbaum); and coho, *O. kisutch* (Walbaum)—spawn in the rivers of North America and eastern Asia. The eggs are laid on gravel redds in streams at the heads of the great river systems and they hatch, yielding larvae, smolts and juvenile fish which migrate down stream out into the Pacific. As they migrate round the North west Pacific gyral they are caught by Japanese drift netters working far out in the ocean, but the Abstention Principle in the International North Pacific Commission prevented a similar exploitation in the Alaska gyral and the Bering Sea. As they mature, the fish leave the gyrals and return to their native streams, the Columbia, the Fraser, the Skeena, the Nass, the Karluk and those rivers that empty into Bristol Bay.

The Indians caught sockeye salmon with shore seines, rough gill nets, spears, and basketwork weirs. Today about 35% of the British Columbian catch is taken by purse seiners, the larger of which work in the Strait of S. Juan de Fuca south of Vancouver Island. Gill netters catch about 40% of the British Columbian catch and about 3500 boats work on the coast and many of them are one man boats that fish in the rivers and estuaries. About 4000 men in 25 000 boats troll for salmon (with hook and line). The sockeye salmon return towards the Fraser river in June and July and pass through the Straits of S. Juan de Fuca and the Straits of Georgia; they are caught by all gears in the area and also by traps in the Sooke area at the mouth of the Fraser river. The Fraser system extends for more than eight hundred miles into the Rockies. The fishery off the estuary of the Fraser and in the Straits lasts for about three or four months, and then the sockeye salmon escape to migrate to their native redds where they spawn and die unless eaten first by grizzly bears, wolves, or coyotes.

Most of the British Columbian catch is canned although a proportion is smoked; in 1741 the German naturalist Steller (who named the now extinct sea cow) discovered some salmon smoked by Indians in an underground storehouse. More than one million cases (each containing 48 lb (22 kg)) of all salmon species leave the canneries in British Columbia each year. This account is taken from Lyons (1969), but equally important fisheries in Southern Alaska and Bristol Bay occur each year exploited in much the same way (Table 1). As salmon is a very valuable fish, the catches from the rivers of North West North America represent a great resource.

The Peruvian anchoveta fishery

The Peruvian upwelling system extends all along the coast of Peru from Punta Aguja to Arica. The anchoveta (*Engraulis ringens* Jenyn) spawns in the southern spring when the upwelling is most intense and as they spawn they feed on the rich phytoplankton patches in which their offspring will

Table 1
Total annual catches (tonnes) of salmon in the Alaska gyral and Bering Sea (Area 67) from 1974 to 1977[a]

	1974	1975	1976	1977
Pink	29 389	35 731	62 070	81 715
Coho	29 124	20 274	26 629	23 440
Chum	31 706	20 719	34 802	32 068
Sockeye	43 819	29 415	50 060	58 181
Chinook	19 101	20 019	22 628	21 607
Total	153 139	126 158	196 189	135 296

[a] FAO *Yearbook of Fishery Statistics* (1977)

grow. They mature between 15 months and two years of age, and although they live naturally up to four or five years they do not survive in the fishery for much more than three years. They are caught by purse seiners working from Callão, the port of Lima, and Chimbote (further to the north) and various minor ports. Most of the catches are taken within 30 km of the coast well within the dynamic boundary of the upwelling system which lies about 50 km offshore. Although catches may be taken at most seasons, the best catches of the largest fish are made during the spawning season in the southern spring. Indeed during the previous four months when small fish recruit to the stock there has often been a closed season to prevent the "peladilla" from being caught.

At the end of the 1950s the catch amounted to about 100 000 tonnes. With increasing demand for fish meal, the fleet expanded during the 1960s and by 1965 the annual catch had reached about 7 000 000 tonnes. In 1970 it reached nearly 12 000 000 tonnes, but the stock collapsed in 1971–2 and has not recovered since. The failure will be examined in a later chapter in more detail as it concerns El Niño, the warm current that floods the Peruvian coast from the north just after Christmas every five or seven years or so. Apart from the peculiar climatic interest, the fishery is itself important. It was the largest single species fishery in the world.

The Peruvian anchoveta fishery was—and still is—the fishery in an upwelling system *par excellence.* Lesser fisheries exploit stocks based on large areas: cod in the Barents Sea, tuna in the subtropical ocean, salmon in the cyclonic gyrals. The coast from the Punta Aguja to Punto Lomas forms a constant angle to the longshore wind which generated the upwelling system. In other regions the coast may only briefly form an optimal angle to the wind and it is possible that the angle of the Peruvian coast is close to the best.

The catches (in tonnes) in the years 1974–77 (FAO *Yearbook of Fishery Statistics*, 1977) were:

1974	1975	1976	1977
3 583 476	3 078 810	3 863 050	792 106

The Gulf of Thailand fisheries

Since 1961, with much assistance from German Technical Aid, the Thailanders have financed the construction of more than 2500 vessels, mostly trawlers. The Gulf of Thailand is nowhere deeper than about 150 m, and up to 300 species caught on the seabed, most in the relatively shallow waters close to the coast. By 1970 more than 660 000 tonnes of demersal fish were landed, half for human consumption and most of the rest for chicken and duck food. The total catches rose from 150 000 tonnes in 1960 to 1 500 000 tonnes in 1970. From observations made by the research cutter *Framong 2*, the catch per unit of effort declined from 249 kg/h in 1963 to 66 kg/h in 1971 (Tiews, 1973).

The catches (in tonnes) in the years 1974–77 (FAO *Yearbook of Fishery Statistics*, 1977; area 71) were:

1974	1975	1976	1977
1 305 000	1 343 001	1 446 917	1 558 020

The Alaska pollack fishery

The Alaskan pollack (*Theragra chalcogramma* Pallas), or more strictly walleyed pollack, has been caught for many years off northern Japan and north-east Korea, but before 1965 catches were less than 100 000 tonnes. By 1970 the total catch in the North Pacific amounted to 3 300 000 tonnes: 700 000 tonnes taken by the USSR, 2 400 000 tonnes by the Japanese, and the remainder by the North Koreans. Although pollack are caught round Hokkaido and Sakhalin, along the north east coast of North Korea, off south-west Kamchatka and the Kurils, the most productive area is the eastern Bering Sea between 165 and 175°W (Fig. 11) (Koslow, 1976).

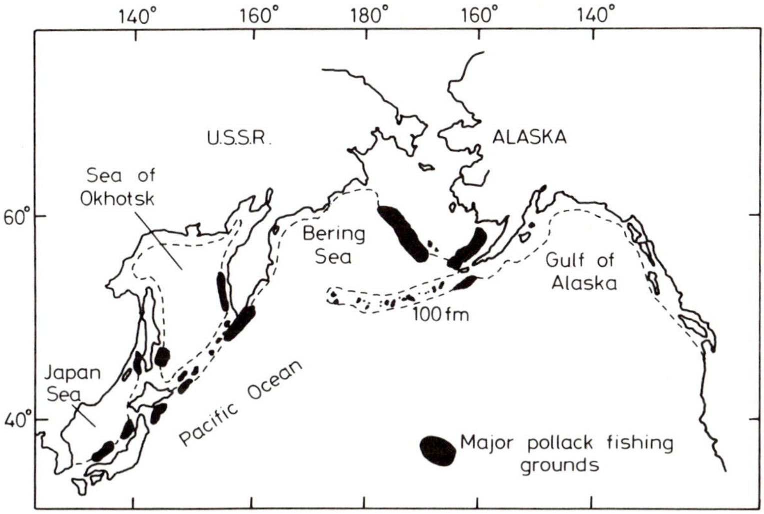

Fig. 11 Distribution of the Alaska pollack fishery in the Bering Sea and elsewhere in the North Pacific (Koslow, 1976) (100 fm = 182.3 m).

Originally the Japanese fished for yellowtail soles (*Limanda aspera* Pallas) in the early 1960s, but when that stock declined they turned their attention to pollack. At the same time a mechanical method of processing pollack into frozen blocks of surimi (minced meat) was devised. Large factory stern trawlers were developed to catch the pollack, together with mothership fleets in the eastern Bering Sea. Medium stern trawlers work off Kamchatka and the Kuril Islands. Although some frozen blocks of pollack are exported to the United States, the major outlet remains surimi and fish meal.

The main gear used is the trawl worked by the stern trawlers, but Danish seines and pair trawls are operated by the mothership fleets; a small gill net fishery takes place. The most extraordinary point about this fishery is that the catch per unit of effort, the index of stock density, increased between 1964 and 1970, whilst enormous catches were being taken. The pollock eats copepods, amphipods and euphausids (Koslow, 1976), and so is abundant as herring in the Atlantic or sardine-like fish in the upwelling areas. Serobaba (1977) suggests that there are four populations: in the Eastern Bering Sea, in the Northern Bering Sea, in the Western Bering Sea and in the Southern Bering Sea along the Island chain bordering the Sea of Okhotsk.

Since 1977, when most nations of the world introduced the 200-mile limit, the major grounds in the Bering Sea have become controlled under the sovereignty of the United States and to a lesser degree the USSR. The catches (in tonnes) in the years 1974 to 1977 in the Bering Sea and the Sea of Okhotsk (FAO *Yearbook of Fisheries Statistics,* 1977) were:

	1974	1975	1976	1977	
Area 61	3 754 807	3 906 067	3 958 928	3 616 861	(East)
Area 67	1 152 565	1 117 858	1 111 445	679 206	
Total	4 907 372	5 023 925	5 070 373	4 296 067	

The Arctonorwegian cod fishery

The Arctonorwegian cod (*Gadus morhua* Linnaeus) stock spawn chiefly in the Vestfjord in northern Norway between February and April; it is a very old fishery, knowledge of which goes back to the twelfth century (Rollefsen, 1955). Purse seines and gill nets have been used in the fjord, but most fish were caught by hook and line from small open boats and were sun dried in the small port of Svolvaer before being exported to West Africa. This trade was a very old one which today has tended to decline.

Offshore the cod are also caught by trawl on the banks of the Norway coast, the Murman coast of the USSR and on the Svalbard shelf between Bear Island and Spitsbergen. Most catches are made by large stern trawlers mainly from the USSR and Norway, but still a few vessels from Britain, France and Germany work this region. Bottom gill nets are worked by inshore fishermen from various ports on the Norwegian coast. Before the recent restraints by the national 200-mile limits, trawlers from many nations worked in the Barents Sea, and often they gathered on dense patches of fish which they detected with their echosounders. Because the trawlers fished quite close together, the skippers call such gatherings "arctic cities". Most of the cod caught in this way was and is frozen, although a few "fresh" trawlers (i.e. the catches are preserved on ice and sold "fresh") continued to fish there in the 1970s.

Large catches were made in the Barents Sea by the Russian fleet in the mid 1960s which may have reduced the stock to some degree later. Since 1975 an internationally agreed quota has restrained catches to about 850 000 tonnes each year; before then, in the early 1970s somewhat larger catches were made, probably more than 1 000 000 tonnes. By 1980, catches were very much reduced to about 350 000 tonnes. The fishery is a very valuable one because the price of cod is high.

The North Sea plaice fishery

Although plaice (*Pleuronectes platessa* L.) have probably been caught near the coasts of the North Sea for a very long time, intense exploitation did not start until the early days of industrialization when the fleets of smacks swept the shallow waters off the coasts of Denmark and Germany to catch large quantities of small plaice; indeed they discarded six times as many as they landed (Cushing, 1972b). The catches were then landed on ice by carriers daily to Billingsgate, the fish market of London. Later when the steam trawlers turned to "single boating", they brought fish back on ice after ten days or so.

Today there are four main fleets fishing for plaice in the Southern North Sea. Side trawlers and small stern trawlers work from Lowestoft in the Northern North Sea west of the Norwegian deep water. Two fleets of Danish seiners work in the North Sea, one from Grimsby which fishes round the Dogger Bank and one from Esbjerg which operates in the Eastern North Sea. Dutch double beamers (which tow large beam trawls quickly, one hauled and one towing) from Scheveningen and IJmuiden catch plaice in the Southern Bight of the North Sea. Between them these fleets have caught between 100 000 and 120 000 tonnes of plaice each year in the last few years.

THE NATURE OF FISHERIES

The demand for fish for human consumption depends on the need for protein in industrialized countries which is slowly being transferred to less developed countries as they approach industrialization. In Victorian Britain and until recently, fish was cheaper than meat. This is no longer so, and in the United States and Western Europe the demand for fish depends on more subtle factors but remains high. The demand for fish for fishmeal depends on the need for the essential amino acids lysine and methionine in animal feedstuffs; the great developments of factory farming require large quantities of feed stuffs which are augmented by fishmeal.

Since fisheries became industrialized, large quantities of fish have been landed from the broad relatively shallow areas on the shelves like the North Sea, the East China Sea or the Svalbard Shelf, from the upwelling areas and to a minor degree from the deep subtropical ocean itself. Cushing

(1973) suggested that the transfer from primary to secondary production was three times more efficient in the subtropical ocean as on the shelves; but as the stock is richer by an order of magnitude there, the production is only three times as great. As fishermen must search enormous areas, they find fish where they are concentrated to spawn, to feed or to gather at an oceanic boundary. This is why the richest fisheries are found on the shelves or in the upwelling areas.

The next chapter, on the biology of fish populations, will reveal how important a part is played by migration in the lives of fishes. A circuit is established between spawning ground and feeding ground within the hydrodynamic structure. This is most important to fishermen because it provides the basis for a regular and seasonal movement of fishermen around their local seas. If they fish on spawning ground and then on a feeding ground, they must shift ground with season sometimes over hundreds of miles.

Chapter Two

The Biology of Fish Populations

The study of fish populations originated in the need to stem the decline of stock under the pressure of fishing and was greatly facilitated by the discovery that fish could be aged readily by annual rings on scales or otoliths (Dahl, 1907; Reibisch, 1899). In recent decades, catches of fish have been properly recorded in many countries, and in some of them samples of fishes have been aged and measured in length on the quays. Since 1966, catches-at-age have been analysed by the relatively new technique of cohort analysis, and matrices of stock numbers (or of fishing mortality) by age and years have been constructed retrospectively for many stocks for some decades. The result is that fish stocks are the best known of all large wild populations, but despite the deployment of such information, many of the world's stocks of fish are overexploited. In the north-east Atlantic alone, 18 out of 43 stocks were considered to be overfished in 1980 (Cushing, 1980).

LIFE CYCLE

Fish live for a long time, grow by several orders of magnitude, and most produce very large numbers of eggs. The life cycle starts on the spawning ground where ripe eggs are released by each individual in batches. They take up water from the sea and increase their volumes by a factor of five (Fulton, 1897). Because the eggs are released in batches, the fish spawn on a number of occasions, perhaps every ten days among spring spawners (which do not feed during the spawning season) and less frequently among summer spawners and those that live in upwelling areas; both of the latter groups feed during the spawning season.

Eggs are usually about 1 mm in diameter, except those of salmon, halibut, lumpsucker and sturgeon which measure about 3 or 4 mm across and those of dogfish and their relatives which are very much larger. The teleostean eggs are stored with transparent yolk and may carry oil globules. They are buoyant and float at or near the surface; exceptions are herring, sandeels and lumpsuckers which lay their eggs on the sea bed, blennies and gobies which spawn on rocks close to the shore, and salmon which spawn on gravelly shallow redds in streams distant from the sea (Cunningham, 1895). Eggs develop in the sea fairly quickly—in two days in warm water and in three weeks in cool water, but those of salmon take five months to hatch in wintry streams.

Except for the few species that spawn on the bottom, fish larvae hatch near the surface of the sea. At first they live on their yolk because the mouth appears only after hatching. The yolk lasts for 10 to 20 days while northern anchovy larvae learn to feed, which may take some time (Hunter, 1972); they learn to attack with a rather slow S-shaped wriggle and indeed plaice larvae may train themselves on the large diatom *Coscinodiscus,* which is found sometimes in the guts at the time of first feeding (Riley, 1966). The larvae are transparent and in general feed on copepod nauplii; indeed cod and haddock larvae in the North Sea may well grow with the nauplii and copepodites of *Calanus* they need (Jones, 1973). Anchovy larvae off Southern California feed on a relatively large gymnodinian, *Gymnodinium splendens* (Lasker, 1975) and plaice and sandeel larvae (*Ammodytes marinus* Raitt) in the southern North Sea feed on the appendicularians *Oikopleura* and *Fritillaria* (Shelbourne, 1957; Wyatt, 1972).

After a period of months, the larvae metamorphose into little fishes. Fin rays and other peripheral skeletal structures are established, and there are changes of body shape, the most spectacular being the well-known twist executed by plaice larvae from a normal dorsoventral posture to one in which they lie on their sides. They stop feeding during metamorphosis, possibly because the processes under endocrine control are considerable.

After metamorphosis, the little fish live during the rest of their first year of life on their nursery ground on beaches and shallow banks. Because bigger fish live in deeper water (an extension of Heincke's Law (1913) which stated that larger plaice in the North Sea live in deeper water) such areas are relatively free of predators on the shallowest part of the beaches where they live. On the nursery ground the little fish are able to grow quickly with reduced chances of predation. As they grow, they move down the beaches and slowly spread into deeper water (described properly as a diffusion (Beverton and Holt, 1957)) until they join the adult stock on the feeding ground.

Anchovies spend a year as juveniles, whereas some cod do not mature until their ninth year; in the final juvenile year the gonad matures slowly during the feeding season. If the fish, for example, herring, reaches a critical length (Burd, 1962), it will start to mature, but if that length is not reached, maturation will be delayed for a year. Consequently, recruitment to the adult stock is spread over two or three years in herring and perhaps between four and nine years in cod.

During maturation, protein that might have gone to somatic growth is converted to gonad growth and feeding slows. During the period of high feeding, much fat is laid down, and this maintains the animal during the period of reduced feeding before spawning. Growth is under endocrine control; but there is no check in growth at maturation as if the reproductive load had been programmed from a much earlier period in the growth pattern (Iles, 1967).

The fish become adult when they recruit to the spawning stock on the feeding ground as they mature for the first time. A most remarkable point about fishes is that, unlike mammals, they grow considerably during their adult lives. A herring increases its weight by 50% during its adult life, but plaice or cod grow by an order ot magnitude during that period of their life cycles. This led Bidder (1925) to suggest that fish were potentially immortal, which is perhaps unlikely, but the suggestion also arose from the increasing realization that fish lived longer than was thought. Table 2 shows how long fish live and the age at which they spawn for the first time. Bigger fish live longer and spawn for the first time at a greater age. In general, fish appear to spawn once a year and so would be expected to spawn many times if the stock was unexploited. If, however, the stock is heavily exploited, the number of age groups is sharply reduced and the population fecundity is much reduced, particularly in larger fish such as tuna.

Fecundity is a function of weight, and the gonad comprises about one-sixth of the body weight in temperate and high-latitude fishes. The eggs of young fish are small, and egg size tends to be greatest in middle-aged fish; subsequently it declines with age, and the eggs of senescent fish are small. During their adult lives, parents produce large numbers of eggs, of which two must survive to replace them in the unexploited population. As the stock is first reduced a little by exploitation, the number of fish recruiting to the adult stock remains constant, so the reduction in stock

Table 2
Ages at first spawning and length of life in some fishes

Species	Age of first spawning (years)	Length of life (years)
Anchovy	1–1½	2–3[a]
Herring	2–5	10–20
Plaice	4	30
Cod	4–9	20–25
Tuna	4	10–15

[a] The history of the 1971 year class suggests that the anchovy off Peru may live as long as five years (Cushing, 1981)

by fishing is automatically compensated. If the stock if reduced too far, the numbers of recruiting fish are reduced and the stock is further reduced—a vicious circle, known as recruitment overfishing.

As fish grow old, the myofibrils in their white muscles, the attack or escape engines, decrease considerably in diameter. Indeed when fishing starts, old thin fish are sometimes found: "watery plaice" off Newfoundland (Templeman and Andrews, 1956), thin and bony plaice on the Dogger Bank (Garstang, 1900–03) and *"praeste flynders"* (priestly flounders) in the Skagerrak (Petersen, 1894). Greer Walker (1970) rationalized such observations with measurements of the myofibrils of old cod and showed that they were much reduced in diameter.

THE CREATION OF BIOMASS

The increment of weight of an individual during its life history from egg to old age is of the same order as the fecundity, 10^3 or so in salmon, 10^4 in herring and 10^6 or 10^7 in cod. If an individual must replace itself, only two need survive from a lifetime's production of eggs or (two per adult lifespan in years). Hence high mortality is to be expected during the life history, and if biomass is to be created after hatching (and it is needed because of the high death rate of eggs), the increment in weight must be greater than the decrement in numbers.

The increment in weight per unit time (expressed in logarithms) is the specific growth rate. It decreases with age in all animals, but in fishes not so sharply as in mammals. The logarithmic growth curves of a number of fish species (Fig. 12) show that small fish grow quickly to their (theoretical) maximum size, whereas big ones attain that objective relatively more slowly; this principle is expressed in the von Bertalanffy equation (Beverton and Holt, 1957). Figure 12 shows that growth is characteristic of the species and expresses its ecological function—filterer, predator or top predator.

The information in Figure 12 is summarized in Table 3 in which four groups are classified by the range of maximal weight within each group in orders of magnitude, by adult life spans and by a description of their niches. The restricted range of growth is found amongst the small filtering clupeids, the herring-like fishes, and the flatfish, isolated benthic feeders. The scombrids and gadoids display a wide range of growth and are predominantly predators, although the smaller ones are filterers. In both groups the smaller fish do not live as long as the larger ones.

Fish die in the sea mainly by predation. They do suffer diseases and carry parasites, but few of the well-documented stocks in the sea are known to be affected in this way; the few well-known examples quoted by Sindermann (1970) reinforce the point. Perhaps this is because fish tend to live through their life cycles in a volume of water rather than on a surface. We therefore believe that the major mortality stored in the gonads of fishes must be represented by predation. Cushing (1975) represented the mortality of plaice

Table 3
Classification of fish groups by adult life span as function of growth pattern, and ecological character

Group	Range of maximal weight (orders of magnitude) 1	3	Ecological character
Clupeids (herring-like)	5–10 yr		Small filterers
Flatfish	10–30 yr		Isolated benthic feeders
Scombrids (mackerel-like)		1–10 yr	Filterer to top predator
Gadoids (cod-like)		5–25 yr	Predators

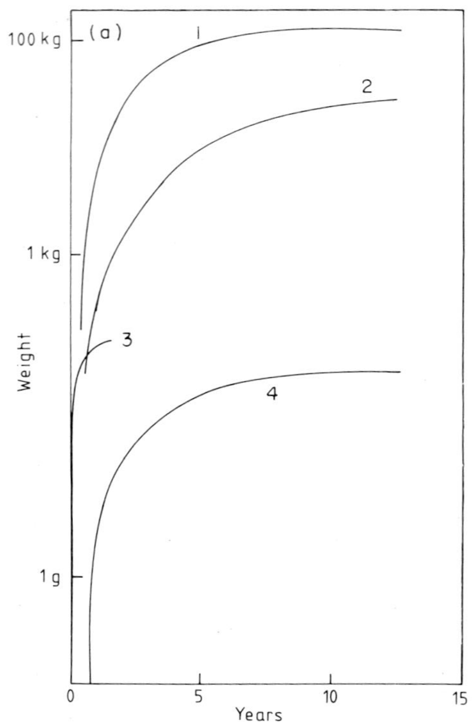

(a)
100 kg
1 kg
1 g
Weight
Years
0
5
10
15
1
2
3
4

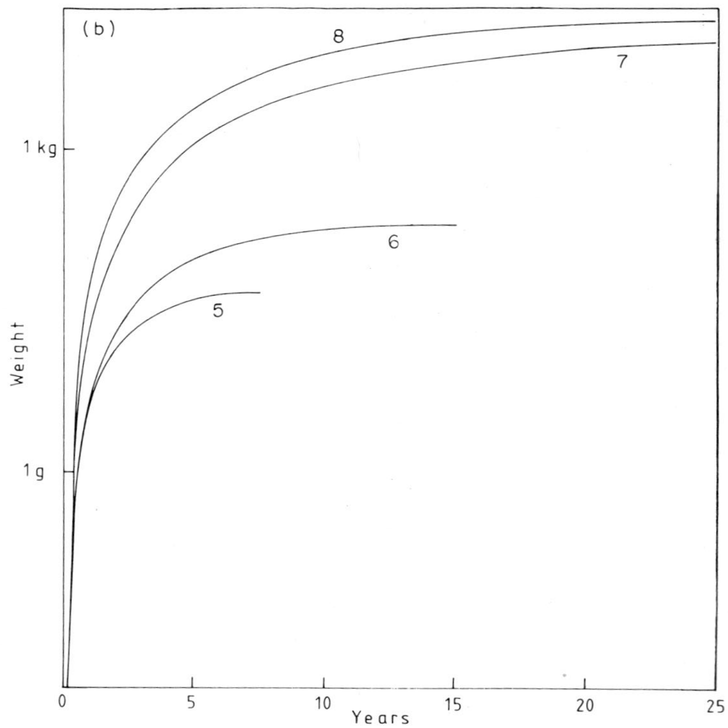

(b)
1 kg
1 g
Weight
Years
0
5
10
15
20
25
5
6
7
8

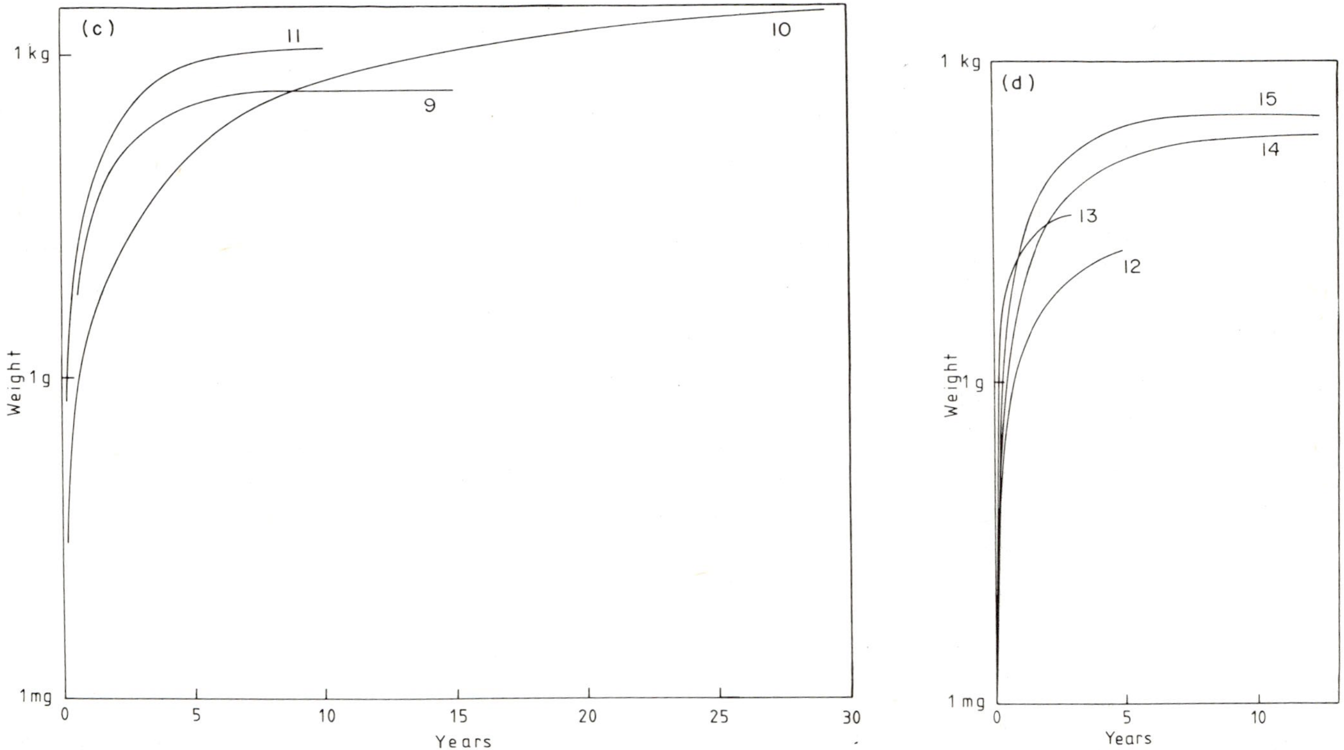

Fig. 12 Logarithmic growth curves of weight on age of a number of fish species, (*a*) (1) bluefin, (2) albacore, (3) *Rastrelliger*, (4) *Pneumatophorus*; (*b*) (5) Norway pout, (6) Georges Bank haddock, (7) Grand Bank cod, (8) Icelandic Saithe; (*c*) (9) sole, (10) plaice, (11) American plaice; (*d*) (12) sprat, (13) anchoveta, (14) herring, (15) Californian sardine.

in a single function. Just as specific growth rate declines with age, so does specific mortality rate. It was assumed that numbers decline as a density-dependent function of age, and the results are shown in Fig. 13; larval mortality rates, nursery ground mortality rates and adult mortality rates, as observed, are matched to the theoretical curve, but the numbers may

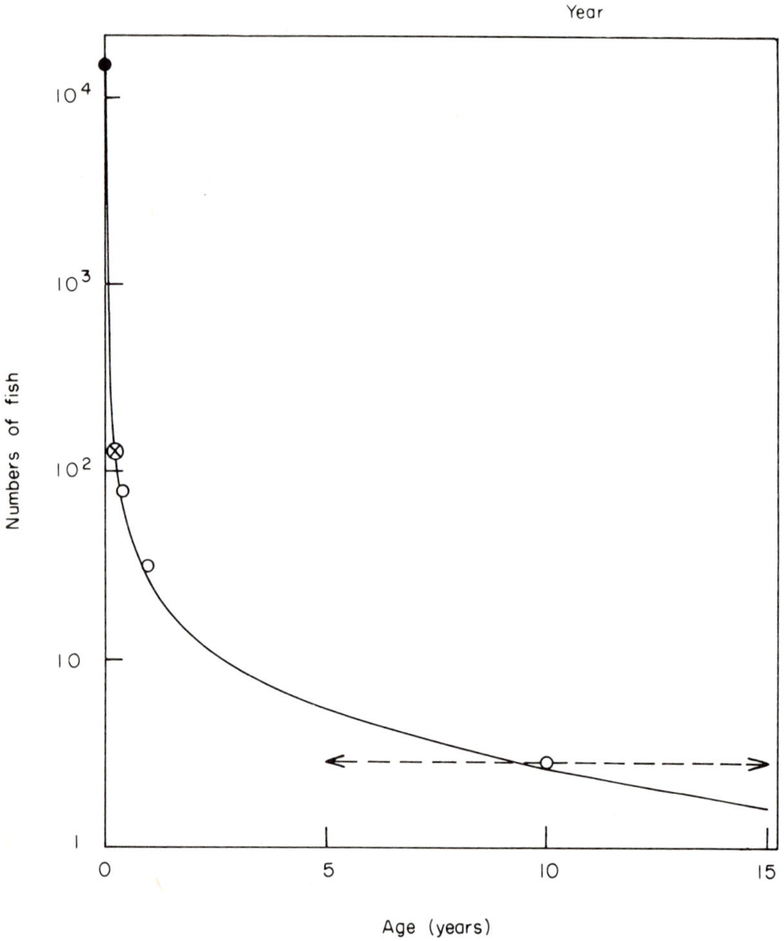

Fig. 13 Natural mortality of plaice represented as a density-dependent function of age. The points represent the observed mortality rates: 5% per day in larvae, 40% per month on arrival on the beaches, 10% per month in the first year, and 10% per year in adult life. The numbers are estimated from initial numbers and two at a critical age of 13 (at which specific growth rates and specific natural mortality rates are equal) (Cushing, 1975).

not be properly represented. If growth and mortality are linked (Ware, 1975), growth may be the governing force, and if growth is density dependent, mortality would be also. If such were the processes in larval and juvenile life, they may well determine the course of mortality during the rest of the life cycle, merely, because the animal grows from one predatory field to the next, and as it grows, the predators are less numerous because they also are larger; hence mortality declines with age.

Figure 14 shows how biomass is created in the plaice, the Lowestoft laboratory animal. Harding *et al.* (1978) have published numbers of Stage I eggs, Stage I larvae and Stage IV larvae for four years of observation at sea. Average numbers of eggs and larvae are derived and are extended to adult life using the curve in Fig. 13, assuming numbers at 100 days are $0{\cdot}0562 \times 10^{12}$ and those at 15 years are $0{\cdot}000071 \times 10^{12}$. Weights of eggs were calculated from their volume, assuming neutral buoyancy; weights

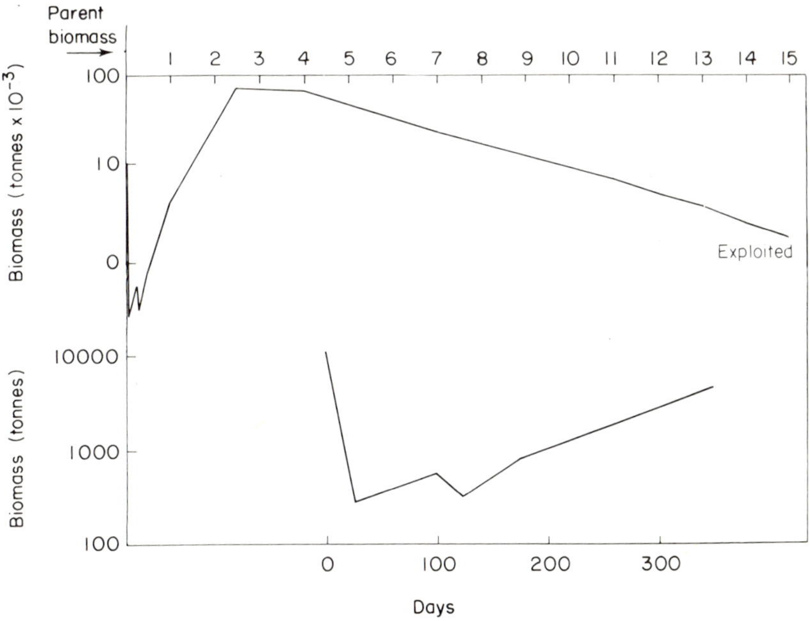

Fig. 14 The creation of biomass; numbers are taken from the curve in Fig. 13, assuming that at 100 days $n = 0.0562 \times 10^{12}$ and at 15 years, $n = 0.00071 \times 10^{12}$. Weights of eggs estimated from their volume, of larvae from Ryland (1966), of 0-groups from P. Walker (pers. comm.) and weights of adults by age from Bannister (1978). The lower figure shows the biomass on an expanded time scale.

of larvae were taken from Ryland (1966) and those of 0-group fish from my colleague Peter Walker (personal communication). Weights of adult plaice were taken from Bannister (1978) (males and females averaged). Figure 14 shows a loss of biomass from Stage I eggs to Stage I larvae of a factor of 40, all of which has to be recovered before the individual matures.

Figure 14 shows the parent biomass in hundreds of thousand tonnes, and the distribution of biomass in age with a fishing mortality of 0.3 exerted from ages 4–15 and a natural mortality of 0.1. A cohort hatched from an unexploited stock would of course have a similar adult distribution but of greater magnitude; because natural mortality may increase to an unknown degree in the unexploited stock, the distribution of unexploited biomass has not been calculated. The distribution of adult biomass under exploitation simulates that in the Southern Bight of the North Sea ' where the eggs were sampled. The sum of biomass from ages 4–15 is 265 000 tonnes as compared with the parent biomass of 183 000 tonnes, which was arranged by setting the numbers at age 15 at the appropriate level. So the cohort throughout its life has roughly balanced the parent biomass in the year of hatching at least well within the range of variability of the recruitment.

The inset in Fig. 14 shows the creation of biomass during the first year. The remarkable point is that recovery starts during the larval stage and continues at about the same rate throughout the year. The specific growth rate reaches a maximum between Stage IV larvae and the arrival on the beaches, after which it starts to decline. Although the larvae take a surprisingly long time to learn to feed, they must do so very effectively in order to start the creation of biomass; or to put it another way, they must do so or metamorphosis would be wasted if fish metamorphose in order to maximize growth with least predation. The greatest biomass is reached at about age 3 when the fish start to mature, which implies that the greatest possible quantity of eggs is produced.

MIGRATION

Figure 15 shows Harden Jones's (1968) "triangle of migration"; its base is the larval drift from spawning ground to nursery ground and the apex is the feeding ground. Feeding ground and nursery ground are quite distinct, and each provides food at a different stage of the life cycle. From the nursery ground, the recruiting year class spreads into deeper water until

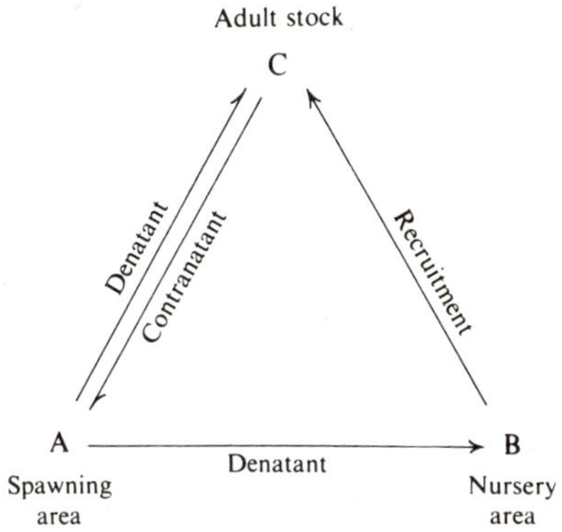

Fig. 15 Harden Jones's triangle of migration. Ripe fish migrate contranatantly to the spawning ground and drift denatantly away from it back to the feeding ground. Larvae drift from spawning ground to nursery ground, which is quite distinct from the feeding ground. Juveniles migrate from the nursery ground and recruit to the adult stock on the feeding ground (Harden Jones, 1968).

it joins the adult stock on the feeding ground. As the adults mature each year, they leave the feeding ground and migrate towards the spawning ground.

In temperate waters, spring spawners spawn on the same grounds each year during a rather extensive spawning season. The Arctonorwegian cod spawn in a narrow strip in 140 m on the northern edge of the Vestfjord in northern Norway (Bostrøm, 1955); the herring in the southern North Sea spawn on small gravel strips about 3600 m in length and 300 m wide which lie along the axes of the tidal streams (Bolster and Bridger, 1958); the haddock in the northern North Sea spawn all over an extensive area of nearly a thousand square miles (Saville, 1959). It is possible that each consists of a group of spawning grounds between which an exchange of spawners can take place from generation to generation; Li (1955) has noticed that such an arrangement would maximize variation in a large population. A broad spawning ground like that of the haddock or a group of small grounds as in the herring both yield large patches of eggs, 50–80 km across, which retain their identity throughout the larval drift; in other words the patches are large enough to remain secure from the diffusive processes in the sea.

The larval drift flows in a regular current, towards the nursery ground (a beach or a shallow bank), and if the spawning ground is fixed in position, so is the nursery ground. Hence the larvae drift along a fixed vector. Spring spawners tend to spawn at the same time each year; indeed the standard deviation of the peak date of spawning for plaice, herring, Pacific salmon and cod is about a week or less (Cushing, 1969). The larvae not only drift between fixed positions in a regular current but they do so at the same season each year in temperate and in boreal waters.

As they drift, the larvae of demersal fishes probably sink in the water column and may drive into highly productive waters. The plaice larvae in the Southern Bight are carried north-easterly in spring into somewhat shallower water where production starts earlier than on the spawning ground (Cushing, 1972). The larvae of the autumn-spawning herring in the central and northern North Sea drift from spawning grounds off the English and Scottish coasts towards the Danish coasts and find themselves in highly productive waters during a pronounced autumn outburst (Cushing, 1967). Indeed the truly autumn spawning herring are found in the north-east Atlantic only where there is a marked autumn outburst. Because the larvae are released into productive waters they grow at the same time as their food organisms, as Jones (1973) has suggested.

Bigger fish live in deeper water, as noted above, so the shallow banks and beaches of the nursery grounds are relatively free of predators. After the initial period of arrival on the nursery ground the growth rate may then be maximal. After six or nine months of successful growth on the nursery ground, the bigger fish of the brood start to move fathom by fathom into deeper water; Riley and Corlett (1965) and Lockwood (1972) have shown that juvenile flatfishes of different species occupy different depth zones each of which increase in depth with age. Later in their life cycle plaice spread offshore up to 160 km from the nursery beaches. The groups of North Sea herring spread away from the coasts in a somewhat similar way but this migration ends in patches of relatively cooler water around the Dogger Bank or on the Fladen ground (Cushing, 1962). The juveniles of the Arctonorwegian cod migrate from the Svalbard shelf or from Skolpen Bank in the Barents Sea into deeper water (Maslov, 1944). Such movements continue until the young maturing fish join the adults on their feeding grounds.

For many years of their lives the adults migrate from feeding ground to spawning ground and back again. They travel for considerable distances, hundreds or even thousands of kilometres, and such journeys may last for two or three months. In temperate waters many spring spawning fish feed little as they migrate to spawn and take the energy they need from fat stored during the feeding season. The nature of fish migration has always been mysterious, but Harden Jones (1968) pointed out that as few have external referents they must usually drift with currents or countercurrents. Harden Jones *et al.* (1979), using the Admiralty Research Laboratory sector scanning sonar, have established that the plaice in the southern North Sea move south in midwater on the southbound tide and remain on the sea bed on the northbound one. They call the mechanism selective tidal transport. After spawning, the reverse procedure was used by the fish in their northern migration. Both southerly and northerly migrations were confirmed by midwater trawling in the region north of the spawning ground (Harden Jones *et al.,* 1979).

A major upwelling area is a biologically self-contained system based on an eastern boundary current drifting towards the equator at the surface and a poleward countercurrent below about 200 m. The equatorward longshore winds cause an offshore drift which drags cool water from below near the coast and this process starts a productive cycle from new. Adults may drift equatorward at the suface to spawn in the most intense part of the upwelling area where production is also intense and after spawning may migrate poleward in the countercurrent. Larvae may sink as they grow, just as they do in temperate waters, until they find themselves in the upwelling water being taken towards the coast where they grow in the richest water.

In upwelling areas the sardine-like fishes apparently eat algae (Longhurst, 1969). Such food can only suffice to maintain the animals, and for growth and reproduction the sardines must use zooplankton (Cushing, 1978). Perhaps the mature fishes eat algae for two reasons (*a*) because spawning may be triggered by algal food, and (*b*) because the algae live in large patches, larvae released in them will grow with their food, the nauplii and copepodites, much as they do in temperate waters. The production cycles in an upwelling area resemble those in temperate waters because both cycles start from very low levels in algal numbers.

Figure 16 (Cushing, 1975) contrasts the dependence of migration on the production cycle in temperate waters and in the tropical seas outside the upwelling areas. Poleward of 40° the production cycle is discontinuous, and in a cyclonic current system the migration circuits are arranged so that the fish release their larvae into the spring outburst; hence the spawning grounds and the nursery grounds are fixed in position and season. In the tropical and subtropical seas the production cycle is continuous and the

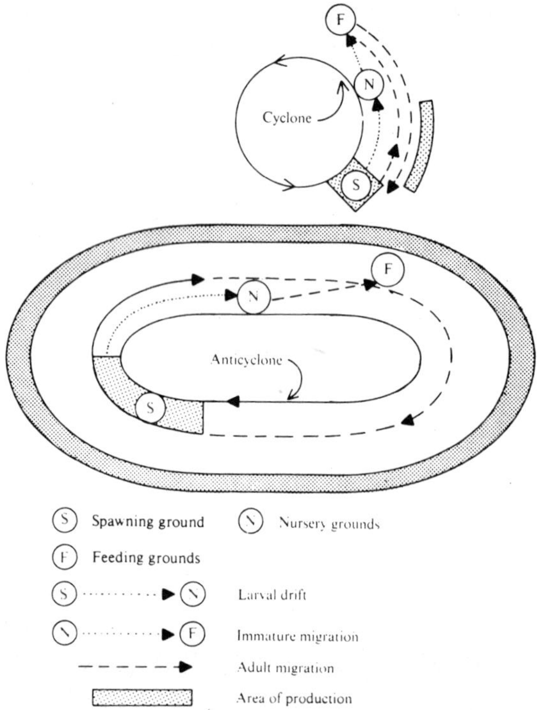

Fig. 16 Dependence of migration on production cycles. In the cyclonic circulation in latitudes > 40°, production is discontinuous, and the adult spawners have to migrate against the current to reach the spawning ground. In the anticyclonic circulation in latitudes < 40°, production is continuous and the spawning ground spreads over thousands of miles, the whole system moving in drift (Cushing, 1975).

whole system is diffuse and much less precise than that in temperate seas; indeed, tuna-like fishes spawn across areas thousands of miles in extent for most of the year.

There are two possible reasons for migration. The first is that larvae are released into the production cycle to eventually grow and avoid predation on the nursery grounds on the beaches and shallow banks. The second is that the adults live in deeper water where the food organisms are also bigger. The food requirements of larval fish are profoundly and obviously different from those of adults, and such differences can only be resolved by migration. The whole phenomenon is an adaptation to the growth pattern of fishes, the increment in mass by several orders of magnitude.

The large adults have to migrate to a spawning ground from which larvae drift to the nursery ground where the juveniles can put on weight rapidly.

THE HYDROGRAPHIC CONTAINMENT OF THE FISH STOCKS

In temperate waters the larvae drift from fixed spawning ground to fixed nursery ground. The larvae leave the regular current in which they drift to find their nursery grounds. In a similar way the adults make use of the same current, or countercurrent, to migrate to the spawning grounds and they employ it to drift downstream to the feeding ground. They join the current on the feeding ground, change their response to it on the spawning ground, and leave it to migrate back to the feeding ground. Thus the stock is contained geographically within the structure defined by larval drift and the spawning migration.

There are groups of spawning grounds between which some interchange is possible in terms of time and distance. Between such groups there may be considerable distances across which the chances of interchange must be very low indeed. As already noted, in large populations (and those of fish often comprise about 10^{10} adults), variance in a genetic sense is maximized, with a number of spawning groups between which there is some degree of interchange. Such a cluster of groups should be isolated in a reproductive sense from any neighbouring cluster which nevertheless lives at a considerable distance. Any stock, or group of stocklets, isolated in this sense from its neighbour is distinct, often in a genetic sense.

Jamieson and his colleagues (e.g. Jamieson and Turner, 1978) with two haemoglobins and seventeen transferrins have shown that the genetic distances between cod stocks in the North Atlantic may be considerable; indeed the chance of mixture between them is as low as 1 in 10^4. Hence the reproductive isolation postulated finds its expression in established genetic differences. Further, such unit stocks are probably contained hydrographically within migration circuits, and although they were established in the distant history of evolutionary processes they are the units with which the fisheries biologists work. Some examples of the dependence of stocks upon the current structure in which they live are given below.

Some of the plaice in the southern North Sea spawn in winter between the Thames and the Rhine. The patches of eggs and larvae drift northeasterly towards relatively clearer shallower and more productive waters (Fig. 17; Cushing, 1972). As the larvae grow, they probably tend to sink in the

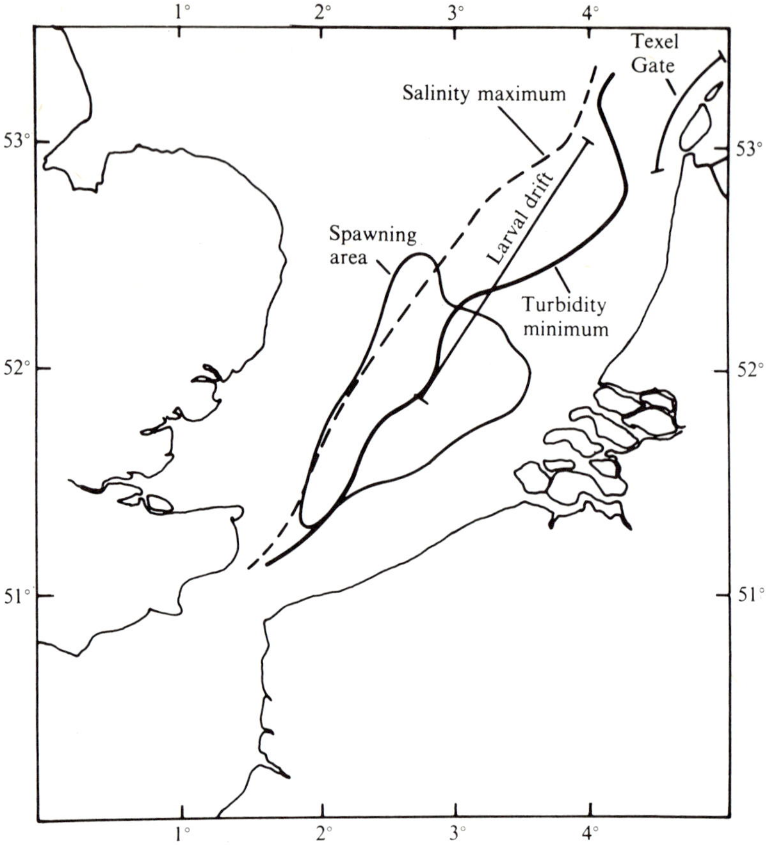

Fig. 17 Larval drift of plaice in the Southern Bight of the North Sea. The spawning ground lies between the Thames and the Rhine. The larvae drift in the most saline and clear water towards the Texel Gate where they migrate inwards towards the Waddensee (Cushing, 1972). The Texel Gate is a boundary between the open sea and the Dutch coastal water through which the little fish can move near the bottom at certain states of the tide.

water column until they reach the sea bed where they start to metamorphose. They are carried through a sharp salinity boundary between the Dutch coastal water and the open sea by a hydrographic mechanism of onshore water movements at certain states of the tide towards the nursery ground on the flats of the Waddensee which lie inside the Friesian Islands (Dietrich, 1954; Zijlstra, 1972). Figure 17 shows the larval drift of the plaice in its regular current from fixed spawning ground to fixed nursery ground. The larvae leave the current merely by sinking to the sea bed as they grow older. Adults alter their migration to the spawning ground to one from it merely by changing the mode of their selective tidal transport. Thus the residual current through the Southern Bight of the North Sea is used in quite different ways by adults and larvae. Not only does the migration pattern serve the differing requirements of larval and of adult fish, but it can be modulated to different purposes within one regular current system. There are two or three groups in the cluster of stocklets with nursery grounds in the estuary of the Rhine, the Waddensee, the Danish coast and on the eastern coast of England, and metamorphosing larvae could be exchanged between them merely in the variability of the wind-driven drift. It is perhaps this character that sustains the variance in a genetic sense between groups in the cluster.

The Pacific halibut (*Hippoglossus stenolepis* Schmidt) live between Oregon and Alaska and are carried through the passages between the Aleutian Islands to the Bering Sea. There are two main winter spawning grounds both in deep water, one off Cape St. James at the southern end of Queen Charlotte Island off British Columbia and the other between Kodiak Island and Yakutat on the southern coast of Alaska. The two egg distributions lie about 1000 km apart, which suggests that the two spawning groups are distinct. The larvae are drifted north-westerly from Cape St. James or westerly from southern Alaska along the course of the Alaska gyral which is a cyclonic circulation, but the juveniles return to nursery grounds in Hecate Strait (between Queen Charlotte Island and the mainland) or on Porlock Bank off Alaska. The metamorphosing animals like any other flatfish must have dropped out of the Alaska gyral to the sea bed and then returned in a compensatory manner to the nursery ground, perhaps in a countercurrent. Figure 18 shows the larval drift and the compensatory movement of the juveniles; this account is taken from Skud (1977). Presumably the adult fish use the currents or the countercurrents of the Alaska gyral to migrate north from Cape Flattery to Cape St. James or east from the Bering Sea to Kodiak Island.

Figure 19 shows where the Arctonorwegian stock of cod lives. They spawn between January and April in 140 m in a transition layer (between cool freshish run off from the glaciers above and the relatively warmer Atlantic water below) near the northern shore of the Vestfjord in northern Norway. The larvae drift out of the fjord into the Atlantic water which streams north into the West Spitsbergen current and the North Cape current, both of which carry the young fish on to relatively shallow banks

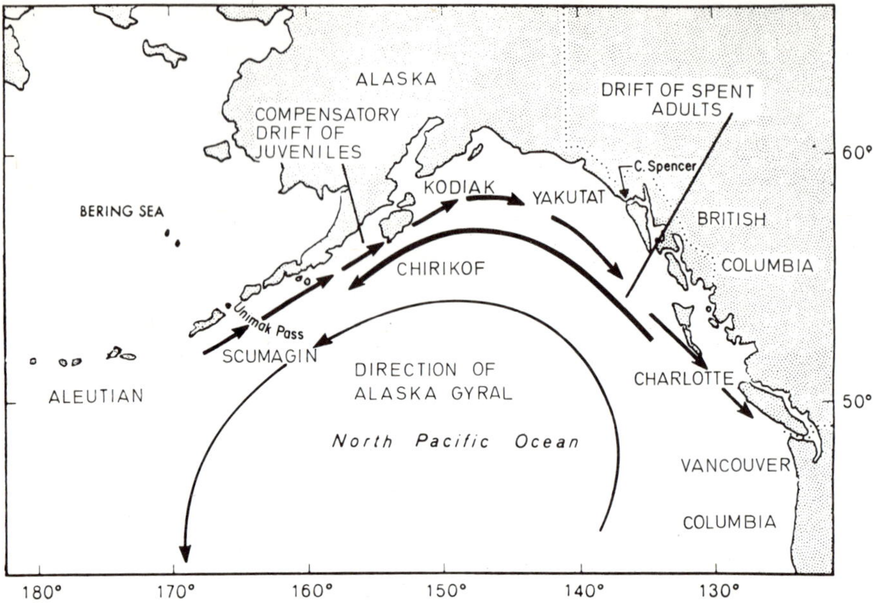

Fig. 18 Dependence of the Pacific halibut upon the Alaska gyral. Eggs are laid in deep water off Cape St James at the southern end of Queen Charlotte Island off British Columbia, and in a more extensive area off southern Alaska. The larvae drift north-westerly and westerly in the Alaska gyral towards the Aleutian Islands. However, the juveniles migrate back in a "compensatory" manner to Porlock Bank off Alaska and to Hecate Strait between Queen Charlotte Island and British Columbia.

(on the Svalbard shelf and on Skolpen bank (Maslov, 1944)). On such nursery grounds the young fish grow until they mature, and the recruiting fish migrate into deeper water where they join the adult stock. The large mature cod migrate from the eastern Barents Sea to the Vestfjord at 9–13 km per day (Maslov, 1944), perhaps in a countercurrent that flows beneath the Atlantic stream (Eggvin, 1964). After spawning, the adults drift back at 2·69–5·90 km per day which is the speed that might be expected from the transport of the North Cape current. The large adults make use of the current structures in the Barents Sea in order to place their larvae in the production cycle off the Lofoten Islands and their juveniles on the rich and shallow banks in the Barents Sea whilst they themselves feed on the shelf edge. But how the fish enter or leave the currents at the right time or place is unknown.

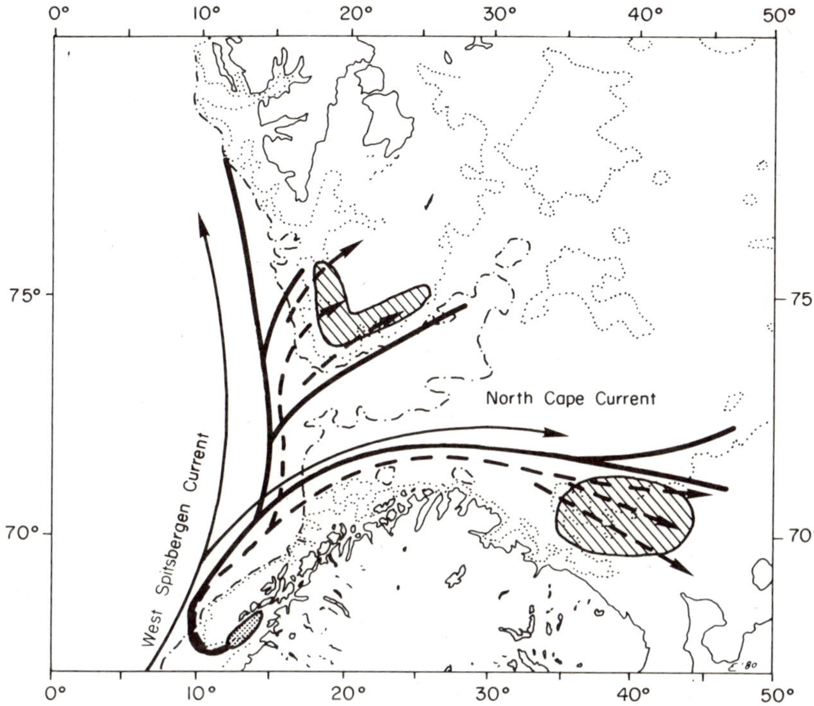

Fig. 19 The current structure in which the Arctonorwegian cod stock lives. The cod spawn in the Vestfjord, and the larvae drift northwards in the West Spitsbergen Current to the Svalbard shelf and in the North Cape Current to the Banks around Skolpen Bank. The adults migrate contranatantly in the same currents back to the spawning ground or perhaps drift in a countercurrent. The young fish recruit to the adult stock by moving off the banks into deeper water (Cushing, 1981). Stippled area = spawning ground; Hatched area = nursery ground; (—) adult migration route; (→) larval drift.

Another gadoid, the Pacific hake (*Merluccius productus* Ayres), lives in the major upwelling area off the west coast of North America. The eggs are laid off southern California, when the upwelling starts early in spring and where it is perhaps the most productive. The adults feed off Oregon, Washington State and British Columbia and presumably migrate south in the Eastern Boundary Current before the upwelling season starts when it is the only current in the region. As the season advances, the upwelling system shifts northwards and with it a deep countercurrent is established

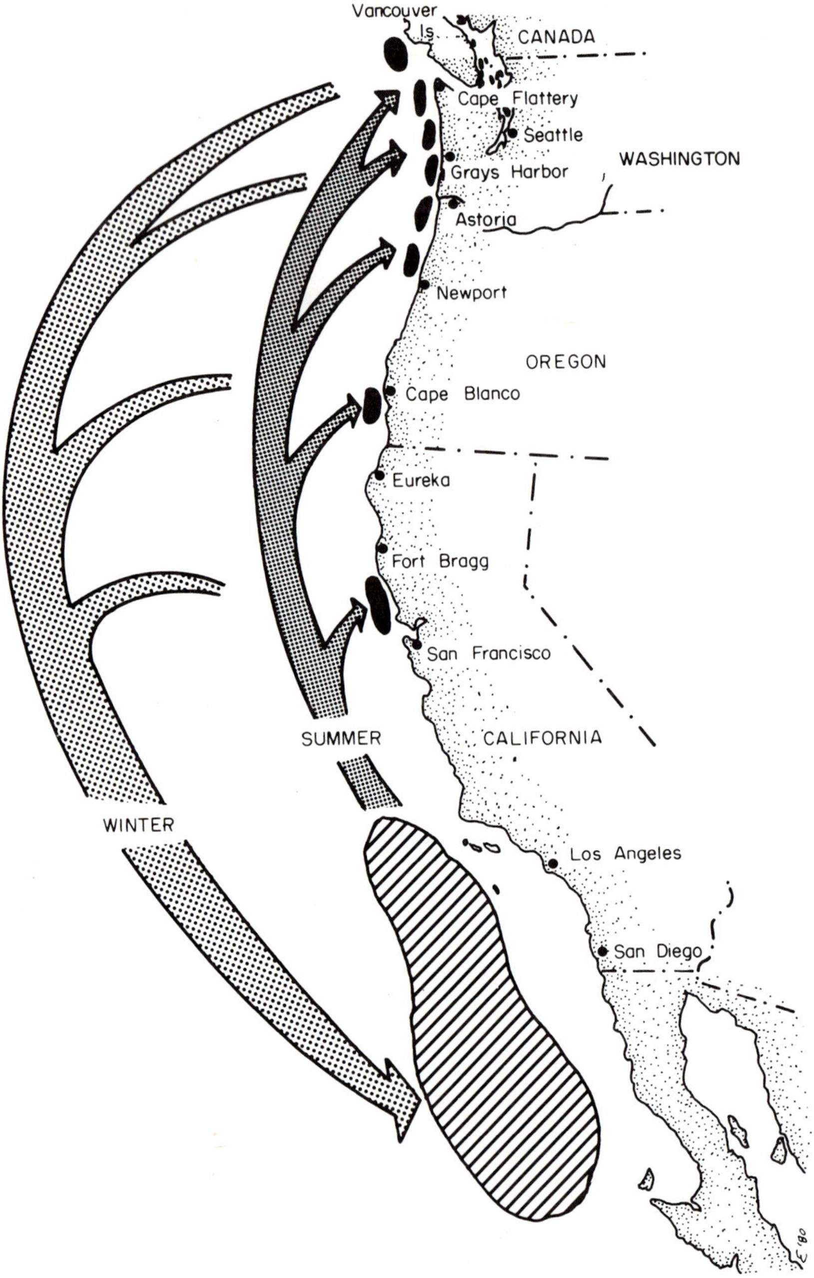

Vancouver Is.
CANADA
Cape Flattery
Seattle
WASHINGTON
Grays Harbor
Astoria
Newport
OREGON
Cape Blanco
Eureka
Fort Bragg
San Francisco
SUMMER
CALIFORNIA
WINTER
Los Angeles
San Diego
E '80

below and it may well slowly carry the spent fish towards Oregon (Fig. 20; Alverson and Larkins, 1966). The position of the nursery ground is unknown, but if the larvae sank into the upwelling water, they would find themselves inshore in the water richest in small organisms. The larvae and juveniles must grow up in the richest part of the upwelling system in the south and the adults maintain themselves near the seabed in the northern part of the region, and they migrate from north to south and back again to put their larvae in the most productive part of the system.

Perhaps the most remarkable and well-documented migration circuit is that of the sockeye salmon. The eggs are laid in the fall on the gravelly redds in small streams in the Rockies and hatch five months later. The larvae drift downstream into lakes, where they metamorphose, and then the smolts migrate downstream in the major rivers like the Fraser in British Columbia. After months, the juvenile sockeyes put to sea and spread across the region occupied by the Alaska gyral and some even migrate into the western Pacific. They are caught in numbers off the Aleutian Islands (Foerster, 1968), and Fig. 21 shows how they might be considered to drift round the gyral (Larkin, 1977). As the fish start to mature, they would have to find ways to leave the gyral in order to edge towards the shore. If they return to the Fraser River they pass through the Strait of San Juan de Fuca and the Strait of Georgia. The fish move into that river in a specific order, first those that have the furthest to go and they swim faster (Killick, 1955). The Fraser network of rivers is very extensive, and as the fish return to their native redds there are many choices for the fish for they meet a succession of tributaries. From the work of Hasler and his associates (Cooper *et al.*, 1975), the salmon must make a succession of chemical choices by smelling the water from each stream and by finally recognizing the parent stream where they spawn and die, if not eaten first by grizzly bears. The larvae depend upon the production in the lakes, and the adults depend upon food close to the surface far out in the Pacific. Strict routines of

Fig. 20 Migration circuit of the Pacific hake. They feed off the Pacific North-West and spawn off southern California and hence must migrate south in the Eastern Boundary Current because it is the only one there before the upwelling season starts. Presumably the spent fish migrate back north in the upwelling countercurrent as the upwelling areas shift poleward in spring and summer. The larvae may find their way inshore in the upwelling streams themselves where they would feed well (Alverson and Larkins, 1966). Hatched area = spawning hake (winter); solid area = feeding hake (spring, summer, fall).

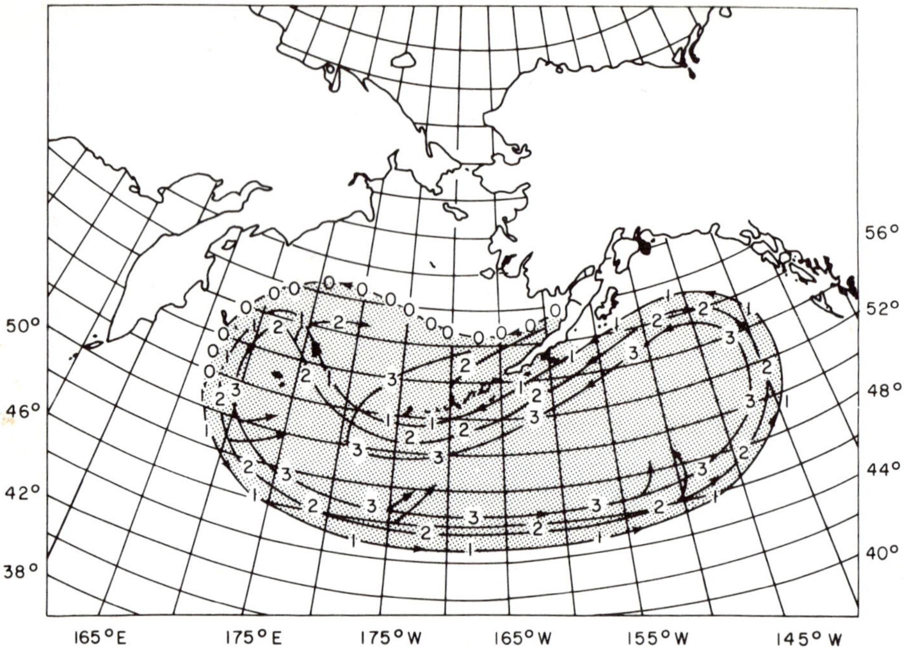

Fig. 21 Migration of the sockeye salmon in the Alaskan gyral. The alevins drift from their native redds into lakes, and smolts migrate to the sea where they drift for months and years at the surface of the Alaska gyral. The maturing adults leave the gyral and find their way to the mouth of their parent river, such as the Fraser, with the use of an unknown mechanism. They smell their way to the river and up it to the parent stream, rejecting many choices until they find the right one (Larkin, 1977). 0 = 0.0 ocean age; 1 = 0.1 ocean age etc. (→) = final herring routes of material fish. Stippled area = general area of distribution.

migration are needed to match two opposing needs, those of the larvae and those of the adults.

The North Sea herring spawn on very restricted grounds on gravel in three groups, Downs (in the Southern North Sea and eastern Channel), Dogger (around the bank) and Buchan (off the Ness). There may be half a dozen little grounds in each group, the products of which are merged together in one large patch as they drift between spawning ground and nursery ground. In the eastern English Channel and in the Southern Bight of the North Sea, the Downs herring spawn in November and December and their larvae are drifted north-easterly towards the productive and shallow waters off the coast of Holland (Cushing, 1967). Off the north-east coast of England and around the western Dogger Bank, the Dogger stock spawns in September and October and the larvae drift eastwards into a productive and late autumn outburst in the eastern North Sea (Cushing, 1967). Off the north-east coast of Scotland, herring of the Buchan stock spawn in August and September and the larvae drift to the same productive area off the Danish coasts. The herring metamorphose on the beaches of north Germany and Denmark five to six months after hatching (Fig. 22).

About a year after hatching the little fish move offshore beyond the 20-m line. Those of the Buchan group, which arrive first, leave first and those of the Downs group last. All spread into deeper water and in the second or third summers of their lives they live in cold water patches east and north-west of the Dogger or on the Fladen ground. When they reach a critical length (for example, 21 cm in the Downs group) they leave the cold water patches and join the adult stocks on their feeding grounds in the northern and central North Sea. The adults migrate south to spawn on the tiny grounds at the same time each year. After spawning, the Buchan and Dogger fish probably drift across the North Sea in an easterly direction to overwinter on the western edge of the Norwegian deep water. The adult migration routes are shown in Fig. 23a and in b are displayed in the tidal streams as drawn by Harden Jones *et al.* (1978); they connected the extreme particle vectors for each tidal stream and if the fishes used selective tidal transport they would follow the tracks shown on the chart. The two figures suggest how the Downs stock may be separated from the other two merely in the structure of the tidal streams.

There are probably two stocks of Pacific albacore, one in the northern subtropical anticyclone and one in the southern. The spawning fish are caught in the North Equatorial current in June and July, and most eggs and larvae are found between 180°W and the Philippines for nearly nine months of the year (Matsumoto, 1966). In the South Equatorial Current, the spawning season extends from October to March, six months out of phase with that in the north. The fishery in each current shifts westerly as if the fish were drifting in that direction (Suda, 1963; Suda *et al.*, 1963). A fishery in the Kuroshio and Kuroshio extension in 40°N exploits immature albacores as if the fish had grown as they drifted from the North Equatorial Current spawning ground into the Kuroshio. In March and May 1954, nuclear bombs were exploded in Bikini Island in the North Equatorial Current in 165°E. Six months later, contaminated fish were being discarded in the Kuroshio extension as far at 176°W, and fish caught off eastern

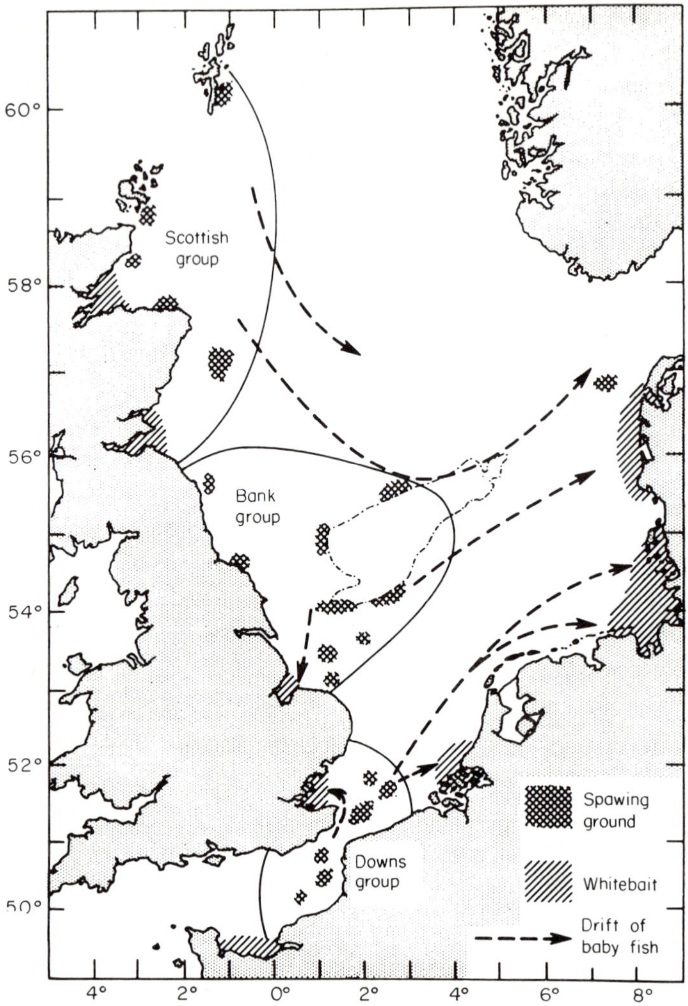

Fig. 22 Larval drifts of three stocks of herring in the North Sea. Each of the three groups of herring, Buchan, Dogger and Downs, spawns on a number of very restricted spawning grounds from which the larvae merge in large patches that drift towards the nursery grounds on the Dutch, Danish and German coasts (Cushing, 1981).

Australia were probably contaminated in the South Equatorial Current (Nakamura, 1969). More detailed models of tuna migration in the Pacific have been constructed (Otsu and Uchida, 1963; Rothschild and Uchida 1968) but they all depend upon drift in the anticyclonic circulation. Tagged albacore have crossed the Pacific in a year or so (Otsu, 1960), but because spawning "ground" and nursery "ground" range across half an ocean for nine months of the year, the migratory circuit cannot be established quite so clearly as in the seas of higher latitudes.

Stocks within tidally dominated systems (as in the southern North Sea) make use of selective tidal transport to make spawning migrations, and the larvae are drifted from spawning ground to nursery ground in a regular

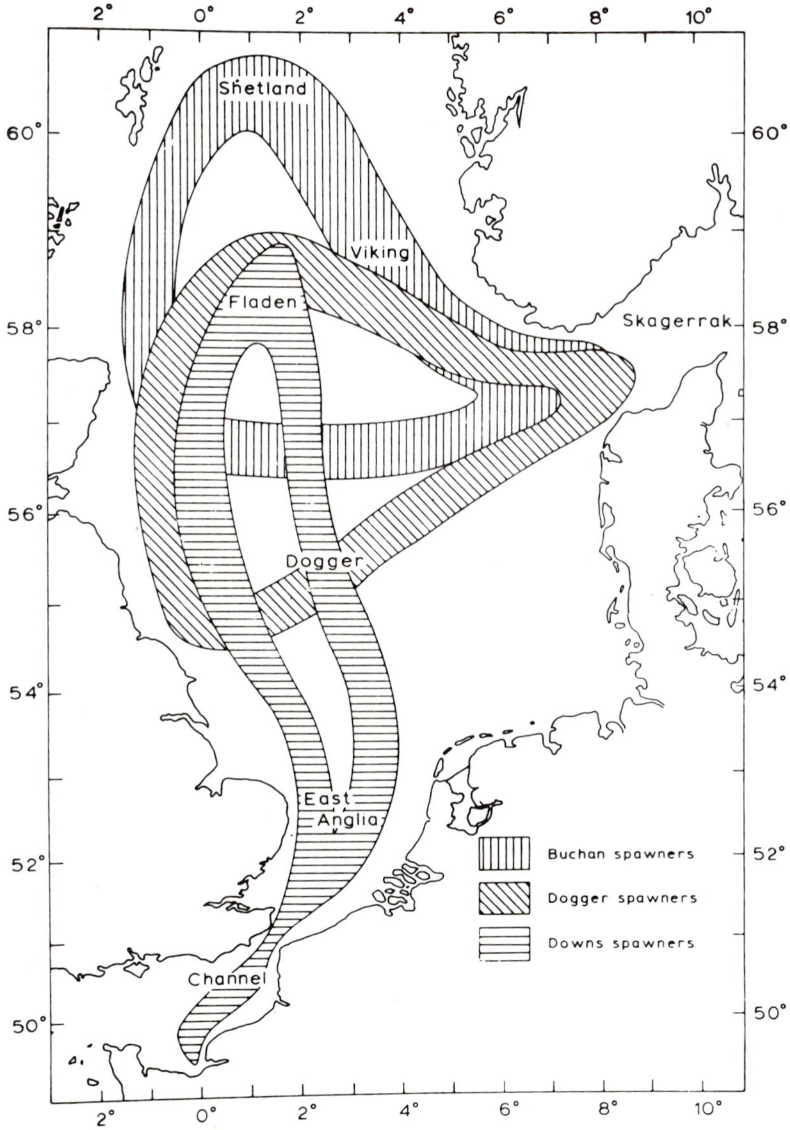

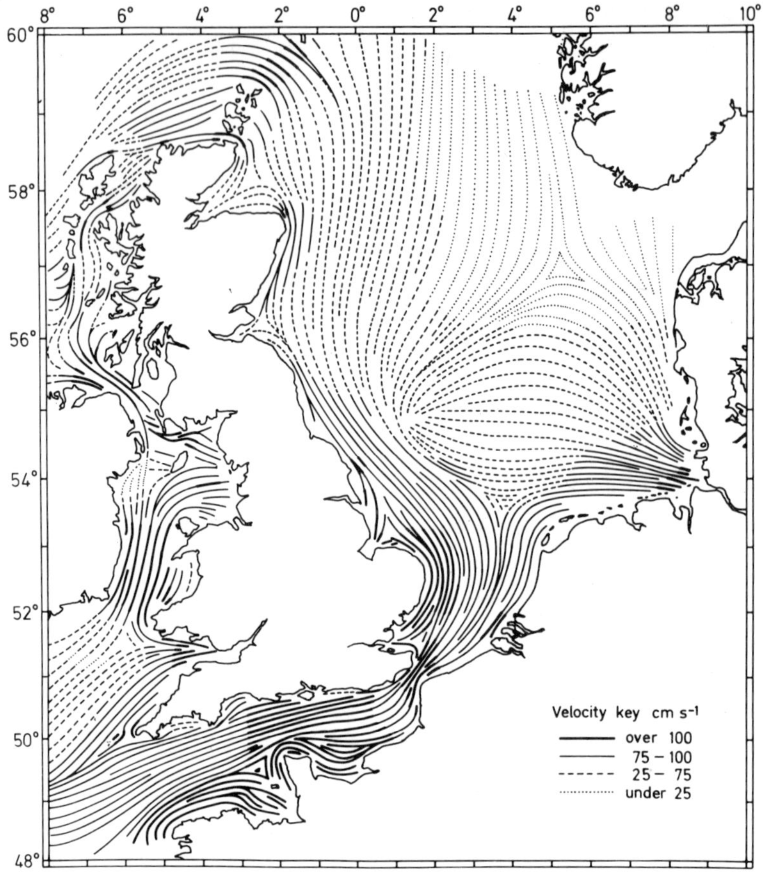

Fig. 23 (*a*) Migrations of the North Sea herring. The Buchan and Bank fish drift away from their spawning grounds off the Scottish and English coasts towards their overwintering ground off the Norwegian Deep Water. In spring they move away to the north and west to the feeding ground and later back to the spawning grounds. The Downs fish drift northerly into the Southern Bight, then north-westerly towards the western Dogger and the northern North Sea where they feed; later they return to the Southern Bight and the eastern English Channel to spawn (Cushing, 1981). (*b*) Tidal streams in the North Sea as a distribution of linked particle vectors (Harden Jones *et al.*, 1978).

residual current. The Arctonorwegian cod employ the West Spitsbergen Current and the North Cape Current to migrate from spawning ground to feeding ground and they may travel in a countercurrent on their journey back to the spawning ground and the larvae drift to nursery grounds in both main currents. Whether the animals use tidal streams or ocean currents, they enter or leave them at particular times and places.

But the most important fact common to all the stocks described is that the single stock is contained within a defined current or tidal system by unspecified behaviour patterns and so the stock remains in the same geographical region. The larval drifts of different stocks of the same species are far enough apart to prevent mixture; for example, there is a second plaice stock in the North Sea in the Moray Firth off the north-east coast of Scotland, and if larvae from the north were drifted south they would sink to the sea bed and metamorphose long before they reached the southern North Sea. Because the two stocks lie far from each other, they are reproductively isolated, a result of the long processes of evolution.

Fish migrate in order to resolve the differing needs of larvae and adults in sizes of food particles; larvae need small and numerous particles for a relatively short time, but the adults need large ones which are not very abundant. Thus Heincke's Law (bigger fish live in deeper water) expresses the progress of individuals through the trophic levels as they grow. The spawning migration must take the adults back to the parent stream or the parent ground even if it is as diffuse as that of the albacore. All animals grow to an enormous degree from the fertilized egg, but fishes are the largest animals that release their eggs into the sea for fertilization. They migrate across great distances in order to grow most quickly in the diverse supermarkets of the sea.

THE STABILITY OF FISH POPULATIONS

The numbers of the fish populations are large and many comprise as many as 10^{10} adult individuals, (and up to 10^{17} eggs), and many stocks are well sampled on the quays where the fishermen land their catches. Fish live for one or two decades and the stock of adults is replenished each year by the annual recruitment; indeed the number of recruits each year to the adult stock replaces the annual deaths. Recruitment is highly variable from year to year, which is not unexpected as its function may be to sample the carrying capacity of the environment. Recruitment varies by a factor of 3,

10 or 20, but of course the variation in stock is reduced by the square root of the number of age groups. In time series the variation in recruitment may trend upwards or downwards or remain steady for long periods. There is a distinction between stocks which resist environmental change and remain steady in time, and those which trend up or down in time and are unable to resist such changes. The former are resilient and the latter are not.

The resilient stocks are the fecund ones, such as cod; those that respond to environmental change, like herring, are much less fecund. Hence the source of resilience lies in the density of eggs and larvae in the sea (Cushing, 1971b). However, the argument is not quite so simple. Darwin wrote that because reproduction is potentially geometric and any increase in food can only be arithmetical, most of the juveniles must die in order that their siblings should survive. Howard and Fiske (1911) pointed out that the control of populations must reside either in the fecundity of the adults or in the survival of the juveniles.

The simplest form of regulation for a fish population is in the density-dependent growth of the adults merely because they may grow in weight by an order of magnitude during their lives after maturation. Alm (1959) showed that the growth of freshwater fish in tanks was potentially density dependent and Le Cren (1965) has demonstrated that the fecundity of perch (*Perca fluviatilis* Linnaeus) in Lake Windermere in north-west England is in fact density dependent. In the sea, however, density-dependent growth has been difficult to detect except in the Gulf of St Lawrence (Lett and Kohler, 1975); but recently my colleague R.G. Houghton (pers. comm.) has shown that for a reduction in stock of three to five times, the mean weight of the North Sea sole (*Solea solea* L.) population has increased by 65%. Similarly in the North Sea cod he found that for a reduction in stock by a factor of three, the mean weight increased by 35%. The degree of density dependence is not very great, and it has emerged as the time series of observations has lengthened.

Figure 24a and b shows two forms of stock and recruitment curve, with weak and strong density dependence. Each curve represents recruitment in numbers (and deaths) plotted on parent stock (as biomass or as number of eggs). Where the recruitment and mortality curves intersect, the population stabilizes; as mortality increases with fishing, the curve of deaths rises towards the recruitment curve and the stabilization point shifts to the left as in the more usual Ricker curve.

However, in the north-east Atlantic many fish stocks are exploited between two and five times natural mortality with no sign of collapse; they included cod, haddock and saithe in the Barents Sea and cod, haddock, whiting, saithe and plaice in the North Sea. The North Sea sole at a ratio of six times natural mortality may be in danger, and North Sea herring stocks collapsed after two decades of exploitation at a high ratio (Anon., 1978).

If total mortality is three times greater than natural mortality, either fecundity or survival must change by a factor of three or more under the influence of density-dependent processes. Hence the changes observed in density-dependent fecundity are not enough to account for the factor needed. Survival can only alter by such a factor either if the fraction surviving the period of density dependence is already less than one-third or the annual instantaneous mortality coefficient must be greater than one. The only period in the lift history when changes in survival occur is the earliest one, the egg and larval stages (Shepherd and Cushing, 1980).

It has been known for some time that recruitment can be predicted at a relatively early age, at the end of the first year for North Sea haddock. However, Rauck and Zijlstra (1978) have shown remarkable correlations between 0-group plaice in the German shrimp fishery and subsequent recruitment. The German shrimp fishery takes place in 15–20 m outside the Friesian Islands (Meyer-Waarden and Tiews, 1965) but the nursery ground lies inside them. Hence the year classes must have been established before the little fish arrived on the nursery ground. Consequently, in plaice the essential processes are completed during the larval stages, including metamorphosis and the period just after.

One of the remarkable points about the studies of egg and larval mortality which have been carried out in the Southern Bight of the North Sea is that egg and larval mortalities in any one cohort are much alike but differ considerably between cohorts (Harding *et al.*, 1978). Hence the predatory mortality may well be common to eggs and larvae. Sheperd and Cushing (1980) suggest that if there is not enough food, growth is density dependent and mortality reduces numbers until there is enough food and then the animals grow at maximal rate. During this critical period it is the cumulative mortality (not the instantaneous coefficient) that is density dependent. In plaice, recruitment may well be determined by the time the little fish arrive on the beaches and if the essential processes are expressed in growth, the critical period must lie between the time of first feeding

(*a*)

(*b*)

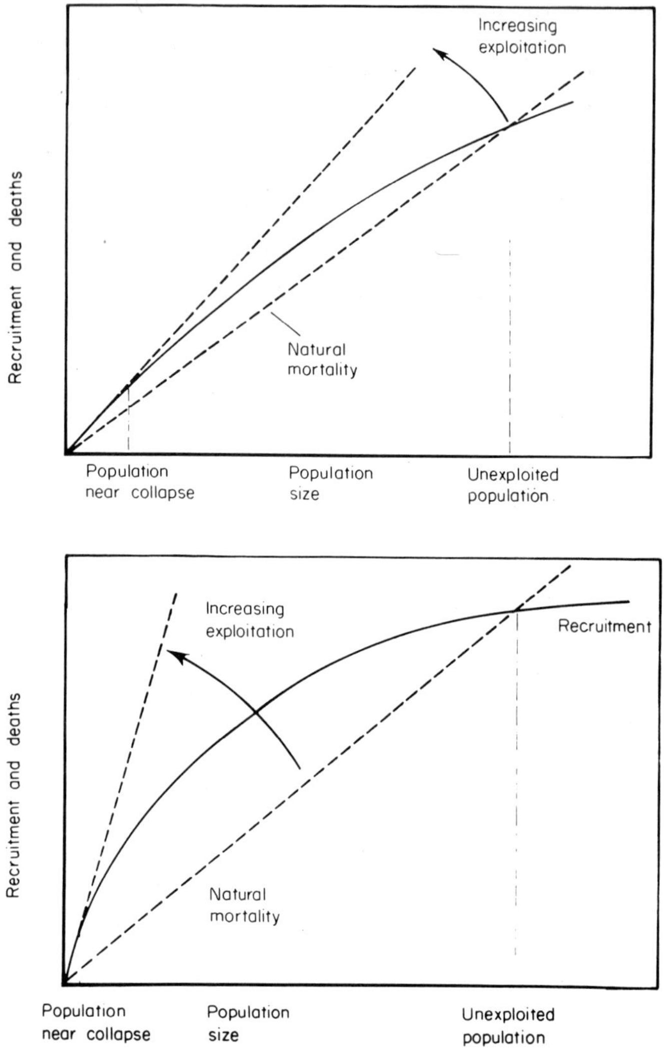

Fig. 24 Stock recruitment curves and mortality curves; the population stabilizes where the recruitment and mortality curves intersect. Figure 24a shows a curve of weak density dependence where $F/M = 1.0$. Figure 24b shows a strong density dependence where $F/M = 3.0$ or more (Shepherd and Cushing, 1980). Figures 24c and d show the relationships with K and A constant.

(c)

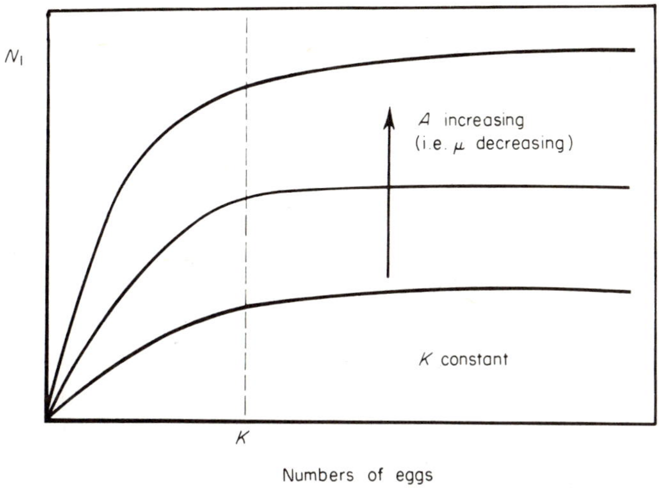

(d)

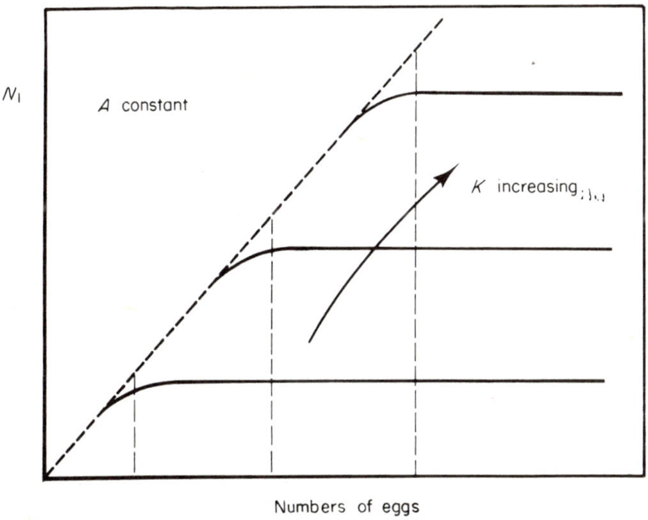

and the time of metamorphosis or just after. Then the predatory field is defined as the size distribution of prey vulnerable to predation during the critical period.

Shepherd and Cushing (1980) have developed a stock–recruitment curve based on the density dependence of growth during a critical period:

$$N_1 = \frac{AN_0}{1 + (1 - A)N_0/k}$$

where

N_0 is the initial number (number of eggs or number of first feeding larvae)
N_1 is the number of recruits at the end of the critical period
A is the fractional survival due to density-independent causes

$$= \exp -\{(\mu/G_{max})(\ln W_1/W_0)\}$$

where

μ is the instantaneous coefficient of predatory mortality,
G_{max} is the maximal growth rate, with no density dependence
W_0 and W_1 are initial and final weights during the critical period
k is the carrying capacity or quantity of food available.

The equation is an asymptotic one in N_1, recruitment. The important point, however, is that recruitment is generated as a function of stock, food and predators—that is, in terms of the trophic levels above and below the stock that generates the recruitment. If the hypothesis put forward above is true, it would not be surprising that the growth of fishes in their early life history should play a prominent part in the stabilization of their populations.

DISCUSSION

Fish live for up to two or three decades, grow throughout their lives, and lay large numbers of eggs. The large populations that support commercial fisheries comprise perhaps a 10^{10} individuals and biomasses of up to 10 000 000 tonnes. Each generation creates biomass in very large quantities and individuals grow from the egg by six or even seven orders of magnitude. Larger animals, such as elephants and whales, grow even more, but they protect their young. Fishes create their biomass within the avid trophic levels of the sea, through which they grow.

Biomass is lost rapidly during the egg stages. But the creation of biomass is started during larval life and continues steadily until the maximum is reached at the start of the mature life of the cohort at least when it is exploited; the age of maximal biomass may be nearer middle age in the unexploited stock. There appears to be a break in the development of biomass at metamorphosis, when feeding is reduced.

The study of migration and of the containment of stocks within the current structure are inextricably linked. Fish appear to migrate to accommodate the conflicting trophic needs of larvae and of adults. The mature fish migrate across considerable distances to release their larvae into production cycles where they can feed and then to nursery grounds where they can grow rapidly. From an array of examples of migration circuits, fish can join or leave currents at particular times and places, with the result that the circuit is completed by a single stock which is isolated in a reproductive sense and is genetically distinct from its nearest neighbour. Briefly, fish migrate across great distances in order to grow quickly most efficiently.

It is possible that the plaice population in the southern North Sea is stabilized during the larval drift and just after, if our hypothesis is right. The process is based on the capacity of the cohort to grow in the face of the predatory mortality, indeed to create the maximal biomass later in the life of the cohort. Reproductive isolation is obtained in a geographical sense by the containment of stocks within a migration circuit. Fish migrate in order to reconcile the very different food requirements of adults, larvae and juveniles. Cohorts are stabilized in numbers very early in their life history, perhaps by the capacity to grow in the face of larval and juvenile mortality. Fish grow rapidly: numbers are probably stabilized by the growth of little fishes which remains density dependent until there is enough food for the survivors; adult fish migrate from feeding ground to spawning ground in order to provide the little fish with the best chances of finding food; because the scale of the migration circuit is considerable, much biomass is created by growth as the cohort ages. Thus the central property in the lives of fishes is their capacity to grow, and by this the large populations are sustained, which is perhaps the most fundamental reason why fish migrate within the current structures and retain their stock identity.

Chapter Three

Changes in the Fisheries during Historical Times

Historians and fishermen have long known that fisheries appear and disappear in time. Such events are sporadic and were often attributed to changes in migration; there are three groups of phenomenon. (1) Some fisheries have appeared and disappeared, and towards the end of last century the possibility of long-term periodicity was discussed, particularly in Scandinavia. (2) In recent years, climatic changes have become well documented both in the "instrumental" period of history and in earlier periods, but the latter evidence is indirect. A dramatic event was the widespread death of tilefish (Collins, 1884) off the eastern seaboard of the United States in 1882, which was attributed to an influx of abnormally cold water from Labrador. (3) Correlations were attempted between catches of particular species and environmental factors, and many have subsequently failed as more information became available. All three groups of phenomena listed above can be interpreted as consequences of climatic change and this chapter provides the historical basis to the thesis that fisheries are affected in this way.

SPORADIC APPEARANCES AND DISAPPEARANCES OF FISHERIES

Some fishes appear regularly at the same place at the same time each year. Such are the sockeye salmon which enter the Fraser River in British Columbia on their spawning migrations and the Arctonorwegian cod as they enter the Vestfjord in Northern Norway in January to March (see Chapter 2). This regular appearance each year is part of the migratory circuit and forms the basis of seasonal fisheries. Some fisheries, however, appear for long periods of time and disappear for equally long periods, and their caprice is familiar to the fishermen, particularly herring fishermen. They attribute such changes to variation in currents or in winds or to differences in weather or in climate, valuable as witnesses of considerable changes. This chapter will describe the variability in historical times which led fishermen to think the way they do. The existence of fisheries may often be determined by economic factors. These have been omitted as far as possible—for example, the rise of major herring fisheries and of trawl fisheries in the North Sea during the nineteenth century was linked to the development of railways and large industrial towns, and the stocks may well then have become more heavily exploited.

Patterson (1905) noted that "It is a curious fact that, for a number of years following Buckland's enquiries (1875), the local fishery [for herring off Yarmouth in eastern England] was hardly worth pursuing, the mackerel [*Scomber scombrus* L.] changing their immigration until the time of the autumnal herring-fishing, when on some occasions they became so abundant that several Yarmouth and Lowestoft boats changed their herring nets for mackerel nets" which have larger meshes. Mackerel were usually caught in very small numbers in the long drift nets used in the East Anglian herring fishery, so the events referred to must have represented a real change in migration. An analogous event on a much larger scale is the appearance of mackerel in very large numbers (representing a local biomass of perhaps 2 000 000 tonnes) in the inshore waters off Cornwall in winter since about 1966 (Lockwood, 1980). We know from echo surveys in that area in the early 1960s that mackerel did not live there before that date in vast numbers (Bridger, 1965). This phenomenon might be part of the complex of changes in the ecosystem in the western English Channel, known as the Russell cycle (see Chapter 5).

Goode (1887) recorded that halibut (*Hippoglossus hippoglossus* L.) were extremely abundant on Georges Bank (off New England) from 1830 to 1850 and that since 1850 they had "partially disappeared from this region . . . into deeper water". Goode and Collins (1887) noted that the winter haddock fishery on Georges Bank dated back to about 1850, when immense quantities of haddock were caught in trawls in Massachusetts Bay. Further, Goode (1887) wrote that "the great schools (of halibut) that were found in the western part of the Grand Bank in February and March 1876 and 1877 appear to be migrating". The earlier disappearance of the halibut from Georges Bank was perhaps a result of fishing, unlike the later discoveries of haddock and halibut on well-known grounds, which were real consequences of changes in migration.

Perhaps the most remarkable invasion was that of haddock into the Baltic in the early 1920s recorded by Johansen (1926). He noticed that sporadic landings had occurred there over a decade or so but that much larger catches were taken in the winter of 1925–6. Earlier, in April to June 1923, haddock larvae had been recorded from all over the western Baltic, the Belts, the Kattegat, the Skagerrak and eastern North Sea. They probably entered the general region of the Baltic in high salinity water during one of Dickson's periods of high salinification (see Chapter 9). In the autumn of 1925 small quantities of this 1923 year class were caught in the Belts, and during the following winter a fishery developed. Such a migration happens rarely—or, more precisely, it has only been recorded on one occasion.

Pilchards (*Sardina pilchardus* Walbaum) were abundant off Great Yarmouth in 1780, 1790 and 1799 (Storrow, 1947), and they penetrated the North Sea during the early 1950s (Aurich, 1953). Cushing (1957) refers to fluctuations in catches off the south coast of Cornwall and off southern Ireland; exports to Italy were greatest in 1831–5, 1845–51, 1871–3 and 1905–6. The fishery was practised by very large seines worked from the beach. They were shot when the shoals were sighted from the cliffs, and the catches must have depended upon the migration of shoals into the bays, among other factors such as the fluctuations in true abundance. From time to time the pilchards have changed their local movements and perhaps even their large scale migrations.

The stocks of herring provide most examples of appearance and disappearance, because the fish has been desired and accessible since the time of the Hanseatic cities and even earlier. It is possible that salted herring was a convenient package of preserved food for travellers and soldiers and for the besieged. The histories of herring fisheries describe the laws to control them and give accounts of their variability. Day (1884) wrote: "Herrings . . . will occasionally disappear from one locality, sometimes reappearing in another. From 1690 to 1709, a very extensive herring fishery occurred at Cromarty; about the latter year an immense shoal was driven ashore near the town; they left the vicinity in a single night and for upwards of half a century no shoals reappeared." Storrow (1947) reports that "in 1780 and 1781, large shoals appeared, accompanied by seven whales of up to sixty feet in length". Such were dramatic events which everybody witnessed.

A number of authorities repeat some reasons for the failure of a herring fishery, such as lowering a copper horse into the Oslofjord in Norway. Howell (1921) lists the following:

The battle of Copenhagen in 1801 drove herring from the Baltic;
Kelp-making drove herring from the Long Island (the Outer Hebrides);
Whaling drove herring from the Shetland Islands;
The church bell at St Monance in Fifeshire in Scotland drove the herring away;
Herring left part of the Irish coast in 1835 when a protestant rector was going to take a tithe;
Trawling ruins the herring trade.

Common to all is an inexplicable disappearance like the failure of the Loch Fyne herring in 1871–3 and at the beginning of the twentieth century. From our present knowledge of those stocks recognizable from the positions referred to, it is unlikely that they were reduced by fishing.

In the Limfjord on the western coast of Denmark a fishery for herring existed twice a year for many centuries (Mitchell, 1864), until the sea broke in and the Limfjord became an estuary; the question then arises why the herring could no longer enter the estuary. A version of Lady Nairne's verses reads:

Wha will buy my caller herring
They're no brought here without brave daring
Some call them lives of men
Wha will buy my caller herring
Fresh drawn frae the Forth.

(Caller means fresh.)

Before 1793 between the Isle of May and the coast of Fife (on the Firth of Forth), spent herring were caught in February and March, but after 1793 the fishery occurred from November to March (Mitchell, 1864). "For centuries the town of Dunbar has been a fishing place of little importance . . . in 1665 . . . yearly about this time there is a great confluence of people to the herring fishing; they told us sometimes to the number of twenty thousand persons". Spring spawning fish were caught in the Firth of Forth during the twentieth century until the late 1940s, when the stock disappeared. Thus if the earlier fisheries exploited Spring spawners they have fluctuated in the Firth of Forth considerably in the last two hundred years.

Storrow (1947) noted that the fishery for herring in the Medway (a river in Southern England) had failed in 1860; today a stock of late autumn spawning herring spawns off Whitstable and in former times may have supplied the Medway fishery. He also noted that in the first decade of the nineteenth century, the early autumn spawning herring fishery at Skinningrove (in Yorkshire in northern England) disappeared. Since the late 1940s a fishery has existed on this now well-known spawning ground. In the 1930s, the Plymouth herring fishery collapsed, as the recruiting year classes from 1926 onwards declined steadily and the fishery ceased in 1937–8 when only older fish were caught (Cushing, 1961). The collapse of the Plymouth fishery due to natural recruitment failure was important scientifically because the natural failure was fully recorded. Since that time many herring fisheries in the North East Atlantic have collapsed under the pressure of recruitment overfishing. Some of the disappearances listed above were due to such recruitment failures, particularly when known spawning grounds appear in the records.

EVENTS ON THE EASTERN SEABOARD OF THE UNITED STATES

The tilefish, *Lopholatilus chameleonticeps* Goode and Bean, is a large fish related to the croakers, native to the Gulf of Mexico, the Caribbean and other subtropical seas. It was first discovered off Nantucket in July 1879 (Goode and Bean, 1895) and very large numbers were found dead at the surface off the eastern seaboard of the United States in 1882. Captain William H. Kirby of the *William V. Hutchings,* a Cape Ann schooner, out of Gloucester, Massachusetts caught 1100 kg of tile fish on codfish trawl lines in 144m in 40°04′N, 70°23′W, which is 130 km S by E of Noman's Land. Captain Dempsey caught tilefish in 150 m in 40°02′N, 70°07′W, 225 km SSE from Beaver Tail. The fish were feeding on the amphipod *Themisto bispinosus,* Boeck. The area of capture became known as *Lopholatilus* bank on the slope inside the Gulf Stream, 4–7 km wide and 105 km in length. In the summer of 1880 the US Fish Commission's vessel *Fish Hawk* found tilefish to be abundant over the whole area between 120 and 230 m. Professor Verrill, aboard the vessel, noticed that there were many species of southern invertebrates on the sea bed (Collins, 1884).

The dead fish were found in March and April 1882 between the Chesapeake and Nantucket; the first record was on March 10. The bark *Sidon* saw many thousands in quite rough weather; this suggests that numbers were underestimated, for it is difficult to see objects on the surface in rough weather. The brig *Rachel Coney* reported dead fish on the same course, but also recorded dead specimens of *Peristedion miniatum* Goode and Bean (a deep sea robin). The schooner *Navarino* claimed that some cod and hake were dead at the surface. The bark *Plymouth* crossed 111 km of dead fish; their gills were red and their blood and eyes were bright, but they were "poke-blown". This means that the swim bladder had expanded greatly and had blown the guts out of the mouth, because they had risen from a fair depth, say 160 m, which represents a pressure change of 16 atmospheres. There was no food in the guts. Figure 25 shows the distribution of dead tilefish constructed from contemporary reports from ships' masters. The area was extensive, about five thousand square miles and if the density were one twentieth of that reported by Captain Lamb, there may have been one thousand million dead tilefish (Collins, 1884). A large population died quickly and rose to the surface with some other species.

In August 1882, the *Fish Hawk* found no tilefish on the seabed and the southern invertebrates had disappeared. No sign of any disease or parasite was found. An extensive seismic disturbance was suggested, but there was no other evidence of such an event. Collins (1884) and Goode and Bean (1895) suggested that the Labrador Current may have intensified, because there was more ice off Newfoundland and it had extended far to the south west in the year 1882.

The tilefish returned in ones and two in the fall of 1892 (Goode and Bean 1895). In the spring of 1915 US Fish Commission vessels caught tilefish in abundance where Captain Kirby had first caught them in 1879. Captain Kirby had also found dead cod floating at the surface on August 1 1876 off the coast of Labrador. A similar record is noted by Cooper Abbs (quoted in Howell, 1921): in 1789 in 68–69°N in the Barents Sea, the surface was covered with large haddock and coalfish, all dead or dying. There are many other records of dead fish at the surface of the sea, particularly from the Indian Ocean during the present century. However, the records cited here are those that had considerable influence in the earliest periods of fisheries science.

The scientists concerned with the tilefish phenomenon—Goode, Bean, and Verrill—believed that the fish were killed on the sea bed by cold water from the north. Although 1882 was a cold year in a cold decade, it remains difficult to understand how they died: the work of Harden Jones and Scholes (1974) suggests that cod adapt themselves to water colder than that in which they usually live. Furthermore, the eyes and blood of the dead fish were bright, which suggests that the fish were freshly dead when they reached the surface. We know that the rates of gas resorption and secretion differ by a factor of three; if the rates diverge with temperature, abnormally cold water could generate high secretion and low resorption. If the fish

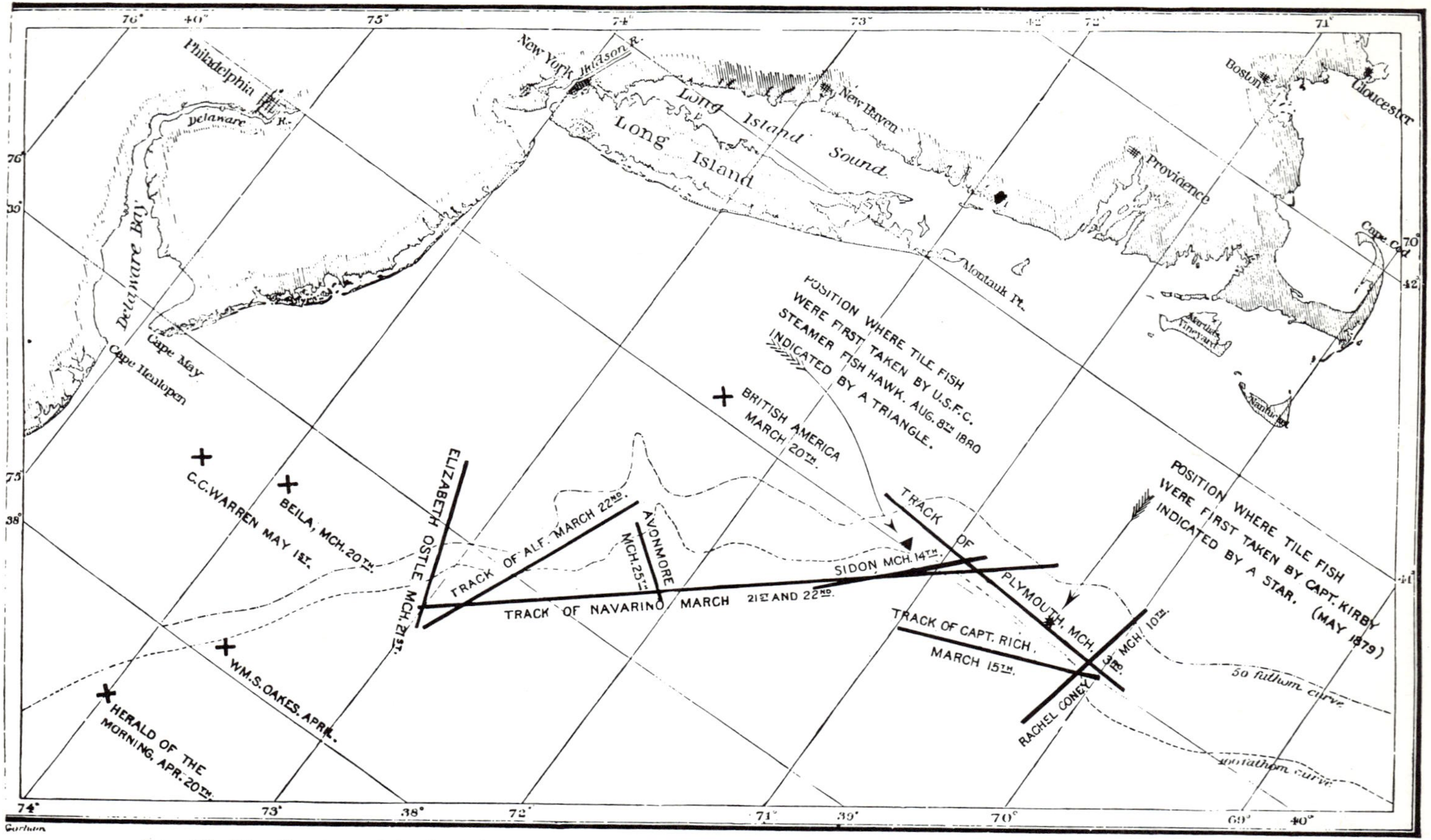

Fig. 25 Distribution of dead tilefish constructed from contemporary reports from ships' masters (Collins, 1884).

rises involuntarily, it may reach a depth at which the swim bladder expands too much and then it rises to the surface and kills the fish.

The region between Cape Hatteras and the Grand Bank is an extensive one where animals of profoundly different origin can mix. In his treatise on the Offshore Waters of the Gulf of Maine, Bigelow (1922) showed that the planktonic animals in the Gulf may originate from the south in the Gulf Stream or from the north in the Labrador Current. The appearance of the tilefish from the south over a broad area, together with an array of invertebrates indicated an abnormal northerly migration, probably from the Gulf of Mexico in the Gulf Stream. Then the death of tilefish, if in fact due to cold water, may have occurred as the Labrador Current pressed and mixed in a southerly direction. A regular fishery for tilefish started in 1915, but the annual catches were very variable, from a few thousand to 5 million kg. Since 1970 fishermen from Barnegat Light NJ have caught tilefish on long lines of up to 64 km in length, and they may take as much as 5000 kg in four or five days. An interesting habit of the tilefish is that they live in burrows on the sea bed, sometimes as long as 1 m (Grunes *et al.*, 1980).

Goode (1884) refers to the bluefish and weakfish (or squeteague) periods off New England. Roelofs (1951) classified the two forms of period in the following way:

BLUEFISH (*Pomatomus saltatres* Linnaeus) (range: Brazil to Nova Scotia)	WEAKFISH (*Cynoscion regalis* Bloch and Schneider) (range: Florida to Long Island)
1672 off New England	
1659–1763 off Nantucket	
1764–1810 absent off New England	Late eighteenth century plentiful in Massachusetts Bay, but disappeared in 1810
1850–65 abundant off Gloucester	
	1867 reappeared in southern Massachusetts
	1870 abundant
	1884 North of Cape Cod
1889 scarce	
	1910–1014 very abundant
	1917 disappeared

The bluefish is a southern species, but the weakfish is resident on the eastern seaboard. Between 1672 and 1810 the southern species predominated and perhaps between 1850–65; the resident species predominated in the late eighteenth century and between 1867 and 1917. Like the bluefish the tilefish came from the south perhaps from the Gulf of Mexico or elsewhere in the Caribbean. Very roughly, the last period of the weakfish was one during which the tilefish was absent. The figures given above are not in themselves evidence of periodicity but they indicate how people started to believe in the temporal variability of fisheries. The eastern seaboard of the United States is flooded normally by the Gulf Stream, but from time to time the cooling influence of the Labrador Current makes itself felt in the region between the Grand Bank and the Chesapeake. The presence of plankton animals and the fishes at different positions on the long coast indicate differences between the two régimes.

LONG-TERM CHANGES IN THE FISHERIES

The great historical changes in the herring fisheries

There are two event of considerable importance in the history of herring fisheries: the Scanian fishery in the western Baltic between the twelfth and fifteenth centuries and the alternation between the fishery for Atlanto-scandian spring herring off the coast of Norway and that for the over-wintering autumn spawning herring off the Bohuslån coast of Sweden which is on the shore of the Skagerrak and Kattegat. The alternation has been recorded since the Middle Ages, and there are indications in the Icelandic Sagas that the periodicity (such as it is) was recognized from earlier times (Devold, 1963); but the Sagas are less historical record than literature.

Scania is the southernmost province of Sweden, and the Scanian fishery provided herring for the cities of the Hanseatic League; the armorial bearings of Lübeck displayed three herrings on a plain gold shield (Samuel, 1918). The Hanseatic League of the early Middle Ages was a commercial network which exchanged furs, salt, fish, flax, timber and tar from northern cities between Bruges and Novgorod for wool, leather and cloth from Venice. The Hansards came to Scania, Pomerania and the island of Rügen (both on the German Baltic coast) for salt from the salt springs in each of the three localities. The herring caught in each were gutted, roused in salt, barrelled and branded (Jenkins 1927), a process used aboard the Dutch and German luggers until the collapse of the North Sea herring fisheries in the late 1960s and early 1970s. As noted above, the use of salted herrings for travellers and soldiers (and others in beleaguered towns) may have been of fundamental importance to the development of the Hanseatic League and its dependence upon Scanian fishery.

The fishery was based on the now small towns of Skanor (*ultima Scaniae ora*) and Falsterbø; undecked boats were used with sails and oars, each of which could carry as much as twelve lasts (about twelve tonnes). The fishery started in late August off Elbogen (the present Malmø) and between the end of September and mid-November the fishermen worked off Falsterbø. No consistent records of catches survive, but 34 000 barrels were cured in 1368 and 50 000 in 1494 (Jenkins, 1927). There were good yields in the second half of the thirteenth century, but in the first thirty years of the fourteenth century the quantity of fish in the Sound was so great that on occasion the rudder "could only be turned with difficulty and fish were taken by hand." Both ports flourished right through the fourteenth and fifteenth centuries; in 1368 no fewer than 160 ships sailed to Scania from Lübeck and about 34 000 tonnes were taken by the whole fleet. In 1463, 20 000 men worked in the two ports, and in 1494, when 501 boats were based on them, 50 000 tonnes were taken between 24 August and 11 November. In 1527 the sea in the Sound was filled with herring from bottom to surface, so they were not only caught with trawls but with wooden shovels at the surface. In 1559 herring were first caught off Bergen in quantity and subsequently at Marlstrand, Bohuslån and Vigen (on the Bohuslån coast of Sweden) and the boats left Scania and Falsterbø by the end of the century. (This account is taken from Schäfer (1887).) An agreement between Lübeck (of the Hanseatic League) and Denmark (which controlled the province of Scania) collapsed in 1569, and the fisheries came to an end (Lundberg, 1886). We do not know whether the Scanian fishery had faded away at this time or whether the source of herring had been taken over by the rising Dutch fishery in the North Sea towards the end of the sixteenth century, but in any case the Hanseatic League collapsed. Mitchell (1864) recorded that in 1080 the fishery had extended from the Sound to Falsterbø, that in 1124 herring were abundant off Pomerania, and that in 1259 the Prussians started to fish for herring in the south-eastern Baltic. Such records and the sporadic records of catches suggest a large fishery on a stock which extended across the western and southern Baltic for perhaps three centuries—if not continuously. Such fisheries were contemporaneous with the Scanian fishery.

The Swedish oceanographer, Otto Pettersson, became interested in the alternation of the Norwegian and Swedish fisheries and later in the period of the Scanian fishery. He pointed out that during the time of this fishery between about 1250 and 1560–1600, cold winters predominated in the Baltic. They started in about 1200, culminated in the fourteenth century and between the thirteenth and fifteenth centuries, the western Baltic, the Sound and the Kattegat were frozen in certain winters. Pedestrians and carriages could cross on the ice. Particularly severe ice winters occurred in 1306–7, 1322–3, 1394, 1407, 1418, 1423 and 1636, after which there was none for 250 years (Pettersson, 1915). The occurrence of cold winters in north-west Europe and in the western Baltic are examined in more detail in Chapter 8, but we cannot substantiate the statement that there was none for 250 years after 1636. However, the records summarized in Speereschneider (1915) and Easton (1928) confirm the severe winters cited above. The Scanian fishery started towards the end of the climatic optimum and ended just before the little Ice Age started (see Chapter 4).

The alternation between the Norwegian herring periods and those of the Bohuslån fishery has been recognized for a long time (Strøm, 1972). Boeck (1871) was one of the first scientists to study the Norwegian spring spawning herring, and from a historical study he believed the fishery to be periodic; a brief summary is given in Watt (1883). The periods of the Bohuslån fishery are given in Ljungman (1882), Pettersson (1921–2), Johansen (1924) and Storrow (1947), but the material in the latter three papers are all derived from Ljungman. He noticed that the great Swedish fishery in the second half of the eighteenth century lasted for 55 years. Then he constructed a series from 1867 backwards in periods of 110 years to the year 915. There were good fisheries during the Swedish periods but "no good fisheries" in Swedish waters during the Norwegian periods. The periodicity is imposed, but the good Swedish fisheries often occurred at some point within the Ljungman period. Hence it is worthwhile examining the material in the detail which was originally presented.

Table 4 gives Ljungman's material in full, together with the dates of the Norwegian fisheries taken from Boeck. Both authors quote historical sources in Sweden and in Norway, but the earlier Swedish records must be less reliable.

Tomfohrde (1914; Table 5) gives the number of vessels southbound from the Bohuslån coast, from customs records between 1556 and 1589. Peter Clausson Friis said that the fishery started in 1556; in 1592 there were no herring, although 4000 lasts were salted in Schonen in 1594.

Usually the dated periods of fisheries have been accepted, although Höglund (1978) suggested that the Swedish fishery between 1660 and 1680 was not typical because the fishermen worked outside the skerries. The dates in the last column of Table 4 are of course not Ljungman's but have

Table 4
Alternation of Swedish and Norwegian fisheries according to Ljungman and Boeck

Ljungman's periods of Bohuslån fisheries	Ljungman's comments	Boeck's dates of Norwegian fisheries
915–970	—	
971–1026 } 1027–1082 }	Good fisheries at least during the reign of Olav the Saint	
1083–1138 } 1139–1194 }	Commercial growth of Konungahalla; probably good fisheries	
1195–1250	Colonization of islands; convents. Probably good fisheries	
1251–1306	No good fisheries at beginnings and middle but probably good towards the end	
1307–1362	Particularly good fisheries at least during the last thirty years of the century	
1363–1418	No good fisheries	
1419–1474	Good fisheries at least during mid-century	
1475–1530	No good fisheries	1500–1568 low abundance
1531–1586	Particularly good fisheries 1556–1587	
1587–1642	No good fisheries	1600–1648
1643–1698	Good fisheries at least between 1660 and 1680	
1699–1754	No especially good fisheries until 1747–81	1700–1784 variable
1755–1810	Rich fisheries 1748–1808	
1811–1866	No good fishery	1818–1870
1867–1922[a]	Rich fishery began in 1877 (and continued to 1906 or 1922)	1896 (or 1910)–1967

[a] Recorded since Ljungman's publication

Table 5
Number of vessels southbound from the Bohuslån coast, from customs records between 1556 and 1589 (Tomfohrde, 1914)[a]

	From Marstrand	All ships
1560	1	42
2	9	53
3	36	67
4	72	104
5	20	35
6	36	58
7	44	75
8	31	107
9	20	113
1574	137	268
5	96	251[b]
6	148	280
7	182	309
8	184	332
9	110	267
1580	152	409
1	86	395
2	78	354
3	51	473
4	34	360[c]
5	118	553
6	106	373
7	67	264
8	81	288[d]
9	34	207
1590	21	163
1	17	148
2	19	195
3	12	188
4	9	195
5	8	169

[a] Scottish, Dutch and North-West German vessels are excluded
[b] Only 598 lasts were brought through the Sound
[c] 50 000 lasts at Marstrand
[d] "*Minderung des Zugangs an Heringengespürt*"

been entered by me; the last Norwegian fishery ended in 1967 probably due to heavy fishing but near the expected end of a period.

The best evidence for periodicity is given in Fig. 26 (Devold, 1963) which shows actual catches from both fisheries since the mid eighteenth century. The alternation of fisheries between 1760 and 1885 is quite clear, but the change from the Swedish period to the Norwegian one between 1890 and 1920 is obscure although it eventually happened. The fact that the fisheries were contemporaneous need not disturb us when we recall that the Swedish fishery was less productive after 1900. Figure 26 shows that Norwegian catches persisted at a low level throughout the most recent Swedish period perhaps because all fisheries were more active with industrialization. Then with the strong and famous 1904 year class the last Atlantoscandian period started, fuelling the fishery from about 1910 on-

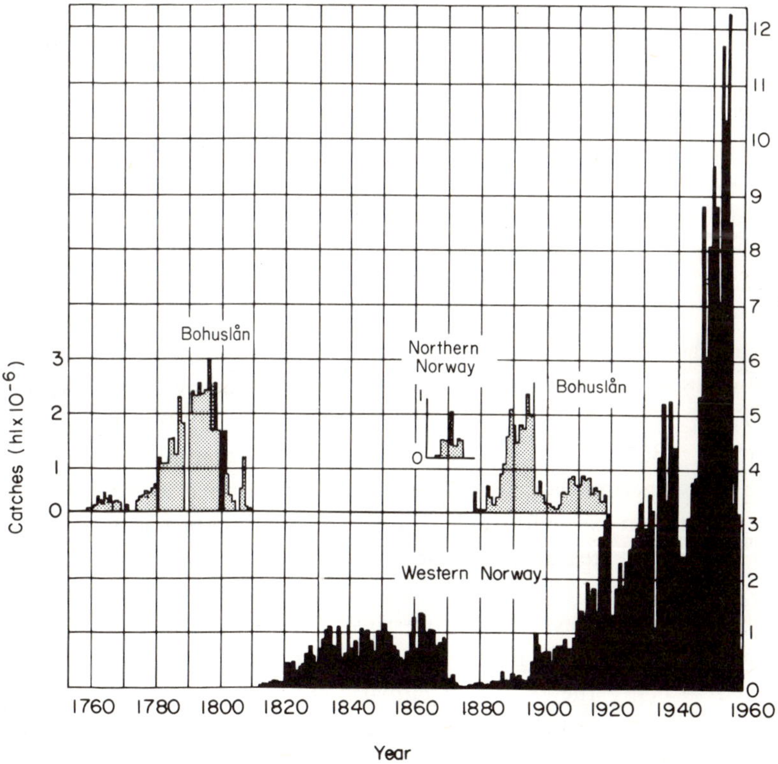

Fig. 26 Catches of winter herring from the Swedish Bohuslån fishery and from the Norwegian fishery from 1760 to 1960 (Devold, 1963).

wards. The fishery was extinguished in 1967 after the adventurous purse seiners discovered the stock in the East Icelandic Current and fished it heavily in the harsh autumnal seas.

From his historical studies, Boeck (1871) suggested that as the Norwegian period approached its end, the fisheries tended to move northwards, first from Utsira to Stadt and then from Stadt to Lofoten. Finally "large herring" appeared off Northern Norway, as Devold shows (Fig. 26), between 1868 and 1874. Devold also showed that in the recent Norwegian period the northward shift took place from Utsira in the early 1950s, and from Stadt in the early 1960s as if there had been a shift from stocklet to stocklet first distinguished by Runnstrøm (1936). There has been some controversy between Devold (1963) and Höglund (1972) on the nature of the alternation. Devold believed that a single stock was exploited by both fisheries which changed its biological characteristics between the end of Norwegian period and the beginning of the next Swedish one and vice versa. Höglund believed that the Swedish fishery exploited North Sea Bank herring; he took vertebrae from the oil pits in Swedish middens on the skerries of the Bohuslån coast and from their estimated age and length concluded that the fish originated from the Bank stock of North Sea autumn spawners.

Petterson (1926) had related differences in internal waves in the Gullmarfjord (on the Bohuslån coast) roughly to the herring catches made during the last herring period. He believed that the Swedish periods were generated by invasions of high salinity water bearing Bank herring into the Skagerrak. He observed that between 1875 and 1897 there were ten ice winters, some so severe that sea traffic was stopped for one or two months. Similarly between 1748 and 1808 there were two groups of excessively cold winters (1780, 1781, 1782, 1784, 1785, 1786 and then 1790, 1800, 1802, 1803, 1804, 1805, 1809). The two decades include the most productive years in this Swedish period (Fig. 26). The association with ice winters depends on the presumption that with high Atlantic influx the depth of Baltic water above it is less and freezes more readily.

The validity of the link between the herring fisheries and the ice winters (and their putative causes) will be investigated more thoroughly in Chapter 8. We are more interested here in the fact that the historical events were interpreted as being the results of climatic change. The previous section gave a short account of the apparent capriciousness of certain fisheries which led fishermen to believe that migration routes were changeable. The work of Boeck and Pettersson—too briefly summarized above—led them to believe that the migration patterns of the herring were profoundly changed through secular periods. As will be shown below, we do not disagree that migration patterns might be affected by climatic changes, but also believe that the greatest changes on a secular scale are due to changes in recruitment, or year class strength, of some orders of magnitude. The 1904 year class of the Atlantoscandian herring stock which started the last Norwegian period was very strong and it increased that stock by perhaps two or three orders of magnitude.

Figure 26 shows the catches recorded in the Norwegian spring herring fishery and in the Bohuslån herring fishery. The stock of the Norwegian herring spawns in spring on three major grounds off Utsira, Stadt and the Lofoten Islands (Runnstrøm, 1936) and lives in the open Norwegian Sea; the adults feed in summer off the polar front near Jan Mayen and migrate in the East Icelandic Current towards the Faroe Islands and thence beneath the Atlantic stream to reach the spawning grounds on the Norwegian coast (Devold, 1963). The Bohuslån herring were probably spawned on the Scottish east coast. The fisheries that alternate as shown in Fig. 26 are based on two stocks which are quite distinct.

Two Swedish periods are shown in Fig. 26, 1760–1808 and 1877–1896; catches continued after 1896 but outside the skerries. The two Norwegian periods extended from 1811 to 1873 and from 1896 to 1967. The stock during the last period was strengthened very greatly by the famous 1904 year class, which remained strong until after the First World War; the high catches towards the end of the period were generated by the strong 1950 year class, probably the strongest during the recent period of warming (see Chapter 5). The end of the Norwegian period of high catches in 1967 was probably premature; the fishery was extinguished by recruitment overfishing partly by a heavy fishery for immature fish in the fjords and partly by an equally heavy fishery for adult fish in the East Icelandic Current during the autumn (Ulltang, 1978). However, catches of the Norwegian stock had been taken further and further north during the 1950s and 60s, as if the Utsira stocklet had given way to that off Stadt and that in its turn to that off the Lofoten Islands (Devold, 1963). Figure 26 shows a period of catches of large herring off northern Norway (1868–1874) which occurred at the end of that Norwegian period in the second half of the nineteenth century. Hence by analogy, we would have expected the end of the Norwegian period in the late 1960s or the 1970s even if recruitment overfishing had not supervened; the reversal of the events of the Russell cycle between 1965 and 1979 may indicate years when the Norwegian herring period might have ended naturally (see Chapter 5).

The last Swedish period (1878 to 1896 or 1918) occurred during a cold period and the last Norwegian period in a warm period if dated from 1920 to 1950 (when the last very strong year class appeared). This is the firm evidence that the Swedish herring periods take place when the climate is cool and when the Baltic freezes more frequently, as Pettersson suggested. Figure 27 shows the distribution in time of sardine catches off Japan, off California, off Spain and in the Adriatic (Zupanovitch, 1968); catches in the three major groups reached their highest levels in the 1930s. The Californian and Japanese sardine stocks were possibly reduced by recruitment overfishing during the 1940s, but the rise in catches may well have been due to common strong year classes in the early 1930s. These events occurred during the recent period of warming throughout the northern hemisphere, common to the rise and fall of the Norwegian stock in about the same period.

Figure 28 shows the long-term changes in the fish catches including the alternation between the Norwegian and Swedish periods of fisheries. The catches of the Bohuslån fishery are expressed as periods of presence and

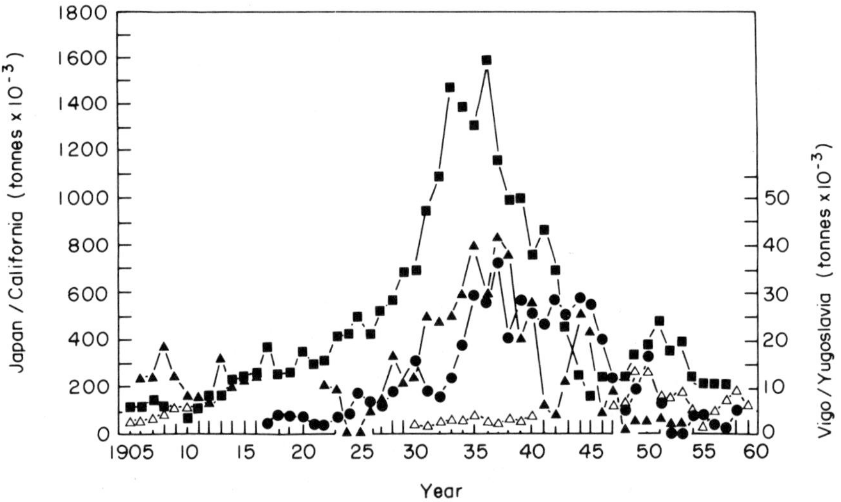

Fig. 27 Distribution in time of sardine catches off Japan, off California, off Spain and in the Adriatic (Zupanovitch, 1968).

■——■ Japan
●——● California
▲——▲ Vigo (Spain)
△——△ Yugoslavia

absence and Ljungman's 110-year period is fitted to the Swedish series in an unconvincing manner, but in a very general way the two sets of catches do alternate. We recall that Höglund (1978) has suggested that the Swedish period towards the end of the seventeenth century did not really exist because that fishery took place outside the skerries, as it did after 1896 in the last Swedish period. However, the difference in position of the fishery is a small one, and we believe that the important point is the presence of Buchan (Scottish) or even Dogger autumn spawners far into the Skagerrak in certain periods. Figure 28 also shows periods of higher and lower catches of the Hokkaido herring and of the Japanese sardine fisheries (Uda, 1957). It will be seen that there are periods of high and low catches of the Hokkaido herring but they do not appear to be related to any of the other three protrayed in the figure. However, the distribution of the Japanese sardine catches in time corresponds fairly well with that of the Norwegian herring and alternates with that of the Swedish catches. Thus extending the argument from the last periods we would expect the periods of the Norwegian herring catches and of high Japanese sardine catches to indicate

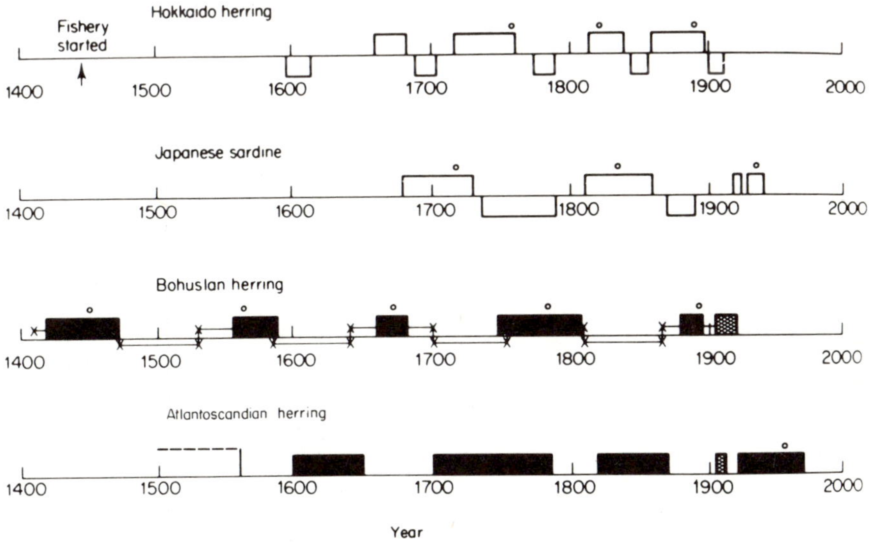

Fig. 28 Presence or absence of catches in the Swedish and Norwegian herring fisheries since mediaeval times. Periods of high and low catches are shown for the Hokkaido herring fishery and the Japanese sardine fishery (Cushing and Dickson, 1976).

periods of warming in the past and to expect the Swedish catches of herring to occur when the Baltic freezes hard.

There is a periodic shift of the wind over the British Isles from westerly to southerly, which may correspond to the recent period of warming (see Chapter 4). During periods of westerly winds, they blow onshore with long fetch on the long exposed coast of Norway, but when the winds are southerly the fetch would be less and more northerly spawning grounds off Stadt and off the Lofoten Islands would be protected to some degree because of the angle made by the coastline to the prevailing southerly wind. If the fish caught off the Bohuslån coast were really spawned off the Scottish coasts, as suggested above, the westerly winds blow straight offshore and the southerly winds would have a relatively longer fetch across most (but not all) of the spawning grounds. If little fish need food from the production cycle in order to survive, then a southerly wind on the Scottish grounds might delay the production cycle, as would a westerly wind on the coast of Norway and vice versa. Such processes will be considered in Chapter 6; all we need note here is that the alternation can be explained in a preliminary sense.

The ice cover north of Iceland in the Denmark Strait has been recorded for centuries. Beverton and Lee (1965) have associated the appearance of Swedish periods with advanced ice cover, and of the Norwegian fisheries during the warmer periods when the ice had retreated far north into the Denmark Strait. Figure 29 shows the alternation of Norwegian and Swedish herring periods associated with the number of weeks of coastal ice recorded on the north coast of Iceland. Koch (1945) has suggested that the polar front lies southerly when the coastal ice off Iceland endures for a long time, but Lamb (1977) has cast some doubt on the validity of some of the earlier data. The Norwegian herring periods are shown as hatched blocks and that in the early eighteenth century is surrounded by a dotted line because Storrow (1947) suggested that this period was variable, a point which has been more recently echoed by Höglund (1978). All the Norwegian periods start when the duration of ice cover started to *decrease,* similarly all of them end as the duration of ice cover *increases*—an observation perhaps independent of Lamb's doubts. Hence the Norwegian periods are associated with reduced ice cover. There is no such association between the degree of ice cover and the Swedish periods, although they tend to occur during long periods of ice cover.

We do not know why the magnitude of the Atlantoscandian herring stock rises and falls so dramatically as the Icelandic ice cover varies. Pre-

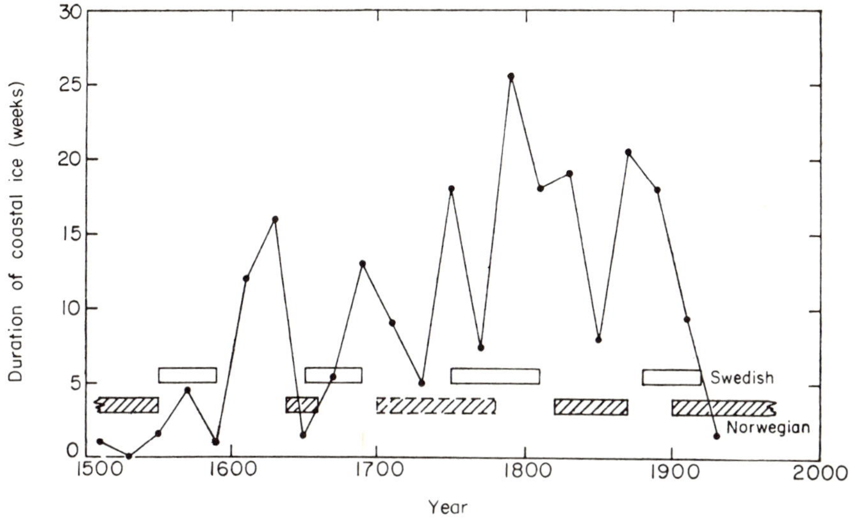

Fig. 29 Alternation of Norwegian and Bohuslån herring and its association with the number of weeks of coastal ice on the north coast of Iceland (Beverton and Lee, 1965).

sumably in cooler periods on a secular scale the ice cover increases and it is useful as an indicator of periodic climatic change. The ice retreated north of Iceland during the recent period of warming, so the association established by Beverton and Lee may indicate an alternation in warm and cool periods extending back across the centuries. It is this evidence that establishes the link between warm periods and those of the Norwegian herring fishery and also between cool periods and those of the Swedish herring fishery. It provides some credence to the belief that the Swedish catches occur when the Baltic freezes. Mason (1976) analysed Manley's (1974) air temperature data in England since 1659 and pointed out that the recent period of warming might have been an abnormal event. If that is the case, the extension of warm and cool periods back into the centuries might not be justified for the periods may be merely periods of different wind directions.

It is usually believed that the pelagic fish species such as herring or sardine respond more readily to climatic change than the demersal species like cod; indeed some theoretical justification will be given in Chapter 11 for this view that cod populations are resilient to environmental factors whereas those of herring are responsive to climatic change. The catches of the Arctonorwegian cod, on their spawning ground off Svolvaer in the

Vestfjord within the chain of the Lofoten Islands in Northern Norway, declined during the period 1898–1925. Before that period (1865–1898) and after it the catches were higher, in extreme three times as high. We assume that the differences in catch from year to year indicate in fact differences in stock; Figure 52 (p. 158) shows the Lofoten catches of cod between 1885 and 1960, and the reduced catches in the first two decades of the present century suggest that stock was really reduced in those years, a period when the cod probably did not visit the Svalbard shelf.

Long-term changes in catches elsewhere

Zupanovitch (1968) has examined the climatic effects upon catches of the Adriatic Sardine, and part of his study includes the records of high and low catches since the middle of the sixteenth century. Zupanovitch compared the periodicity which he obtained with that found by Uda (1957) for the Japanese herring shown partly in Fig. 27, as shown in Table 6.

More information on the Japanese sardine (*Sardinops melanosticta* Temmicnk and Schlegel) is given than in Fig. 28, enough to relate the Japanese sardines a little more closely with those of the Norwegian herring. From Table 3 it is clear that the periods of high and low catches in the Japanese sardine fishery correspond in a rough and ready way with those in the

Table 6

Comparison of periods of high and low catches of the Japanese sardine (Uda, 1957) and the Adriatic sardine (Zupanovitch, 1968)

Japanese sardine		Adriatic sardine	
1500–1600 (start)		1553	+ +
		1588	+ +
1680–1730	+	1670–1730	+
1716–1724	+ + +	1718–1725	+ + +
1736–1789	–	1730–1780	–
1760–1780	– – –	1775	– – –
1830	+ + +	1830–40	+ + +
1884–1880	– – –	1882–1878	– – –
1917–21	+	1919–21	+
1929–39	+ + +	1929–40	+ +
1941–57	–	1946–55	–

+, high catches; – low catches

Adriatic. Common to both is a poor period between 1941 to 1957 before the end of the full Norwegian period, but it is when the recent period of warming ended and when the air and sea temperatures started to reduce. If the time series do correspond, as suggested by Figure 28 and Table 3, and are really linked back in the centuries to climatic events, as suggested by the Beverton-Lee correlation, then the times of high catches of Norwegian herring, Japanese sardine and Adriatic sardine indicate periods of warming, whereas those of the Bohuslån herring are signs of cool periods.

Long-term changes in fish stocks as indicated by scales recovered from anoxic sediments

Cores from two basins filled with anoxic sediments off California (Santa Barbara and Soledad) have been examined by Soutar and Isaacs (1974), and they have compiled records of fish scales extending back as far as 2000 years. Examination of such time series requires that the cores really do represent a continuous and uninterrupted set. The authors suggest that from 1785 onwards it is in fact a continuous series. They have established high correlations between the numbers of scales/cm^2/year in the cores and billions of two-year-old sardines and millions of tons of total anchovy biomass as sampled on the fish markets or in the Calcofi cruises. They showed that ratios of sardine to anchovy displayed two periods of high sardine numbers, 1820–1865 and 1890–1945, which correspond roughly to the Norwegian periods. Smith (1978) converted the rates of scale deposition to biomass in five-year periods from 1785 to 1970 for sardine (*Sardinops caerulea*), anchovy (*Engraulis mordax* Girard), hake (*Merluccius productus* (Ayres)) and mackerel (*Scomber japonicus* Houttuyn).

Changes in biomass on a logarithmic scale are shown in Fig. 30 and the logarithmic mean for each stock is marked as a horizontal line, so positive and negative anomalies are effectively indicated. Two sardine periods are shown, 1815–1870 and 1896–1940, which correspond rather roughly to the two Norwegian herring periods in Fig. 26; the Californian sardine fishery collapsed in 1949–50, probably owing to recruitment overfishing, shown by Murphy (1966). Hake and anchovy declined after 1925, and there is some suggestion that the anchovy had decreased somewhat before then. During the 1960s the stock of anchovies recovered and there is perhaps some indication of recovery by the hake. The change from sardine to anchovy during the 1950s and early 60s has been well documented from the Calcofi surveys (Ahlstrom, 1966) and it probably took three generations

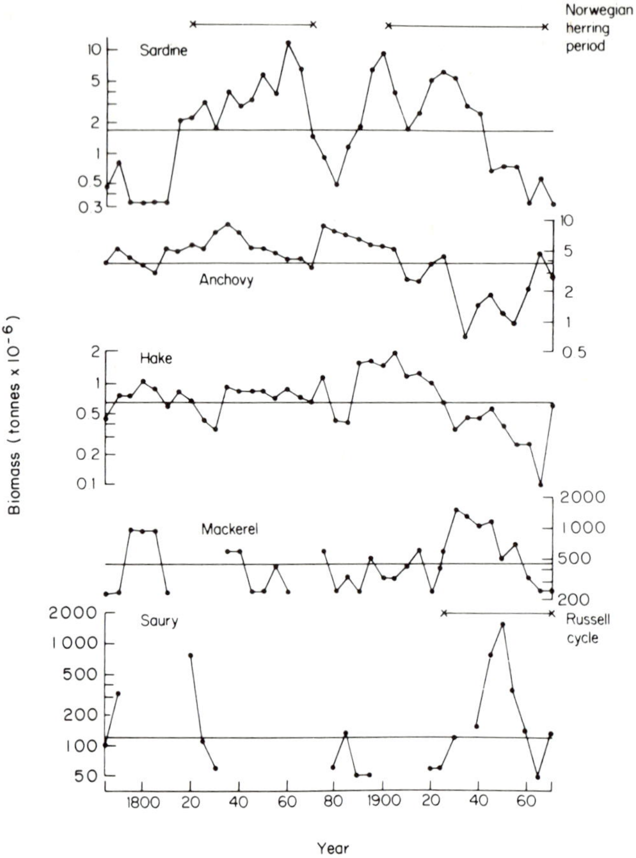

Fig. 30 Changes in biomass off California on a logarithmic scale of sardine, northern anchovy, hake, mackerel, and saury (Smith, 1978).

for the anchovy stock to fully occupy the niche in numbers formerly held by the sardine. Between 1925 and 1960 mackerel and saury became more abundant than before. This period, which was one of reduced hake and anchovy abundance, is roughly that of the Russell cycle in the western English Channel and of the recent period of warming in the north-east Atlantic (see Chapter 5). The sardine periods correspond with the Norwegian herring periods and in terms of the North-east Atlantic would be called warm periods. This suggests that off southern California, sardines indicate warmer periods in the northern hemisphere, but the hake and anchovy indicate cooler periods but it remains only a suggestion.

An upwelling system depends upon the direction of the longshore wind, and Yoshida (1967) has shown that there is an optimal angle made by the longshore wind to the coastline. Obviously the wind direction deviates from the optimal angle in nearly every year even if the deviation may not be of great magnitude; similarly one might expect deviations to exhibit trends with time analogous to the trends in the proportions of westerly winds over the British Isles. Such differences will generate different rates of upwelling or different degrees of intermittency and hence different rates of production are possible; indeed Cushing (1971) suggested that the slower the rate of upwelling the greater would be the production, because a high rate would bring the algae to the grazers too quickly. Cushing (1981) has shown that differences in zooplankton off California are correlated with upwelling indices; Bakun and Parrish (1981) correlated the recruitment of Pacific mackerel (*Scomber japonicus* Houttuyn) with the same indices. Thus there is capacity within the system to change the ratio of, for example, sardine to anchovy across the decades. This approach requires that the two systems of coastal upwelling and of offshore southerly transport (or offshore divergence) distinguished by Bernal and McGowan (1981) have common effects upon the fish stocks spread across them.

Conclusion

There are two sets of material on the long-term changes in the fish catches. The first comprises the records of periods of presence or absence, or periods of high and low catches (Fig. 28); statistical records, of course, do not exist, so the evidence is historical rather than statistical. The second set is the statistical record of herring catches in the Norwegian and Swedish fisheries (Fig. 26) and the record of sardine catches during the recent period of warming in many parts of the world. The two sets are linked by the Beverton-Lee association of the periods alternating between the two Scandinavian fisheries and the extent of ice cover north of Iceland. With such evidence we conclude that periods can be detected in the past, but it must be admitted that from instrumental records the recent period of warming might have been exceptional.

However, our purpose is not so much to understand climatic changes in the past than to disentangle the effects of climatic factors upon the fish stocks. What is clear is that some pelagic fish stocks, herring and sardines, can be augmented or reduced by several orders of magnitude. Further, such enormous increments or decrements take place under the influence of climatic factors, including temperature, irradiance, wind strength and direction; however, all that need concern us here is that the numbers in certain fish stocks change considerably in response to physical factors.

One of the most remarkable points is that some fish stocks respond very sharply to climatic changes which are themselves a little difficult to detect. Of course the changes in air or sea temperature or in wind direction are detected with instruments working continuously for long periods. However, the very strong year class of Norwegian spring spawning herring that started the last Norwegian period was hatched in 1904 and its magnitude was probably determined within months. It is very difficult to detect from the instrumental records what was climatically significant about those critical months. Indeed this is almost certainly the wrong way to tackle the problem; it is much more likely that the choice of a particular year may well be randomly determined within ten years or so. When the large year class appears or when a stock collapses, the slow and very variable climatic changes appear to be rectified—if that is not too absurd a metaphor.

EARLY CORRELATIONS BETWEEN FISH CATCHES OR YEAR CLASS STRENGTH AND CLIMATIC FACTORS

The events recounted by Boeck and Pettersson in Scandinavia and by Goode, Bean and Collins in North America led fisheries biologists to attempt more direct relationships between fish catches or year class strength and climatic factors. Probably the first of such correlations were those between catch anomalies on Schultz's ground in May; inverse correlations between catch anomalies were found from 1903 to 1924 and temperature anomalies on Skagens Rev and Anholt Knob (Johansen, 1925). The correlations were significant, and the input of high salinity and low temperature may be the effects of a salinification of the type discussed in Chapter 9.

Carruthers (1938) demonstrated that herring year classes in the Southern North Sea were correlated with wind strength in the area. He also showed that haddock year classes in the Northern North Sea were correlated with wind strength from 1918 to 1937. The argument behind these correlations was that fish larvae might be blown by prevailing winds away from the regions where food was rich. The correlation between haddock year classes and westerly winds on the same argument was repeated by Rae (1957). Saville (1959) showed that Carruther's correlation failed with more information but that another could be established between haddock year classes and the area in the northern North Sea free of oceanic or neritic indicators. Cushing (1966) correlated the haddock brood strength from 1933–61 with the quantities of *Calanus finmarchicus* as sampled on the Flamborough Line for the same period of years (the Flamborough Line was a line of six stations between Flamborough Head and the Dogger on which a Hensen net was hauled every month from 1933–1963). With the 1962 year class of haddock, which was 25 times larger than the preceding average, all the haddock year class correlations failed; we shall return to this point in Chapter 7.

There are other examples of simple correlations which break down as more information becomes available. This suggests not a multiple correlation of which only one component is used but a simpler correlation which changes its nature. In Chapter 11 it will be suggested that differences in year class strength depend partly on differences in wind strength and direction. Very often the effect of wind strength in fact depends upon the direction from which it blows particularly with reference to the shore.

CONCLUSION

Any historical study is limited by the quality of documents used and these become scarcer and harder to interpret as we go further back in time. Indeed, early history is a description of what material is available. The history of fisheries are often based on economic, legal or commercial records which yield indices of some value, and it is usually impossible to detect all the changes in stock necessary for our understanding because all we have are sporadic records of catches, sometimes many decades apart. Also climatic change must be detected against the noise of economic pressures and the effects of fishing. In other words for the effects of a climatic change to be detected properly it must be a profound change, and if the effects survive the long attenuation of time there is a sense in which they acquire credibility like a legend.

The evidence on the appearance and disappearances of fisheries (often herring fisheries) is often taken from the testimomy of fishermen at enquiries of a public nature. The collapse of a herring fishery remains in men's memories longer than its advent because the disaster has destroyed their livelihood. We cannot help being struck by the fact that herring stocks appear to have been more vulnerable than those of other fishes. Line fishermen must have taken roundfish steadily during the long periods of herring variation but records of rise and collapse are virtually non-existent. Indeed, the shifts of migration of haddock and halibut off the east coast of the United States were only referred to when scientists started to keep records. In Chapter 2 it was noted that the low-fecundity species such as herring were more vulnerable to recruitment failure (including recruitment overfishing as well as natural failure) than high-fecundity species such as cod. The herring that disappeared from the Medway, from Skinningrove and from Loch Fyne may well have suffered recruitment failure as did the Plymouth herring in the 1920s and early 1930s.

The rise of the 1904 year class to start the last Norwegian period in the alternation between Norwegian and Swedish fisheries is the obverse of recruitment failure. The alternation suggests linked causes (see Chapter 11), but nobody has suggested that the Scanian fishery might have been an extended fishery for Bohuslån herring further into the Baltic. It is not surprising that the interest in the fluctuations in fisheries originated in Scandinavia and that the problem continues to be studied there. Johansen's early correlations between catches and environmental factors were a first step in this direction. As will be shown in Chapter 11 this concern persists. The eastern seaboard of the United States is another region where dramatic events inflicted themselves upon the fishermen and fisheries scientists: the conflict between the Gulf Stream and the Labrador Current may well have killed the tilefish in such large numbers in 1882. From a historical point of view four scientists laid the foundations of our study, Boeck and Pettersson from Scandinavia and Goode and Collins from the United States.

Chapter Four

Climatic Change Relevant to Fisheries

Much of the study of climate concerns events on land or in the air and derives from the need to forecast the weather for airline pilots, ships' captains (and fishermen), and farmers. The ocean does not play a predominant part, although its function as a temporary store of heat, or as agent of poleward heat transport, for many months is of great importance. Climate represents changes in weather averaged for periods of time and so is largely a record of past events, although some events, such as an El Niño, can now be forecast.

In studies of the ocean, the physical oceanographer and the biologist have often worked independently of each other because they operated on different scales. Until recently the physical oceanographer was happier describing a large structure, such as the Alaska gyral or the Somali Current, rather than the smaller structures of interest to the biologist; for example, the stock of Pacific halibut within the Alaska gyral may depend upon a countercurrent which would be a small and indistinct part of the main structure, in which case the interests of the two disciplines would lie apart. However, the work on upwelling since the early 1970s and the pan-oceanic study of El Niño have shown how the physical oceanographers are starting to match the physical processes to a biological scale. In both instances the meteorological components of the processes are of predominant importance.

There is an analogous problem of scale in the study of climate with respect to fish populations. In one sense the study of climate is that of great changes in time, from ice age to ice age, or of the subperiods of climatic change in the last millenium to be described below. The recent period of warming from the end of the nineteenth century to the Second

World War is a small event on such climatic scales but a large one to the fisheries biologist, extending much further than the two or three decades of his usual time series. Fisheries arise and disappear on the scale of often many decades, primarily owing to changes in the recruitment, changes in the incoming year classes to the stocks. This chapter gives a layman's description of the essential physical mechanisms, a short history of the changes since the time of Duke William, and a brief account of some of the important and possible factors effective during the last millenium.

THE ESSENTIAL MECHANISM

The dominant feature in the weather and climate of the earth is the band of westerly winds in middle latitudes at the top of the tropopause, that is, at a height of about 16–17 km in low latitudes and at 11 km in middle and high latitudes. The westerlies are generated by the considerable differences in heating between the equator and the pole. The air rises more against gravity in the tropics because it is warmer and it endows the atmosphere with more potential energy, with the consequence that there is a pressure gradient between low and high latitudes. The gradient is steepest in middle latitudes, and this is where the upper westerlies blow. Because the pressure difference between the warm and cold zones increases with height, the maximum wind speed is high in the atmosphere at 11–17 km. In the southern hemisphere, because the rough land masses are smaller, with less surface friction, and because the western boundary currents in the ocean carry less warm water towards the pole, the pressure gradients are sharper and the westerlies stronger; because of greater angular momentum the southern hemisphere westerlies are a little closer to the equator than those in the northern hemisphere (Fig. 31)).

A parcel of air displaced either poleward or equatorward tends to retain its original vorticity (or spin, which increases with latitude), which means that the parcel tends to be returned towards its original latitude. Each excursion towards the equator forms a cold trough and each excursion towards the pole forms a warm ridge. Such excursions on a large scale form

---

Fig. 31 Average height of the 500 mb pressure level (*a*) in the northern hemisphere 1949–53; (*b*) in the southern hemisphere, 1952–4. The westerly wind blows along the contours, counterclockwise around the low pressure areas in the northern hemisphere and clockwise in the southern hemisphere (Lamb, 1972).

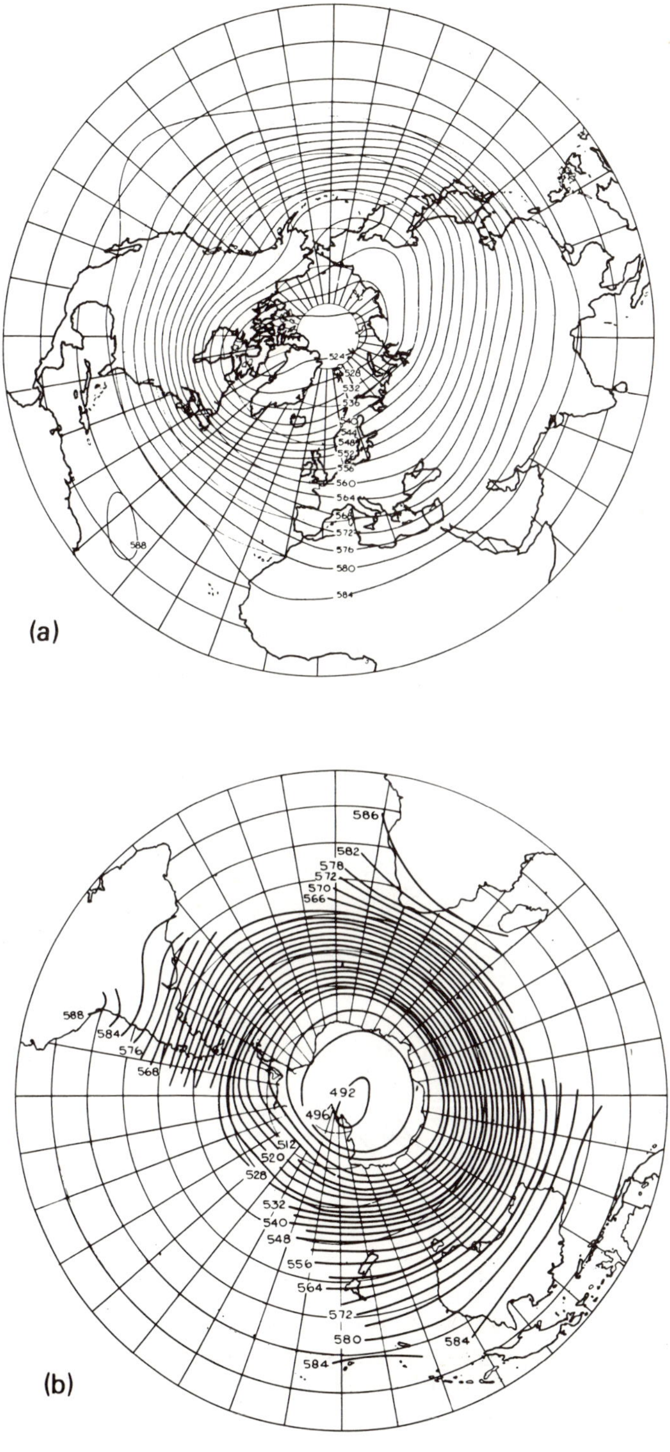

524
528
532
536
540
544
548
552
556
560
564
568
572
576
580
584
588
(a)
586
582
578
572
570
566
588
584
576
568
492
496
512
520
528
532
540
548
556
564
572
580
584
584
(b)

waves within the system of westerlies. These Rossby waves, are shallow in amplitude if the westerlies are strong and vice versa. There may be as many as five or more waves within the circumpolar vortex and standing waves are of some importance; with a wavelength of about 5000 km (Fig. 32) they spread right across a broad region.

The system is modified by geographical features. In early autumn the spreading snow cover intensifies two large troughs over north-east Asia and north-east Canada, and both tend to persist until the snow melts; in summer the Canadian trough retreats to the Arctic and the Asian one moves over the cool ocean currents of the North West Pacific. A barrier to air flow is formed by the Rockies so a semi-permanent ridge lies over western North

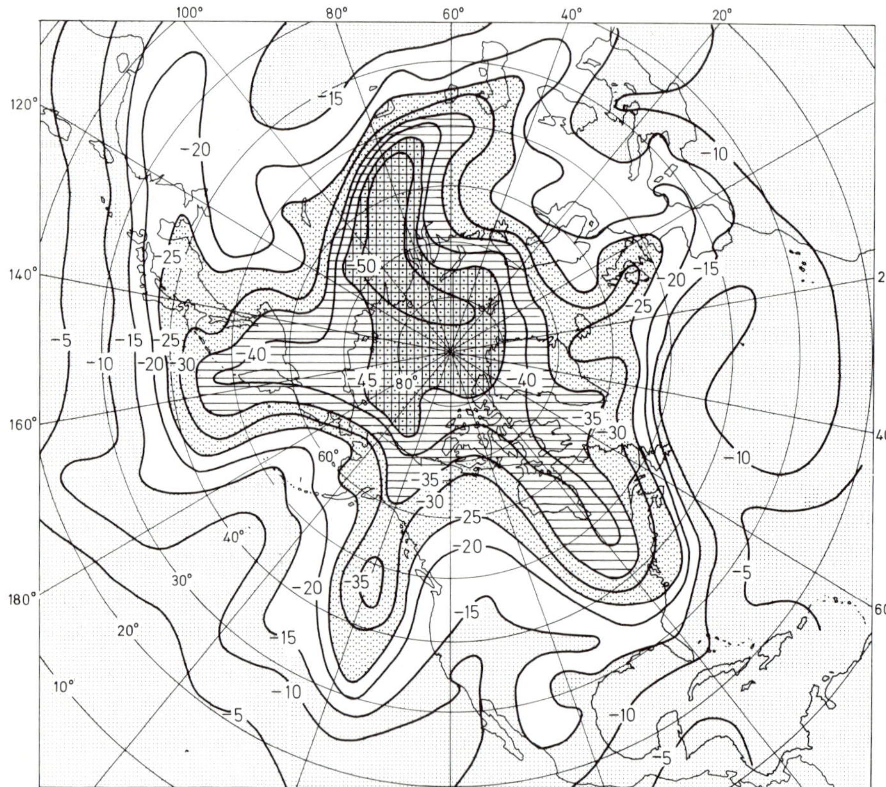

Fig. 32 Rossby waves in the southern hemisphere described by the trajectories of meteorological balloons at a constant level of pressure (200 mb) during 5–6 December 1971 (Cushing and Dickson, 1976).

America and this reinforces the trough over north-east Canada, because vorticity must be conserved. The two major troughs and the ridge might vary in position and intensity, but they are usually present in the same positions from year to year in charts of the upper air pressure distribution. In the southern hemisphere, ridges are observed over the Andes and the southern Alps of New Zealand, and the troughs develop downstream of these ridges. A major trough develops in the Indian Ocean where the Antarctic continent lies closest to the equator.

The weather system at the surface of the earth is dominated by the upper westerlies high above it. Surface cells form in the westerly wave train where flow lines tend to converge or diverge. Where the westerlies encounter parts of the planetary wave in which the pressure gradients are locally sharper than average, they turn from their normal path along the pressure contours. The atmospheric mass is redistributed, generating cyclones or anticyclones at certain characteristic positions on either side of the axis of the westerlies. Anticyclones usually appear on the warm side of the stream and cyclones on the cool side, and in the subtropical anticyclones belt, the high pressure systems on the warm side and the subpolar low pressure belt on the cool.

A common method of study is to compare anomalies over a region with the long-term average for that period; for example, a persistent high pressure system may be described in a particular region. This represents a persistence of short-lived systems in that region as anticyclones pass through or become delayed there. A sequence of such time-averaged charts shows how systems develop and die away. Figure 33 shows the track of the Great Pacific Anticyclone of the winter of 1949–50 for a period of five months (Namias, 1951) with 30-day charts (charts of shorter time intervals show a much more complicated picture). This illustrates the contrast between the day-to-day differences in weather and the month-to-month or year-to-year changes in what we call climate.

Sawyer (1965) examined the possible causes of such spatially extensive and temporally persistent phenomena lasting for some months. Because the total circulation energy of the atmosphere could be dissipated by friction and viscosity within about a week, the source of persistent regularity must lie outside the atmospheric system. Any external factor must be larger than 1000 km, persist for a month or more and must release heat at a rate of at least 50 langleys per day. Of the possible factors, sea ice remains in a normal position to within 100 km and soil temperature differences generate only shallow anomalies and are dissipated too quickly. Variations in snow

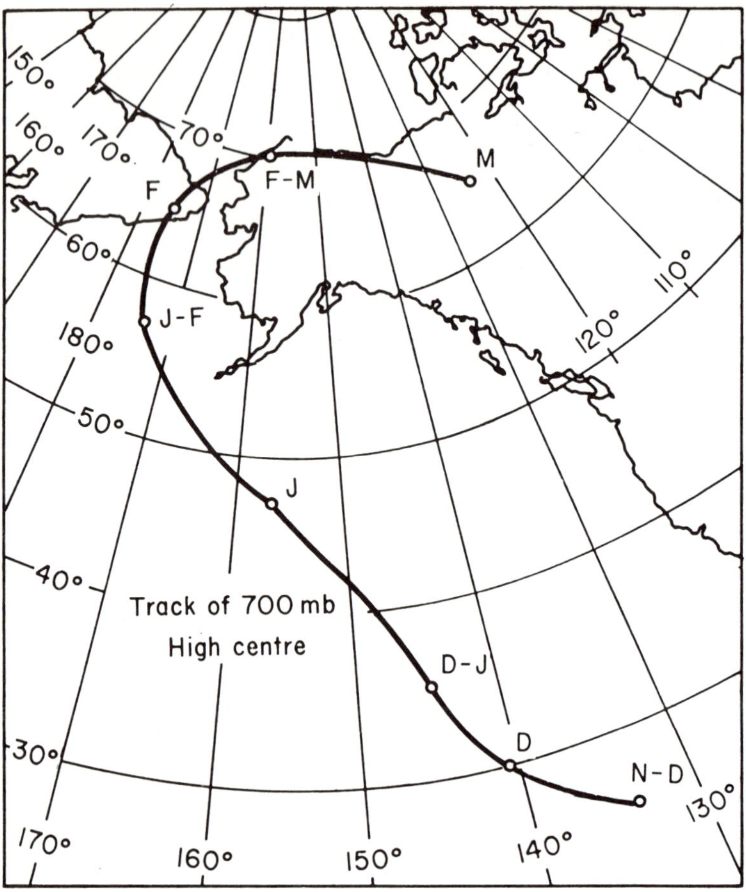

Fig. 33 Track of the great Pacific anticyclone in the winter of 1949–50 from 30-day mean charts at a height of 700 mb. Months are indicated by the initial letter, D for December etc. (Namias, 1951).

cover are confined to limited areas of middle latitudes in spring and autumn. But sea surface temperature (SST) anomalies of over 1°C are widespread, move slowly and may extend to about 100 m or more in depth and as such fulfil the three conditions given above. Figure 34 shows large SST anomalies in the North Pacific. Namias (1972) correlated the 20-year record of SST anomaly from a central point with similar records from 154 surrounding points; Figure 34 also shows the phenomenon in atmospheric observations at 700 mb and the two events, atmospheric and oceanic, are of the same scale. Namias and Born (1970) autocorrelated the monthly pattern in SST at discrete lags of 1–12 months and found a decay in the correlation coefficient with time, as might be expected, from 0.55 at one month to ~ 0.15 at 6–12 months. In the atmosphere and in sea level pressure the correlation coefficients were very much lower, although they were highest at one month: 0.19 and 0.14, respectively. But in the ocean the phenomenon is persistent.

Namias (1973) showed that SST anomalies develop most in the cold season and Fig. 35 shows the position of greatest surface temperature variability during the winters from 1959 to 1974. There are three sites at which the standard deviation of mean winter surface temperature is greatest: off Japan, in the east-central Pacific, and along the western coast of North America. Although the depth of such SST anomalies may reach 100–150 m, they are in general rather shallower and more variable.

The areas of SST anomaly receive heat from the atmosphere and the system of processes allows feedback between its component parts. For example, if a warm SST anomaly lies to the east of a cold water pool, the thermal gradient may be transferred to the air masses above and so depressions which enter this region linger and develop. Northerlies on the west of the depression tend to maintain the cold pool, and southerlies on the east will reduce the heat loss from the warm one, so the thermal gradient is reinforced and maintained by the feedback, at least for a short time. We shall return to this mechanism in discussing the events controlling the inflow of high salinity water into the Baltic (see Chapter 9).

From the quasi-persistent character of the Rossby wave system and the analogous persistence of atmospheric structures for short periods, links between distant regions might be expected. Figure 36 (O'Connor, 1969) shows the probability of *sign* of an anomaly at other 10° grid intersections when a positive or negative centre lies at a given position. With a negative centre at 60°N 40°W in winter (about Cape Farewell, in Southern Greenland), there is a 98% probability that it will be accompanied by a positive anomaly of 700 mb height over northwest Europe. Such a link, or teleconnection, is partly a consequence of the conservation of vorticity but may also depend upon the form of feedback relationship described above. Namias (1981) has linked recent climatic events in North America, such as the 1975–6 drought in California (and the hot dry summer in Britain in 1976) or the severe winter in North America and the anomalies of 1978/9, to the SST anomalies in the North Pacific in somewhat analogous ways.

Within this system, which generates structures of short-term persistence,

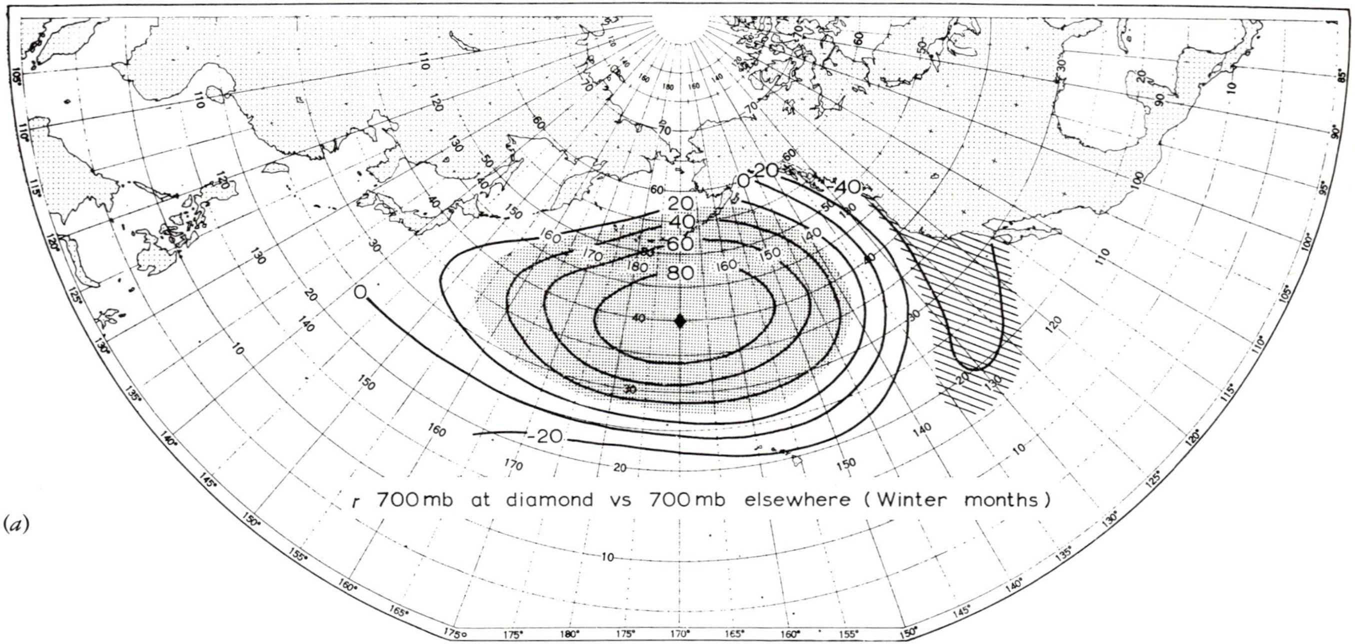

(*a*)

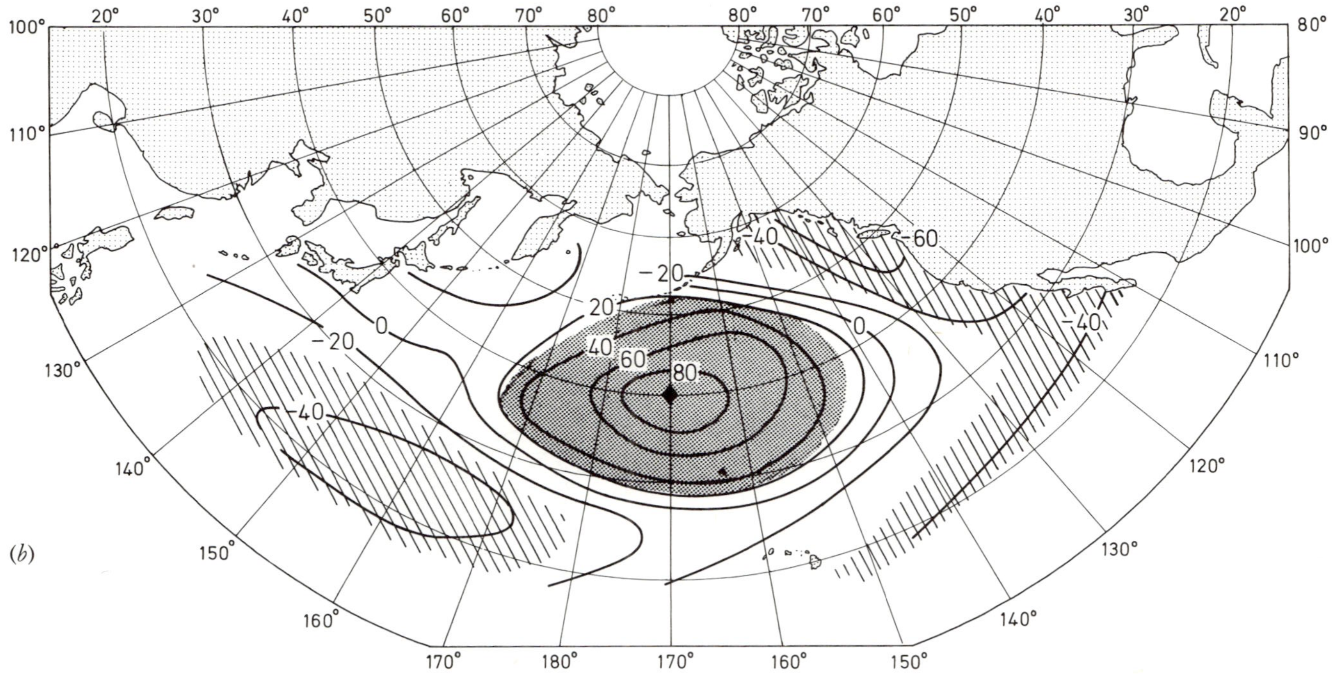

Fig. 34 Contemporary correlations in winter between 40°N and 170°W (position at the diamond) and elsewhere for (*a*) 700 mb heights (*b*) sea surface temperature. Shaded areas represent correlations at 1% significance; positive correlations stippled, negative correlations hatched (Namias, 1972).

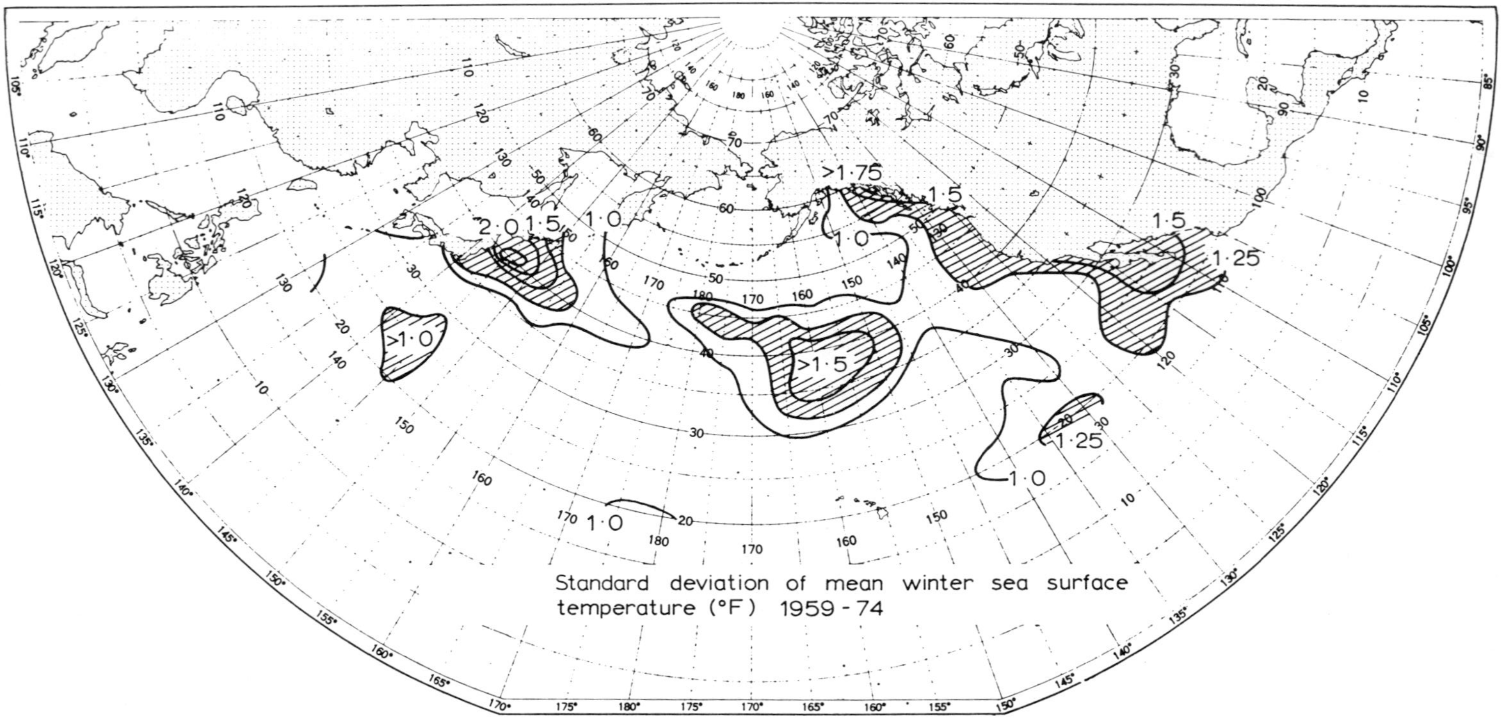

Fig. 35 Standard deviation of mean winter sea surface temperature (°F), 1959–74 (Namias, 1973).

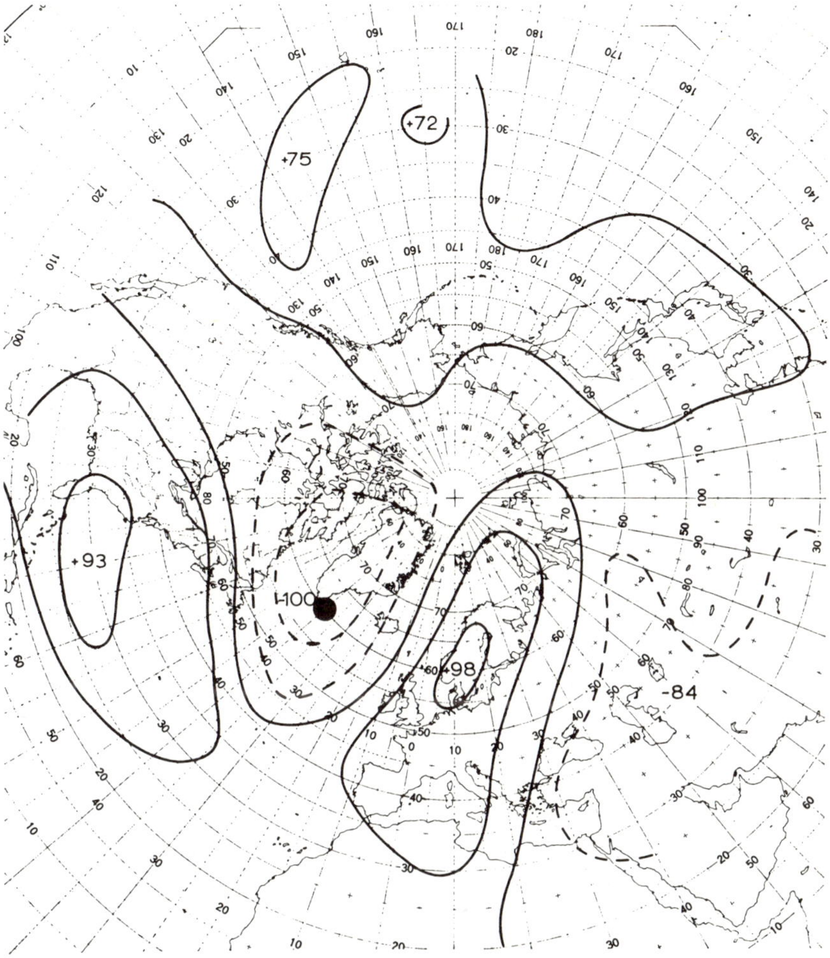

Fig. 36 Isopleths of probability of sign of 700 mb high anomaly when a negative anomaly centre lies within the 10° square centred at 60°N, 40°W in winter (shown by the black circle) (O'Connor, 1969).

are two factors of importance to fisheries. First, the wind may blow over a given region of the sea with some consistency, like the predominance of westerly or even south westerly winds over the British Isles (see below). Secondly, links across the oceans need not surprise us, if differences in wind strength and direction on average are linked across 5000 km or more in the Rossby wave system (see Chapter 5). This account is largely taken from Cushing and Dickson (1976).

TEMPORAL CHANGES

The longest records of fish catches in numbers or in weight extend back to the eighteenth century (for example, see Fig. 26), although there are historical descriptions of fisheries from mediaeval times (for example, the Scanian fishery during the period of the Hanseatic League, and the alternation of Norwegian and Swedish herring periods; see Chapter 3). Hence the very long term changes across tens or hundreds of thousands of years that are probably caused by changes in the earth's orbit are of no direct interest. The description of climate during the last millenium does concern us because it can possibly be linked to historical changes in the fisheries.

Figure 37 gives estimates of temperature in central England expressed in 50-year averages. It comprises observations since 1659 standardized by Manley (1974) and earlier indirect estimates derived from botanical information. There are also unadjusted values based on early meterological evidence. An analogous description (Fig. 38) shows changes in the lengths of glaciers during the last thousand years or so (Ahlmann, 1953), which must be related to changes in the temperature régime. Based on independent methods both figures show the Little Optimum in mediaeval times (a warm period) and the Little Ice Age from 1550–1700 (or perhaps a little later). Figure 39 shows the Manley record in greater detail from 1659 to 1973

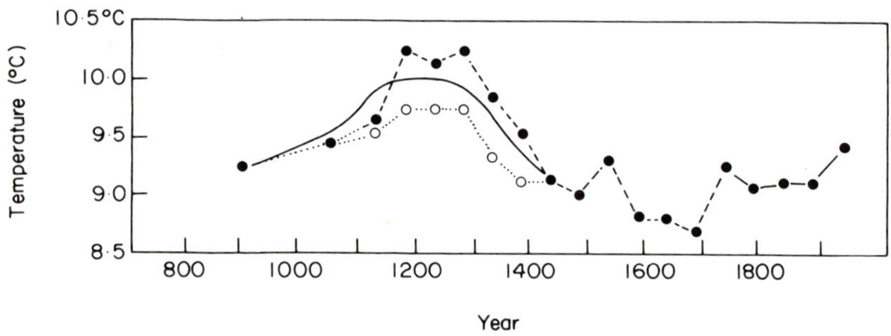

Fig. 37 Temperatures in central England from 50-year averages (Lamb, 1965).

● —— ● observed
● - - - - ● estimates from botanical indicators
○ · · · · ○ unadjusted values based on meteorological evidence
—— analyst's opinion
● · · · · ● connects points with sparse data

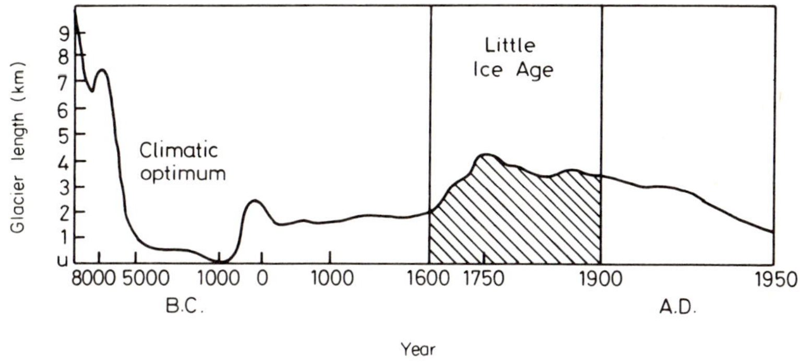

Fig. 38 Variation in the length in km of the glaciers in Iceland and Norway (note changes on time scale) (Ahlmann, 1953).

(Mason, 1976) which may show the last years of the Little Ice Age at the end of the eighteenth century but also displays clearly the period of warming in the first half of the twentieth century.

Lamb (1977) distinguishes four subperiods since the twelfth century: (1) the Little Optimum from about 1150 to about 1300; (2) the climatic shift to a cooler and more variable regime towards the end of the Middle Ages, 1300–1550; (3) the Little Ice Age from about 1550 to about 1700, although it may have started earlier and ended later; and (4) recovery during the nineteenth century. The Little Optimum is so-called to distinguish it from the Climatic Optimum which lasted for about four thousand years between about 5500 BC to about 1500 BC (see Fig. 40b). The evidence for the Little Optimum is indirect; for example, the tree limit, the upper limits for tillage, for vines, for fruit and grain were all 100–200 m higher than in the subsequent period of climatic shift in Europe, south of 55°N. There were Viking colonies in West and South West Greenland and they

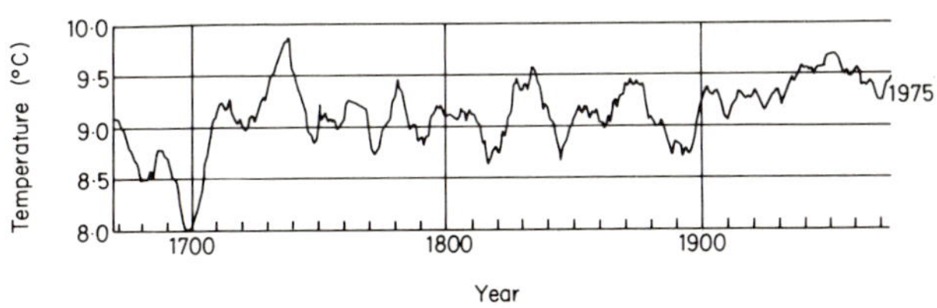

Fig. 39 Manley's observations of air temperature in Central England from 1700 to 1975 (Mason, 1976).

buried their dead which suggests soil not permafrost. Lamb's indices of winter mildness or severity (see Chapter 8) imply a period of mild winters and dry summers in Britain and in North West Europe. The anticyclones in the atmosphere were possibly displaced somewhat towards the poles during this period.

When the climate shifted in the late Middle Ages, summers became wetter (as indicated by records of floods) and winters cooler. The glaciers advanced in Iceland, Norway and in the Alps, the tree levels and the fruit and cereal lines were lowered, and the sailing routes to Greenland were abandoned. The proportion of $^{18}O_2$ to $^{16}O_2$ in ice or in wood indicates the temperature at formation, because more of the heavier component in $H_2{}^{18}O$ means that it condensed at a lower temperature (because of its lower vapour pressure) and vice versa; there are latitudinal, seasonal and altitude biases, but at one position a record of the isotope ratio in time is a reasonably consistent index of changes in temperature. Figure 40A (Hammer *et al.* 1980) shows the oxygen isotope ratio record compared with ring-dated trees in California, with the Manley record of Central England temperatures and differences in acidity in the Greenland ice cores. Figure 40B shows a longer time series of $\delta^{18}O_2$ for the last 13 000 years taken from ice cores in the Greenland ice cap (Dansgaard, 1981). The four subperiods can be seen in the observations for the last thousand years and even the recent period of warming can be detected; as noted above the main Climatic Optimum of some thousand years ago is also shown. During the Little Ice Age, temperatures in Europe were cooler at all seasons, 1°C lower in winter and 0.5°C lower in summer (Schuurman, 1981a), and the weather was more variable. Rivers were frozen in Holland thirty to forty times in each century, two to three times more frequently than in the nineteenth century. The depression tracks probably lay further south as if the polar cap were more extensive.

After the end of the Little Ice Age in the eighteenth century, the climate ameliorated and the period of warming in the early decades of the twentieth century ensued. A valuable and specialized data set is the record of sea ice observed on the coasts of Iceland in 20-year averages (Koch, 1945); there are some historical misgivings on the quality of material from mediaeval times, but since the eighteenth century records are probably reliable (Lamb, 1977). Much of this account is taken from Lamb (1977), where a mass of detailed observation is given together with interpretations of the changes.

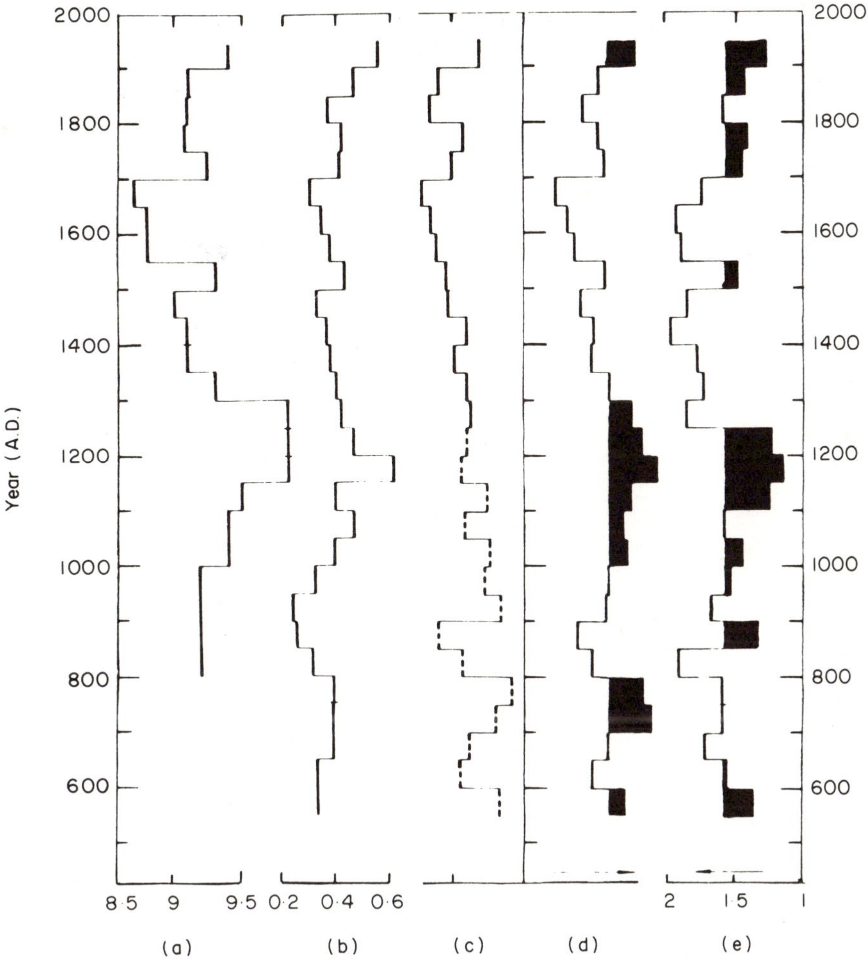

Fig. 40 (A) Record from the oxygen isotope ratio anomaly (*c*), compared with central England temperatures (°F) (*a*), differences in temperature deduced from tree ring diameters (m) in the White Mountains in California (*b*), a combined index of temperatures (*d*), and differences in acidity (μEq/kg) in the Greenland ice core (Dansgaard, 1981).

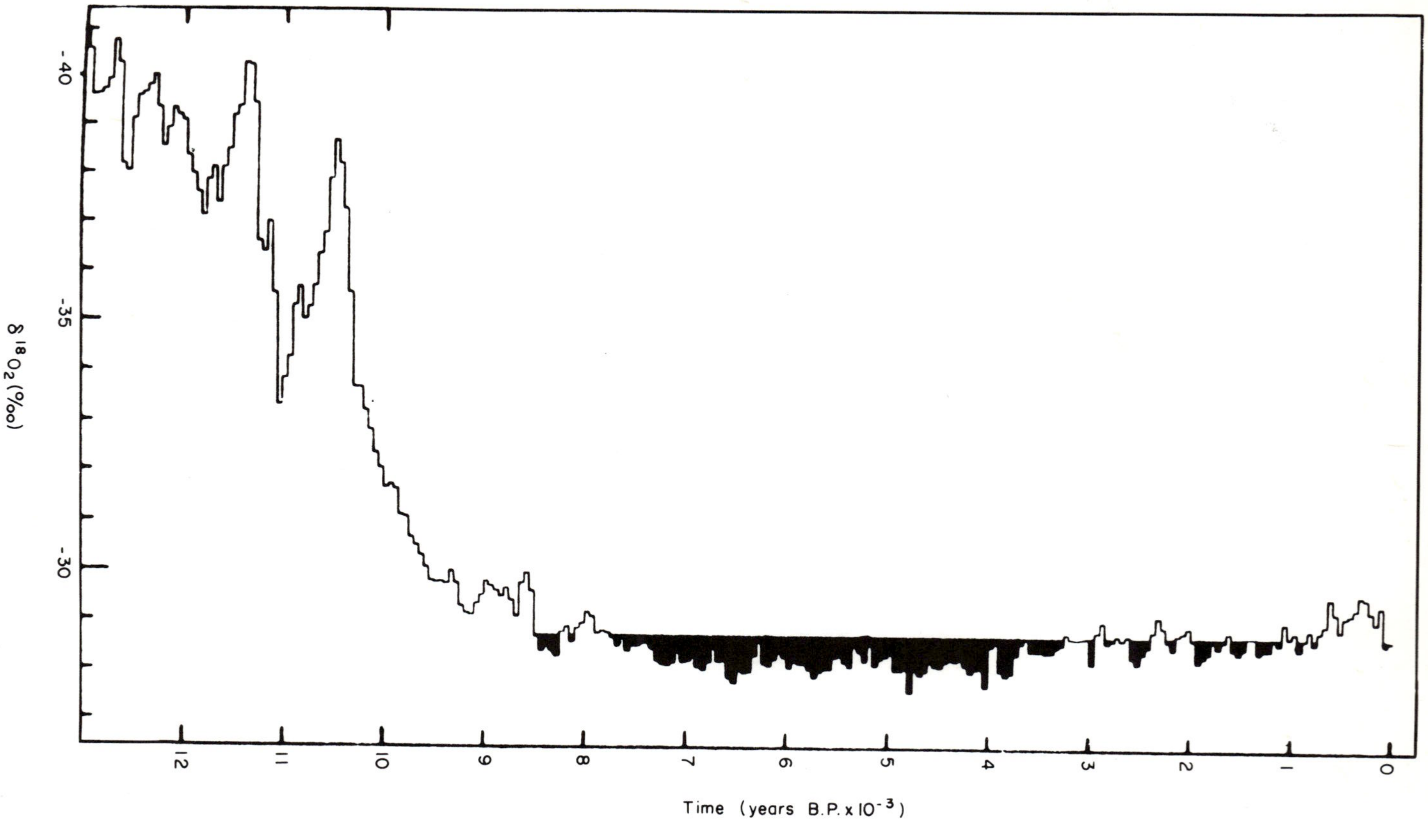

Fig. 40 (B) The last 13,000 years (Dansgaard, 1981).

Schuurman (1981a) summarizes the information in the following way:

	Winter	Summer	Year
1700–1900	Cold dry	Variable temperature + rain	Cold dry
1550–1700	Very cold dry	Very cool, very wet	Very cold and wet
1300–1550	Cold dry	Variable temperature + rain	Cold, dry
1150–1300	Mild wet	Very warm, very dry	Cold, dry
1000–1150	Cold		

The amelioration of climatic conditions in the early twentieth century extended right across the northern hemisphere. Between the periods 1900–19 and 1920–39, the mean annual air temperature in the region of Spitsbergen increased by more than 7°F and by 2–5°F in the European arctic more generally. In this region, sea ice was reduced to a minimum in 1938 and in Siberia the permafrost boundary retreated northwards by about 40 km between the 1830s and the 1940s. Figure 41a shows the change in global mean surface air temperature, as deviation from the long term mean in five year averages (Mitchell, 1963; Cushing and Dickson, 1976) and as annual recorded temperatures in the northern hemisphere north of 30°N for the period 1891–1978 (Gruza and Ran'kova, 1979) (Fig. 41b). The warming progressed from the first decade of the twentieth century, or perhaps a little earlier, until the 1940s after which there was a period of cooling, which persisted till the end of the 1960s; during the 1970s there has been a tendency to be a little warmer. The increment amounted to 0.5°C at the maximum, which is considerable. The biological consequences of this event are discussed in detail in Chapter 5.

This prominent event, warming the high latitudes was linked with an increase in the strength of the general circulation over the North Atlantic (Defant, 1924; Wagner, 1940; Scherhag, 1950); in this sense it probably resembles the Little Optimum, a period when the anticyclone tracks lay closer to the poles. Presumably, the oceanic circulation was also affected.

Figure 42 shows the number of days per year of general westerly circulation over the British Isles which is also averaged as 10-year means at 5-year intervals (Lamb, 1965, 1977; Anon., 1980). From the 1880s and 90s the proportion of westerlies increased until the 1920s and 30s, after which there was a decline. Dickson and Lamb (1972) showed that the circumpolar vortex may have contracted with a northward shift of zonal wind, at both high and low levels. The Iceland low lay further north, and the retreat of sea ice may well have been associated with an invasion of the Arctic by depressions. Mitchell's figure (Fig. 41), which shows the global

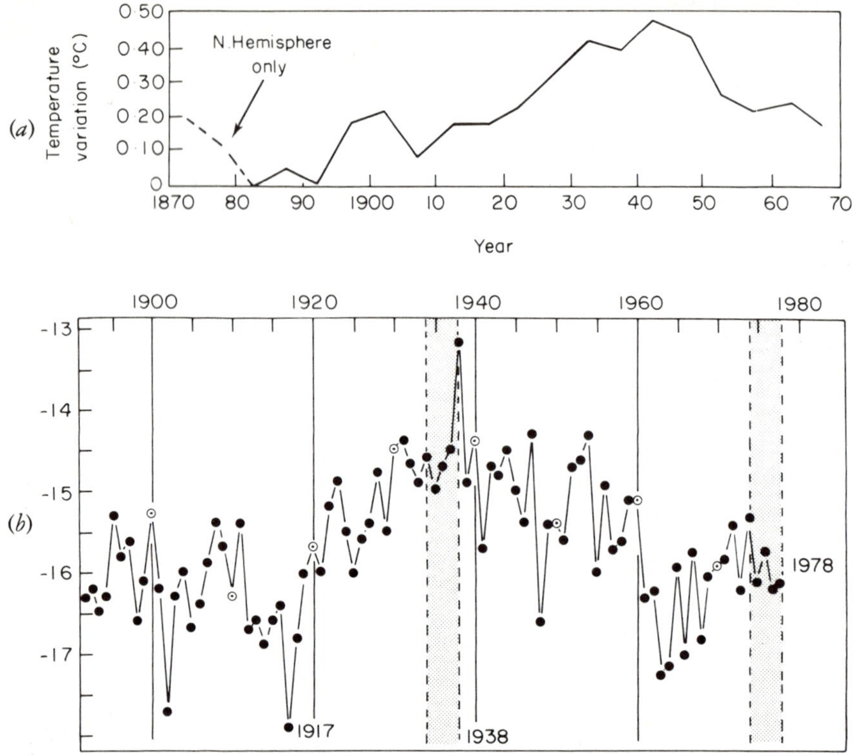

Fig. 41 (*a*) Variation in the global mean surface air temperature as deviation from the mean in successive five-year averages from 1880–84 (Cushing and Dickson, 1976) and (*b*) as annual records north of 30°N (Gruza and Ran'kova, 1979).

increase in temperature, implies that the intensification of atmospheric circulation was not limited to the North Atlantic. Thus the increasing prevalence of westerly winds over the British Isles was probably associated with intense circulation of air throughout the northern hemisphere and perhaps all over the globe. The subtropical anticyclones in the Pacific and Atlantic oceans also intensified and broadened their circulation (Lamb, 1977).

Since the 1920s and 30s the number of days of westerly wind over the British Isles has decreased. This decrease has been accompanied by four events on a world-wide scale. (1) The axes of winds and pressure belts have shifted slightly towards the equator with latitudinal shifts of the tropical zonal rain belts (Winstanley 1973a,b; Lamb, 1977; Wahl and Bryson,

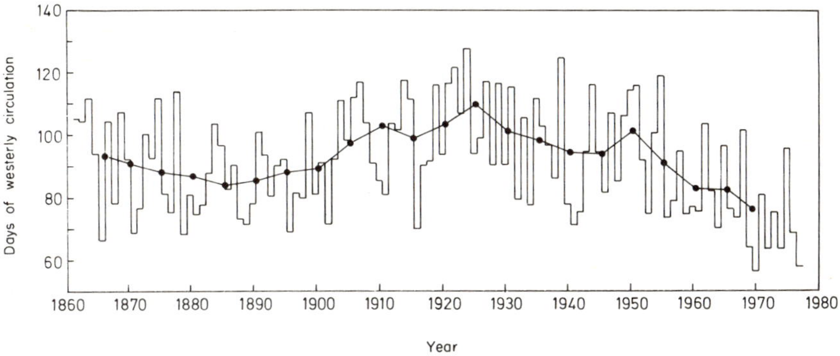

Fig. 42 Number of days per year of general westerly circulation over the British Isles as ten-year means at five-year intervals (Lamb, 1977; Anon., 1980).

1975). (2) The upper westerlies have shown a greater tendency to meander; the Canadian trough has tended to regress westward towards its position of the mid mineteenth century (Wahl and Lawson, 1970; Eichenlaub, 1971) and an extra trough has tended to develop over Franz Josef Land and Novaya Zemlaya. (3) The global atmosphere has cooled by 0.3°C in pentade mean surface temperature since the 1920s and 30s (Fig. 41); the cooling was most pronounced at high latitudes, maximal cooling in the Arctic where the greatest increase had taken place, which resulted in a southerly extension of snow and ice cover. (4) Budyko (1968) and Pivovarova (1968) have demonstrated a decline of about 4% in the strength of the solar beam on clear days from actinometric measurements since the 1940s. As the global atmosphere cooled, the upper westerlies meandered more and they have shifted on average towards the equator. The general circulation of air has presumably slackened, and this may also be true of the oceans.

Other, and perhaps associated events have taken place over the North east Atlantic since the 1950s. A high pressure system was established over Greenland and maintained in the 1950s and 60s; this increased the frequency of northerly winds in the Norwegian Sea and as far south as the British Isles. Figure 43 shows that in the months December to February the increase in mean pressure at sea level amounted to 7 mb between the period 1900–39 and 1956–65 and that this increment continued in the late 1960s. When the periods 1956–65 and 1966–70 are compared (Fig. 44), it will be seen that the shape of the system altered, with a steeper pressure gradient between Greenland and the eastern Norwegian Sea. Consequently areas to the north and west of the atmospheric and oceanic polar fronts became

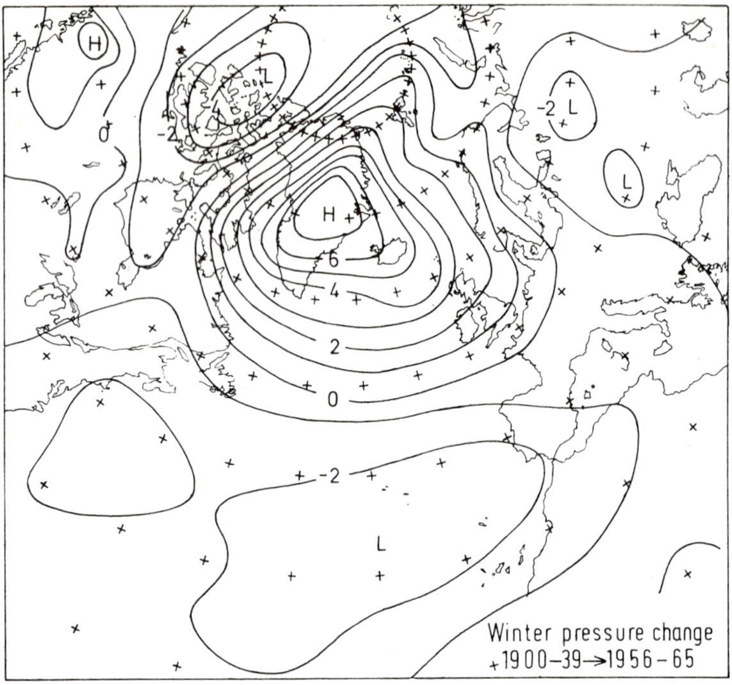

Fig. 43 Distribution in increments of mean pressure at sea level in the North Atlantic between 1900 and 1939 and 1956 and 1965 (Cushing and Dickson, 1976).

much cooler in winter; for example between 1949/50 and 1967/8 mean winter air temperatures declined by 5.54°C in the Isford in Spitsbergen, by 5.60°C at Bear Island and by 4.06°C at Jan Mayen. Decade mean air temperatures in Franz Josef Land dropped from between −10.8°C and −10.0°C for the period between the 1920s and the 1950s to −14.5°C during the 1960s (Rodewald 1972). There were two important consequences:

(a) The North Sea temperatures decreased sufficiently to be associated with enhanced recruitment to the cod stock (Dickson *et al.*, 1973). In Chapter 7, other factors will also be associated with this event, suggesting that the decrease in temperature and a delay in the production cycle were linked.

Fig. 44 Distribution of increments of mean pressure at sea level in the North Atlantic between 1956 and 1965 and 1966 and 1970 (*upper*) and between 1966 and 1970 and 1971 and 1974 (*lower*) (Cushing and Dickson, 1976).

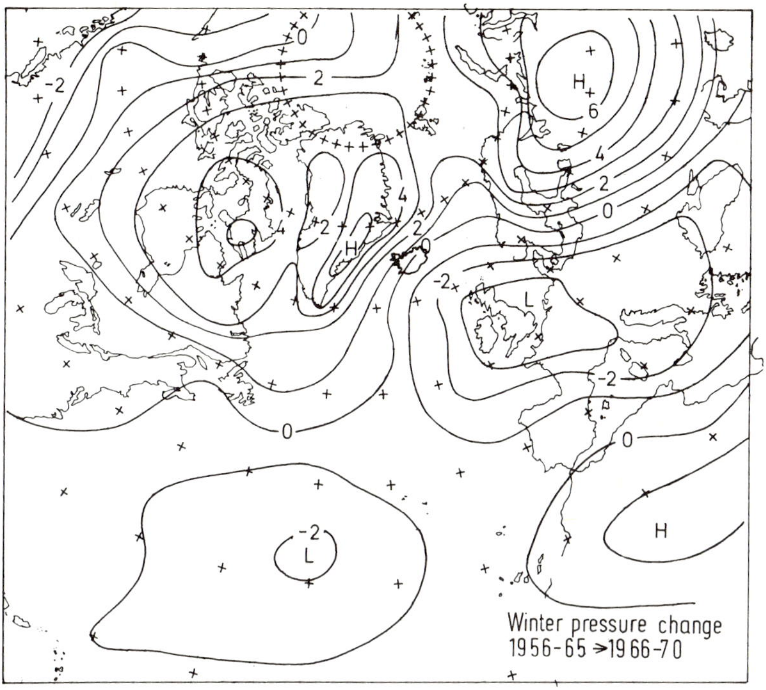

Winter pressure change
1956-65 → 1966-70

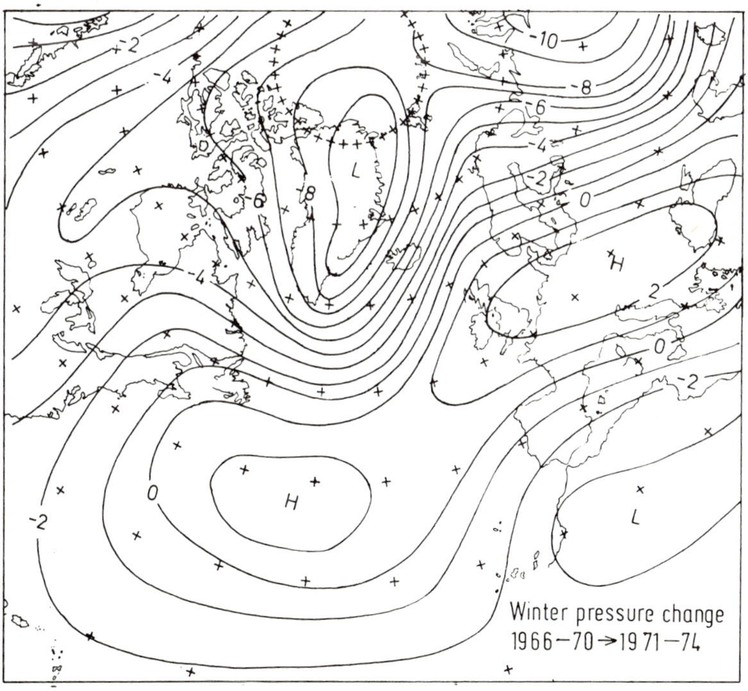

Winter pressure change
1966–70 → 1971–74

(b) The easterly winds across the Denmark Strait were replaced by northerly winds, which meant that the transport of the eggs, larvae and 0 group cod from Iceland to West Greenland would be inhibited; this atmospheric event may have been the prime cause of collapse of the West Greenland cod fishery. The easterly winds which may have strengthened the Irminger Current would also have sustained relatively warm water on the West Greenland banks where the adult cod live.

A further consequence of these events was a southerly extension of ice off Labrador, south of Cape Farewell, south of Bear Island and particularly on the north and east coasts of Iceland. Hence the polar front was shifted south and east, and Jakobsson (1969) suggested that the migration route of the Atlantoscandian stock of herring was shifted away from the north coast of Iceland, where there had been a good summer fishery before the stock collapsed due to recruitment overfishing. In the winter of 1970–71, the high pressure cell over Greenland collapsed, but northerly winds were sustained in the Denmark Strait. Part of this account is taken from Cushing and Dickson (1976).

THE RESOLUTION OF CLIMATIC FACTORS

Much of our interest in climate originates in the need to know when an Ice Age may recur. The Milankovitch thesis states that the periodicity of Ice Ages with their subsidiary events depends upon changes in the Earth's orbit due to changes in the relative positions of the planets. Berger (1981) simulated the appearances of main and subsidiary events with an astronomical model and predicted that the next Ice Age will appear in about 6000 years. He cannot resolve differences of a thousand years, so the historical events of the last millennium lie below the noise level of such a system.

However, the four subperiods since the eleventh century are real and their causes are unknown. Lamb (1970) developed a dust veil index (DVI) which estimates the relative quantities of volcanic dust in the upper levels of the atmosphere. He suggested that some of the variability in climate in recent centuries might be a consequence of volcanic activity: an eruption is followed by a cold summer, with reduced air temperature, increased rainfall and more snow over the icefields. Bray (1974) showed that the advances of the glaciers since the sixteenth century were correlated with Lamb's dust veil index DVI. Kelly (1977) related the Index to Central England temperatures (CET) and to the frequency of south westerly winds over the British Isles (SW), which indicates the strength of the circumpolar upper air flow, the upper westerlies:

	DVI	CET (annual)	CET winter	
SW (annual)	−0.85*	+0.61	+0.86*	*$P < 0.05$
SW (winter)	−0.76*	+0.55	+0.90*	
DVI		−0.67*	−0.64	

Thus DVI is inversely correlated with the frequency of south westerly winds over the British Isles and inversely with Central England temperatures on an annual basis. Mitchell (1975) shows how the recent period of warming and the subsequent cooling might be inversely related to the quantity of dust in the northern hemisphere from the major eruptions. However, the dust is assumed to remain in the stratosphere for not more than a year, the eruptions occur once or twice in each decade, and Mitchell assumes climatic response time of 6.5 years. Thus it is possible that the cool periods since the Little Optimum of early mediaeval times are associated with the dust from volcanoes, but other factors may be involved. Dansgaard (1981) has compared a Northern Hemisphere temperature index based on the Manley record, the Greenland isotopes and on White Mountain tree rings (Fig. 40a, as indicated above); these indices were related to a Greenland ice acidity index linked to acid gases from volcanic eruptions and were significantly correlated from 600 AD to the present day.

On shorter times scales (decades) attempts have been made to resolve periodicities in various sets of data. For example, Mason (1976) examined the Manley record of Central England temperatures (1659–1973) with spectral analysis using the maximum entropy method; he found significant peaks at 2.1, 5.2, 7.6, 14.5, 23 and 76 years. The 23-year cycle is the double sunspot one when the Hales effect occurs (the single sunspot cycle is hardly noticed by this spectral analysis). Johnsen *et al.*, (1970) found two predominant periods of 78 and 181 years (± 5%) from Dansgaard's oxygen isotope material. Lamb (1977) quotes Siren's (1961) analysis of tree ring data from northern Scandinavia in which the following periods occurred:

11·3, 18, 30 and 90 years. However, no estimate of noise, red or white, was given.

Records of catches by species have been established for decades and in one or two instances for centuries, but estimates of recruitment extend back for somewhat shorter periods—two, three or four decades. It is unlikely that spectral analysis can resolve periodicities within the relatively short time series of recruitments. The longer series of catches are inevitably somewhat noisier. However, it will be shown in later chapters that the recruitment to the fish stocks under some circumstances depends upon atmospheric events which are perhaps not distributed periodically in the strict sense. For example, the salinifications in the Baltic, the invasions in the North Sea of *Calanus finmarchicus,* or the appearances of El Niño all occur once every five years or so, as described below, and are all events which may be described as intermittent but not periodic.

CONCLUSION

There are different scales of climatic change within the last millennium during which changes in the fisheries have been recorded. The shortest scale is that of the weather, which can be forecast for a few days. It is of no relevance to studies of fisheries because catch data are usually reported by months. Sometimes catches are reported by days across a grid of statistical rectangles but differences from day to day would be hard to link to the broad scale of weather changes. There are anomalous features that persist for months in the atmosphere: events which generate the cold winters in north-west Europe (Chapter 8) or the periodic Baltic salinifications are of this order.

On a somewhat more extensive scale in time and space are events such as El Niño with its special origin in pan-Pacific phenomena. But differences in the recruitment to stocks of fish appear to become pronounced from decade to decade. Sometimes this phenomenon is expressed in an upward or downward trend and on other occasions a profound change in the magnitude of recruitment occurs in a given year which persists for many years, for example, the 1904 year class of Atlantoscandian herring and its successors for six decades, or the gadoid outburst in the North Sea which started in 1962 and has persisted since. Such phenomena extend in time beyond the duration of the mechanisms described in this chapter.

On the longest scale are those changes of the order of centuries, for which the information is necessarily meagre. There appear to be events in the fisheries which are rather roughly linked to one of the four or five subperiods in the last millennium. The last—the recent period of warming from 1890–1940—is well documented and many biological events are associated with it (see Chapter 5). Earlier subperiods in antiquity are decorated with features in the history of fisheries, but it is very hard to decipher more than sporadic indications of what actually occurred.

Chapter Five

The Recent Period of Warming

Four groups of biological events occurred between the 1920s and 1960s: (1) the invasion of the waters round western France and the British Isles by animals from more southerly waters and a corresponding invasion of high latitudes by boreal animals; (2) the spread of intertidal organisms (as indicated by barnacles), a migration of a southerly species during the period of warming, and a retreat during the period of subsequent cooling; (3) the rise and fall of the West Greenland cod stock (and the fishery which exploited that stock) between the twenties and the sixties; and (4) the Russell cycle, the profound changes in the ecosystem that took place between 1925 and 1935 and which reversed between 1965 and 1979.

THE NORTHWARD MOVEMENT OF ANIMALS BETWEEN THE 1920s AND 1940s

"Indicator" organisms are large conspicuous animals that live long enough to survive in seas far from their native waters. They are seen by fishermen in their gear and by naturalists on the shore, classes of people ubiquitous enough to ensure that the appearance of such animals is a real event. The simplest form of such an invasion is that of the Portuguese Man of War, the siphonophore (*Physalia physalis* L.), which is blown by the wind at the surface of the sea. Woodcock (1944) suggested that *Physalia* is drifted to the left of the wind in the northern hemisphere and to the right in the southern hemisphere. Wilson (1947) recorded the many appearances of *Physalia* between July and October 1945 on the coasts of Devon, Cornwall and South Wales and suggested from a meteorological study that they were blown ashore by sustained winds from the Azores and mid-Atlantic region.

Associated with this invasion were other conspicuous animals such as *Lepas fascicularis* Ellis and Solander and *Lepas pectinata* Spengler, species of goose barnacle, *Velella,* the by-the-wind-sailor and the loggerhead turtle (*Caretta caretta*), all of which are found at the very surface of the sea. Wilson also examined earlier records and found that in the British Isles, northern and western France and in Belgium, *Physalia* appeared in the years 1834, 1852, 1862, 1884, 1907, 1912, 1913, 1919, 1921, 1934, 1935, 1945 and 1946. In 1912 and 1913 there was a big invasion in March and April, the animals appearing in the Scilly Isles, Cornwall, Sussex, North Devon and in Cardigan Bay. If Wilson is right, the appearance of *Physalia* in south-west Britain indicates persistent winds from the Azores and they occurred in particular years. It is assumed that the animals are always abundant enough in the region from which they originated. In the last hundred years, *Physalia* strandings have been more frequent since 1912 than before that year. The increase in frequency may indicate the period of warming in rather general terms.

Another manifestation of exotic invaders is the infrequent plague of *Octopus vulgaris* Lamarck in the South Western Channel. There was one in 1899–1900 and another in 1950 (Rees 1951; Rees and Lumby, 1954). The plague originates on the southern side of the Channel between the Cotentin peninsula and Ushant, including the Channel Islands, where they spawn in May if the temperature is $<$ 11.5–12.0°C. Rees and Lumby show that the plague is associated with the highest temperatures in the Western Channel. The parent animals are presumably drifted in from the south, and if the temperature is high enough, they can breed successfully on the southern side of the Western Channel. There are two components to this invasion: the northerly migration and the subsequent successful breeding in the warm waters. If the animals are carried by the water, drifting with the streams, the simplest explanation is that warmer water appeared from the south carrying *Octopus* and it was warm enough to sustain subsequent reproduction.

A third form of migration is shown in Fig. 45 from a study of the benthos on the Svalbard Shelf (Blacker, 1957; 1965) which compared the records of presence or absence of conspicuous benthic animals in the period 1878–1931 with records taken by R.V. *Ernest Holt* between 1949 and 1956. In the earlier period, Atlantic species such as *Lithodes maia* (Linnaeus), *Hippasterias phrygiana* Parelius, *Geodia barretti* Bowerbank and *Sabinea sarsi* Smith were found between 100 and 200 m west of Bear Island as far north as the North West Gully. In the later period between 1949 and 1956, the

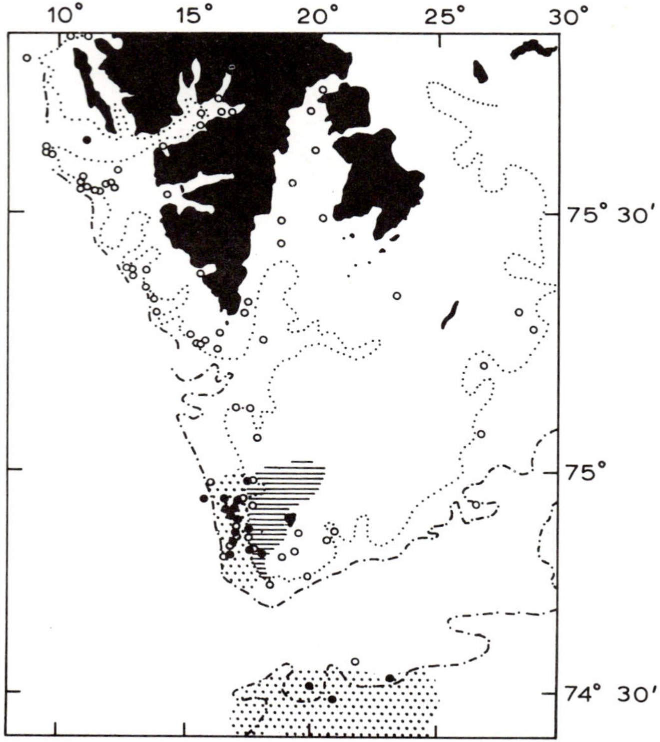

(a)

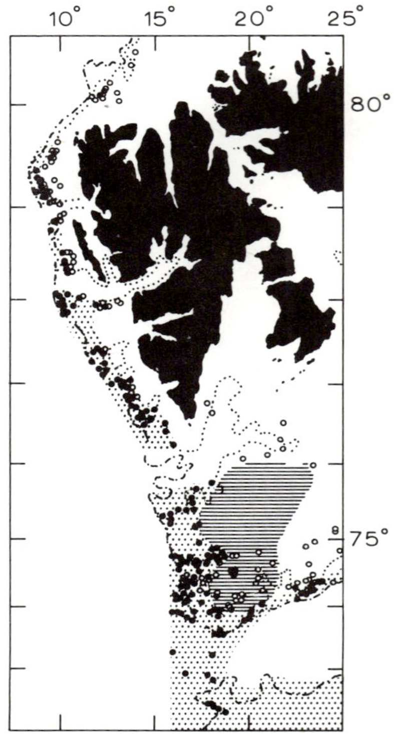

(b)

Atlantic species were recorded in the same depths as far north as Prince Charles' Foreland off the Svalbard Archipelago; some were also recorded in shallower water on the top of the shelf. Similarly Atlantic species extended their range from Bear Island towards the South East Gully and towards Hope Island. Blacker plotted the records on temperature and depth and showed that the Atlantic species were associated with a particular temperature range. The northerly invasion of the Atlantic species implies that the warm West Spitsbergen current flowed at shallower depths and may have overtopped the shelf, thus allowing the benthic larvae to survive a northerly drift. The West Spitsbergen Current may have flowed on a shallower course, which may have been a consequence of intensification. Nesis (1960) demonstrated the spread of boreal (or Atlantic as opposed to Arctic) benthic forms in more temporal detail in the Murman current:

Years	1921	25	35	38	45–46	47	48	49–51	56–57	58
Atlantic forms (%)	42	44	72	92	64	72	85	88	86	73

Thus the spread of benthic animals of more southerly origin into the Murman current took place from 1935 onwards, probably most in the late 40s and early 50s.

The three examples illustrate three forms of invasion, the purely wind blown one which deposits *Physalia* on the coasts of Western France and southwestern Britian in certain years, the dependence of *Octopus* breeding on warm water in May in the southwestern part of the English Channel and the spread of Atlantic benthic species northwards in the West Spitsbergen Current and the North Cape Current by the early 1950s as that current spread into shallower water. As noted in Chapter 2 Harden Jones (1968) suggested that most fish must migrate by following (or drifting in) the oceanic currents or countercurrents because they usually had no external referents.

The animals may well be drifting with the tidal streams when in midwater. If that is the general rule for animals in the sea (in current or in tidal stream), then the appearance of an exotic invader at a particular point means that the currents between that point and the point of origin might have strengthened.

Table 7 lists appearances of subtropical invaders off the western coasts of France and off the British Isles during the 1930s. They included not

Fig. 45 Northerly movement of Atlantic benthic animals (black circles) on the Svalbard shelf (*a*) 1878–1931 and (*b*) 1949–56; positions of Arctic benthic animals are shown in open circles (Blacker, 1957).

Table 7

Records of conspicuous surface-living animals between the West coast of France and Fair Isle between 1930 and 1935

Region	Depth[a]	Species		Remarks	Date	Reference
W. coast of France	s	*Balistes* sp.	trigger fish	25 recorded, 17 in 1930–1	1930–1	
	s	*Dermatochelys coriacea* L.	turtle		1932	
	s	*Callianthus ruber* (Rafinesque 1810)	medusa		1932	(Desbrosses, 1932, 1935, 1936)
Plymouth	s	*Velella* sp.	siphonophore		1932	
	s	*Ianthina* sp.	pelagic mollusc		1932	
	s	*Salpa fusiformis* (Cuvier)	tunicate		1932	
Fastnet	p	*Cyclosalpa bakeri* Ritter	Tropical tunicate		1932	
	p	*Oxynotus paradoxus*	shark		1932	
	p	*Carcharinus glaucus* (L.)	blue shark		1932	

Isle of Man	p	*Scomberesox saurus* (Walbaum)	saury pike		1933	
Yorkshire coast	p	*Scomberesox saurus* (Walbaum)	saury pike		1932	
Scottish waters	p	*Cyclosalpa bakeri*		Fair Is. passage	1932	
	s	*Ianthina*				
	s	*Velella*		Skye	1933	
	s	*Lepas*				
	p	*Carcharinus glaucus* (L.)	blue shark		1933	
	p		albacore	Fair Isle	1933	
				Holy Loch	1934	
				West of Scotland	1934	
	s	*Ianthina britannica* Jefferies		Skye	1935	(Stephen, 1938)

[a] s = surface layer
p = pelagic

only surface-living animals such as *Velella,* which may well have been driven by the wind on to our coasts, but also pelagic animals such as sharks, albacores and saury pikes which migrated in shallow depths, although not at the very surface. The animals at the surface itself, like *Physalia,* may well have been blown by persistent winds from the tropical mid-Atlantic, whereas those a little deeper in the surface layers may well have been drifted by an intensification of the surface currents from southerly regions. Most appearances occurred between 1930 and 1935 (especially 1931–3), and although some of these animals were sporadically recorded in later years (see for example the records of "Rare Fishes" in the *Annales Biologiques* of the International Council for the Exploration of the Sea), the records in earlier years in the region were few. We conclude that there was a considerable invasion of subtropical animals in the region during the first half of the 1930s.

Table 8 shows the evidence of a more extensive migration of boreal animals into high latitudes in the North-east Atlantic (Cushing and Dickson, 1976); the swordfish is really a subtropical animal but it appeared off the Faroe Islands and Iceland together with many truly boreal animals. They lived at all depths, surface, pelagic, mesopelagic, demersal and benthic. Hence the migration extended to all depths and spread from Faroe, Iceland, Greenland, Jan Mayen and the Barents Sea. The most extensive invasion occurred at Iceland, where mackerel, bluefin tuna and saury pikes, among other pelagic species, were observed. Three new mesopelagic species were recorded as well as about a dozen demersal animals, including mullet, conger, two rays, pollack, and a forkbeard. The Irminger Current bathes the southern and southwestern coasts of Iceland before it swings westward towards East Greenland. The witch and turbot spread to the northern and eastern coasts, whereas the capelin, a high-latitude species, disappeared from the southern coast. If the animals were carried by the Irminger Current, it must have spread further north than it had and may have warmed the northern coasts to some degree. This considerable invasion occurred between 1924 and 1949.

Three demersal species appeared at East Greenland during the 1920s and 30s. The salmon, the Ca'ing whale (*Globicephala melaena* L.), haddock, coalfish (*Gadus virens* L.), spurdog (*Squalus acanthias* L.), ling (*Molva molva* L.) and dealfish (*Trachipterus arcticus* (Brünnich)) appeared at West Greenland between 1925 and 1934. As the Irminger Current swings south, it mixes with the cold East Greenland Current and the mixed stream turns round Cape Farewell and floods the offshore banks of West Greenland. If boreal

Table 8

Records of appearance in the Northeast Atlantic between the 1920s and 1940s.

Region	Depth[a]	Species		Remarks	Dates	Reference
Faroe Is.	p	*Alosa finta* (Cuvier)	twaite shad			
	s	*Xiphias gladius* L.	swordfish			
	d	*Pollachius pollachius* L.	pollack		Tåning (1953)	
Iceland	s	*Xiphias gladius* L.	swordfish		1924–1949	
	p	*Alosa finta* (Cuvier)	twaite shad			
	p	*Callionymus maculatus* (Rafinesque)	dragonet	new records at Iceland		
	d	*Pollachius pollachius* L.	pollack			
	d	*Raja nidarosiensis* Collett	ray			
	s	*Cetorhinus maximus* (Gunner)	basking shark			
	p	*Scomber scombrus* (L)	mackerel			
	p	*Thunnus thynnus* (L)	bluefin tuna			
	p	*Scomberesox saurus* (Walbaum)	saury pike			
	mp	*Paralepis rissoi* Kroyeri		increased abundance		
	mp	*Paralepis brevis* Zugmayer				
	mp	*Scopelus elongatus*	myctophid			
	d	*Trachurus trachurus* (L)	horse mackerel			
	d	*Conger conger* (L)	conger			
	d	*Raja clavata* (L)	ray			

Table 8 *(cont'd)*
Records of appearance in the Northeast Atlantic between the 1920s and 1940s.

Region	Depth[a]	Species		Remarks	Dates	Reference
	d	*Crenimugil labrosus* Risso	mullet	increased abundance		
	d	*Phycis blennoides* Brunnich	forkbeard			
	d	*Raja lintea* Fries	ray			
	?	*Centrolophus britannicus* Günther	rudderfish			
	d	*Argentina silus* Ascanius		extended their distribution		
	d	*Somniosus microcephalus* (Bloch and Schneider)				
	d	*Glyptocephalus cynoglossus* (L)	witch	spread to the northern and eastern coasts		
	d	*Scophthalmus maximus* (L)	turbot			
	p	*Mallotus villosus* (Müller)	capelin	disappeared from southern coasts		Fridriksson (1949)
East Greenland	d	*Sebastes marinus* L.	redfish	in small shoals in twenties; abundant in thirties	20s	
	d	*Reinhardtius hippoglossoides* (Walbaum)	Greenland halibut		30s	
	d	*Anarrhichas minor* Olafsen	catfish			Jensen (1939)
West Greenland	s	*Globicephala malaena* Traill	ca'ing whale		1926	
	s	*Salar salar* L.	salmon		end of twenties	
	p	*Halopsis ocellata* A. Agassiz	craspedote medusa		1928	

	d	*Sebastes marinus* L	redfish	South of Julianehaab	1925	
	d	*Melanogrammus aeglefinus* (L)	haddock	South of Julianehaab	1929	
				Nanortalik	1932	
	d	*Pollachius virens* (L)	coalfish	1st spawning in present century	1932	
	d	*Squalus acanthias*	piked dogfish		1931–6	
	d	*Trachypterus arcticus* (Brünnich)	dealfish	7 records (only 1 before 1923)	1923–34	
	d		ling		1928	Jensen (1939)
Jan Meyen	p	*Clupea harengus harengus* (L)	herring		1930–1	
	d	*Gadus morhua* (L)	cod	Some had Icelandic hooks in them	1930–1	Jensen (1939)
Barents Sea	d	*Gadus morhua* L.	cod	plentiful in 1925, having perhaps absent since 1883		
	d	*Melanogrammus aeglefinus* (L.)	haddock	plentiful in 1925, having perhaps absent since 1883	1925	Jensen (1939)
	d*	*Gibbula tumida* (Montagu)	gastropods	benthic animals which appeared on the Murman coast for the first time in the thirties		
	d*	*Akera bullata* (Müller)	gastropods	benthic animals which appeared on the Murman coast for the first time in the thirties		
	d*	*Eupagurus bernhardus* L.	hermit crab	benthic animals which appeared on the Murman coast for the first time in the thirties	30s	
	d*	*Cerastoderma edule* L.	cockle	benthic animals which appeared on the Murman coast for the first time in the thirties		Berg (1935)

[a] s = in the surface layer
p = pelagic
mp = mesopelagic
d = demersal
d* = benthic

animals appeared during the dramatic decade (1925–35), they were carried by the Irminger Current and the new appearance of animals—surface, pelagic and demersal—suggests that the current may have strengthened and more warm water appeared at West Greenland.

Cod and herring were recorded at Jan Mayen for the first time in 1930–1, and cod and haddock returned to the Barents Sea in numbers in 1925, having perhaps been absent or in low abundance since 1883. Four relatively well-known benthic animals appeared on the Murman coast for the first time. If cod, haddock and herring drifted in the North Cape Current, that current must have been somewhat stronger in the early 1930s; if benthic animals reproduced on the Murman coast in these years, their parents must have been carried there.

In 1929–30, cod appeared at Novaya Zemlaya, and in 1932, salmon, cod and *Clupea harengus pallasi* appeared in the River Kara which enters the Kara Sea. Haddock were caught in the White Sea where they had not been recorded before. Catches of the Murman coast herring (in tonnes) then increased:

1920	381	1925	164	1930	1000
1921	270	1926	16	1931	22928
1922	910	1927	89	1932	29296
1923	73	1928	654	1933	67939
1924	189	1929	23		

The big catches in 1931–3 must have arisen in year classes in 1926–7.

Invasions were also recorded in the Northwest Atlantic, in the Bay of Fundy, and off Nova Scotia. Arctic species, such as the catfish and the sea tadpole *Careproctus longipinnis* Burke, appeared there between 1932 and 1936, presumably carried south by the Labrador current. During the same period, southern animals such as the Greater silver smelt, *Leptagonus decagonus* Bloch and Schneider, the sea snail, *Neoliparis atlanticus* Jordan and Evermann, the sucker, *Remora remora* L., the mousefish, *Histrio histrio* L., and the tunicates, *Salpa zonaris* (Pallas) and *Salpa vagina* Tilesius, appeared in the region (McKenzie and Homans, 1937). The latter presumably rode the Gulf Stream until they were flung into the Gulf of Maine. The only connection with the spread of animals, subtropical and boreal into the Northeast Atlantic, is the fact both Northeast and Northwest Atlantic are fed by the North Atlantic Drift and Gulf Stream and that the two invasions were contemporaneous.

In summary, (1) surface and pelagic subtropical animals reached western France and the British Isles; (2) boreal (and one or two subtropical) animals at all depth ranges reached further north, Faroe Islands, Iceland, East and West Greenland, Jan Mayen and the Barents Sea; and (3) northern and southern animals were found in the Bay of Fundy and off Nova Scotia. The North Atlantic circulation is common to all areas, although southwesterly winds and the Labrador Current must have also played a part. The most important point is that all events occurred in the same period, 1924–49, with most between 1930 and 1935 as if the circulation had intensified then. This is the simplest form of explanation if it is realized that animals in the sea must in general be drifted by the oceanic currents. A strange invader in waters distant from its own can only reach them if the migration circuit of its population has been distorted or extended.

Table 9 gives some records from the Pacific during the decade 1925–35. There are two groups, exotic invaders in the seas off the Pacific Northwest which are Californian natives and subtropical migrants off the coast of California. The observations were recorded in 1926 and 1931. There are two groups, the northward migration of animals native to Californian waters to the Pacific Northwest and of animals native to tropical seas of the waters off California. The migrations are unlike those in the Atlantic, which are really based on the circulation, even if the proximate cause of *Physalia* invasions in British waters is persistent southwesterly wind. In 1958–9 warm water flooded Californian waters from the south as a consequence of the periodic flooding of the eastern tropical Pacific with warm water which is associated with El Niño. A major El Niño occurred in 1925–6 when the invaders reached the Pacific Northwest; in 1931 *Pleuroncodes planiceps* Stimpson appeared off California. Walsh (1976) has associated the appearance of this pelagic crab off California with El Niño and there was a minor one in 1931 (Miller and Laurs 1975). El Niño appears periodically right through the recent period of warming and subsequent cooling and the two events are unrelated in any time series. A coincidence remains, that the years 1925–6 and 1930–1 were notable years in both the north-east Atlantic and in the north-east Pacific.

THE SPREAD OF INTERTIDAL ORGANISMS

Two coastal barnacles, *Balanus balanoides* L. and *Chthalamus stellatus* (Poli), may compete on the littoral of the British Isles and northwestern France. Southward and Crisp (1956) showed that the range of temperature between which the beat rate of the feeding cirrhi of *Balanus* is optimal extends from 0–17°C; that for *Chthalamus* lies between 5 and 30°C. Further, a high

Table 9

Records of Southern species off the west coast of North America in the decade 1925–35

Region	Depth[a]	Species		Remarks	Date	Reference
Pacific Northwest	s	*Velella*	by-the-wind sailor	British Columbia	1926	
	s	*Mola mola* L.	sunfish	British Columbia	↓	
	p	*Engraulis mordax* Girard	northern anchovy	North Columbia R	↓	
	p	*Thunnus alalunga* (Bonnaterre 1788)	albacore	Oregon	↓	
	p	*Pneumatophorus planiceps*	mackerel }	Coos Bay	↓	
	p	*Atherinopsis californiensis* (Girard)	jack smelt }	Oregon	↓	
	p	*Seriola dorsalis* (Gill)	yellowtail	Washington State	1931	
		Palometa simillemma (Ayres)	pompano		1931	(Hubbs, 1948)
California	p	*Sphyrna zygaena* (L)	hammerhead shark	California	1931	
	p	*Pleuroncodes planiceps* (Stimpson)	pelagic crab	S. Catalina Is.	↓	Hubbs (1948)
	p	*Etrumeus micropus* (Temminck and Schlegel)	Japanese herring	California	↓	
	p	*Manta birostrus* (Walbaum)	giant ray	↓	↓	
	p	*Fodiator acutus* (Cuvier and Valenciennes)	a flying fish	↓	↓	
	p	*Verrunculus polylepis* (Steindachner)	a trigger fish	↓	↓	
		Chaetodipterus zonatus (Girard)	spadefish	↓	↓	(Walford, 1931)

[a] s = surface
p = pelagic

summer temperature affects the fecundity of *Chthalamus* and survival of their larvae; perhaps low winter temperatures delay embryonic development until the larvae can benefit. Southward and Crisp compared the distributions of the two barnacles around the British Isles between the 1920s and 1930s. *Balanus* retreated eastward from the south coast of Cornwall; *Chthalamus* spread eastwards from St Alban's Head, Swanage, off the Isle of Wight and southwards along the east coast of Scotland from Nybster to Wick, Lybster and Latheron.

Crisp and Southward (1958) compared the distributions of intertidal animals on the French and English coasts of the Channel. *Chthalamus* increased in the Gulf of St Malo between the 1930s and 50s, and during the same period *Balanus perforatus* Bruguière increased along the coast between the Scillies and St Catherine's Point. There is a boundary between St Catherine's Point and the Gulf of St Malo and the southern forms, *Chthalamus stellatus, Balanus perforatus, Patella aspera* Lamarck, *Patella intermedia* Jeffreys, *Gibbula umbilicalis* da Costa, *Monodonta lineata* da Costa and *Littorina neritoides* (L.) lie to the westward on both sides of the Channel. The boundary is in fact tidal and westward the water stratifies in summer and eastward it does not. During the recent period of warming *Chthalamus* spread eastwards to the tidal boundary.

Southward (1967) reported on the regular watch on the relative abundance of barnacles between 1949 and 1966. He showed that between 1960 and 1966 *Balanus balanoides* spread back along the coasts of Dorset, both north and south coasts of Devon and Cornwall and occupied the sites recorded in the 1930s; further there was a considerable increase in abundance, in Cornwall of up to three orders of magnitude. The "barnacle front" extended westwards in 1962, reaching Lyme Bay in 1963, Plymouth in 1964, St Austell Bay in 1965 and the Fal estuary in 1966. On the north coast of Cornwall the front extended from an outlying population in St Ives Bay to Portreath in 1963, Newquay in 1964 and Trevoze in 1965. A barnacle index was developed which is the proportion of *Chthalamus stellatus* in the total intertidal barnacle population; it is positively correlated with temperature. Between 1949 and 1965 the barnacle index at two sites near Plymouth remained above 0.7; in 1964 it fell to 0.4 and 0.6 and in 1965 to 0.2 and to 0.55. In 1934 at the same stations it was about 0.3.

Between the 1930s and 1949 the boreal barnacle shifted eastward along the coasts of the English Channel and the southern one replaced it. Between 1962 and 1965, such changes were reversed quite sharply. Although we cannot determine when the first change took place, because there was only one observation in 1934 before the main series was started in 1949, the second change started in 1962 and was completed by 1965; the first year was that when the gadoid outburst started in the North Sea and the second that when the first signs of the reversal in the Russell Cycle appeared (see below). More generally, the recent period of warming continued until the late 1940s and subsequently a period of cooling set in. Both the Russell Cycle and the spread and return of the intertidal barnacles obviously are linked to the climatic change in a very general way and also perhaps to significant years.

THE RISE AND DECLINE OF THE WEST GREENLAND COD FISHERY

It was noted above that a number of animals appeared for the first time at West Greenland between 1925 and 1934, salmon, haddock, coalfish, spurdog, ling and dealfish. Cod appeared on the offshore banks between 1912 and 1923. Tåning (1950) noted that in former times they were numerous as far north as Disko Bay between 1840 and 1850. Meinardus (1934) showed that there was little ice off Iceland during the period 1840–55 (but there was no reduction in the Davis Strait). They lived in the fjords of southern Greenland, but after 1850 they were probably absent from the offshore banks which were visited from time to time by North American and Faroese cutters (Jensen and Hansen, 1931). The Tjalfe expedition found cod in a few places in 1908 and 1909 (Jensen, 1925) and some experimental fishing in 1906 and 1914 yielded the same result (Hansen, 1949). Annual catches of cod increased from 23.5 tonnes in 1912 to 243.5 tonnes in 1917 and 600 tonnes in 1919–23 (Jensen, 1925). In detail the catches (in tonnes) in the crucial early period were as follows:

1912	23.5	1916	124.5	1920	396.5
1913	89.5	1917	243.5	1921	413.5
1914	95.5	1918	428.0	1922	522.5
1915	82.5	1919	612.0	1923	681.5

The table suggests that the build up in stock was gradual. Because the fish were not found on the offshore banks, the first increase may have occurred in the fjord population. Figure 46 shows the international catches from 1930 to 1972 in thousands of tons (Beverton and Hodder, 1962, ICNAF Statistical Record).

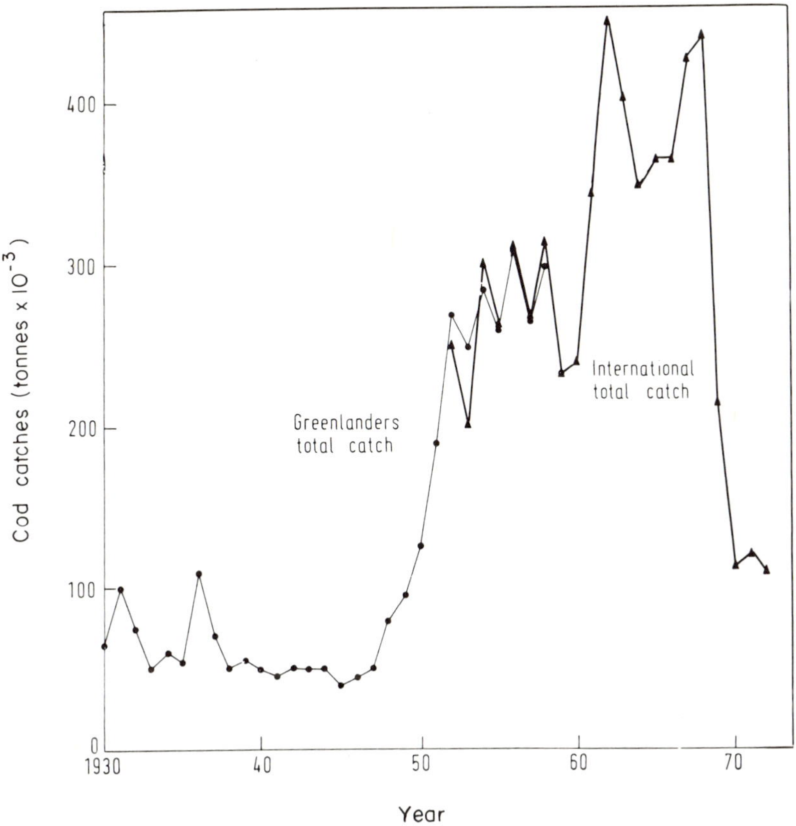

Fig. 46 International catches of cod at West Greenland (Beverton and Hodder, 1962; ICNAF Statistical Record).

Hansen (1954) showed that the rise in catches followed a series of good year-classes, 1917, 1922, 1924, 1926 and, later, 1934 and 1936. Further good year-classes hatched in 1942 and 1945 and were probably the strongest of all; those of 1956, 1961 and 1963 were the last good ones, although there was a significant one in 1968 and a small one in 1974 (see below). The trend of catches in Figure 46 is the fairly obvious consequence of this parade of year-classes. In the early days of the fishery, the positions of catches progressed northwards along the coast of West Greenland: Julianehaab in 1917, Godthaab in 1919, Sukkertoppen in 1922, Holsteinberg in 1927, Disko in 1931 and 72°45′N in 1936 (Jensen, 1939). Off East

Greenland, cod were recorded at Angmagssalik in 1912 and by 1930 were distributed all over the inshore area (Hansen, 1949). In 1952, German fishermen started to fish on the rough grounds off East Greenland and in 1955 they started to work the Dohrn Bank which had been discovered by Dr Arno Meyer.

The 1917 year class probably spawned on the southern coasts of West Greenland in 1925, for eggs and fry were caught on Fiskennes Bank, Størehellefiske Bank and Godthaab fjord. In 1963 on the Norwestlant Expedition, cod eggs were caught off West Greenland, East Greenland, Iceland and right across the Denmark Strait, a much more extensive distribution than that published by Tåning (1937) and Schmidt (1909). In 1903–06, Schmidt found no cod eggs on the north coast of Iceland and they were distributed only on the south and south-west coasts. Jensen (1939) found cod eggs and fry on the north and north-east coasts. Figure 47 contrasts Schmidt's distribution of cod eggs and larvae at Iceland in 1903–6 with that observed on the Norwestlant expedition (Hansen, 1968). From these charts we conclude that cod eggs and fry existed right across the Denmark Strait in 1963, but not in the first decade of the century. Although eggs and fry were found off the southern coasts of West Greenland in 1925, the question still arises whether the appearance of cod on the offshore banks at W. Greenland was due to a drift of eggs, larvae and subsequently 0-group fish from the Icelandic spawning ground. Meteorological evidence on the difficulty of crossing the Denmark Strait during the collapse of the stock in the 1960s (see below) suggests that it was perhaps initiated and then sustained from Iceland; if so, a corollary to the argument would be that the Iceland stock also increased during the recent period of warming. We shall return to this point later.

The link between the Iceland and West Greenland stocks of cod was investigated in a remarkable series of tagging experiments reported by Hansen *et al.* (1935) and Tåning (1937). Between 1924 and 1936 mature and maturing cod were tagged at West Greenland and from 1929–36 they were also tagged on the southern coasts of West Greenland. Fish were recaptured on the spawning ground south and southwest of Iceland during the spawning season (none were taken between October and January). Hence there was a migration of adults from West Greenland to Iceland. The first recapture at Iceland was taken in 1927, but they did not appear in numbers until 1930 and they originated from northern and central areas of West Greenland. During the same period, 4939 fish were tagged at Iceland on the spawning ground and 443 were recaptured there, but 17 were recovered

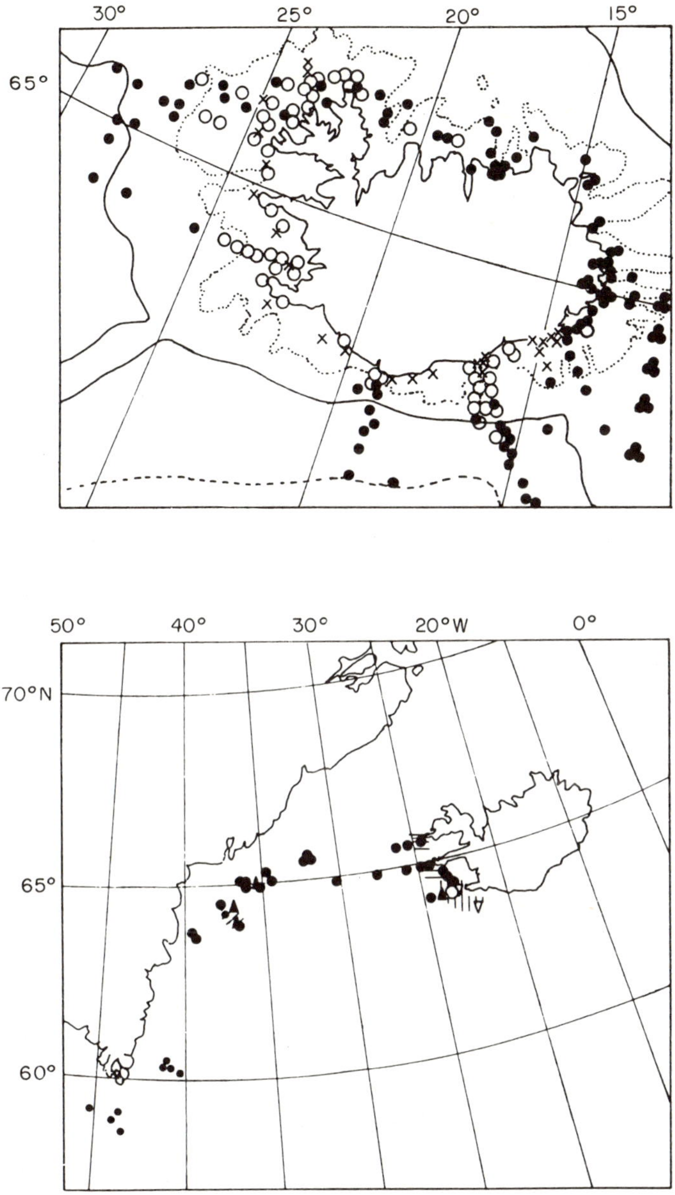

Fig. 47 (*upper*) Distribution of cod eggs and larvae 1903–06 (Schmidt, 1909).

○ Pelagic fry × Fry absent but eggs present ● Nothing

(*lower*) Distribution of cod eggs and larvae in 1963 (Hansen, 1967).

Cod larvae 1–9 ▭ 10–99 ▭ 100–999 ▭

Cod eggs 1–100● 11–100 ▲ 101–100 ▽ > 1000 ○

at West Greenland, but whether the latter were of Greenland or of Icelandic origin is of course unknown. Such tagging experiments were continued after the Second World War, but no fish tagged at Iceland were recovered at West Greenland; further, of cod marked in the northern part of West Greenland, none migrated to Iceland (Rasmussen, 1959). Between 1929 and 1944, 45.1% of all the recaptures were taken on the Icelandic spawning ground, but between 1945 and 1952 that proportion declined to 4.4% (Hansen, 1954). In later years, the proportion remained low, most being recovered from the central region of West Greenland (Horsted, 1973).

The detailed pattern of recaptures by years is shown in Table 10. The most remarkable point is that in 1931, 1933 and 1934 more fish were recaptured at Iceland than at Greenland. Yet the return migration did not start until 1930, although tagging of "mature and maturing" fish started in 1924. Further the proportion of Icelandic recoveries fell after 1949, never to recover. Thus the most intense period of migration lasted from 1931–1944, perhaps when the period of warming was most intense.

Thus there were three parts to the migration: the dominant movement of mature cod from West Greenland to Iceland to spawn, the journey of a few Icelandic spawners back to West Greenland, and the decline in the proportion of migrants from West Greenland to Iceland after 1945. At the same time the stock at West Greenland must have become more self sustained, although as will be shown below, the Greenland stock still contributed a fair proportion to the Icelandic stock. Relatively warmer water could reach the offshore banks at West Greenland when the Irminger Current was strong enough to mix with the cold East Greenland Current and warm it beyond Cape Farewell on the offshore banks at West Greenland. If warm enough, the water might sustain a population there throughout its life history.

After the discovery of the East Greenland fishery, it was noticed that a good spawning fishery occurs off East Greenland eight years after the hatching of a good Iceland year class. Subsequently in their life history some of these fish probably emigrated to Iceland never to return, as do the West Greenland migrants. The good East Greenland year classes were hatched in 1945, 1956, 1958, 1961, 1962, 1963, 1964 and 1968. According to Dr Arno Meyer this succession of good year classes followed a relative decline of those at West Greenland in the 1960s. Two questions arise. Was the East Greenland stock sustained by the drift of eggs, larvae and 0-groups from Iceland and if so did that drift reach East Greenland

Table 10
Recoveries of cod tagged at West Greenland

The 1929–36 experiments			
	Recaptures		
Year	Greenland	Iceland	Reference
1924	1	0	Tåning (1937)
5	15	0	
6	16	0	
7	16	1	
8	6	0	
9	9	0	
1930	16	7	
1	32	47	
2	44	35	
3	22	57	
4	48	53	
5	58	44	
6	31	12	

Year	Percentage recovered at Iceland	Reference
1924–8	1.8	Hansen (1956)
1929–33	54.5	
1934–39	42.1	
1940–4	38.7	
1945–9	3.2	
1950–2	5.7	
1968–9	5.8	Horsted (1973)

but not West Greenland? Did the East Greenland stock become abundant before that at West Greenland during the first two decades of the century?

Figures 43 and 44 showed the change in mean winter sea-level pressure (mb) between the periods 1956–1965 and 1966–70, compared with a similar change between the periods 1900–39 and 1956–65. The pressure over West Greenland increased between the first and second periods but "a remarkable difference of pressure" occurred "between Greenland and the eastern Norwegian Sea; it has increased since 1950 in every month of the year, and the annual mean value of the differences between 40° and 0°W at 70°N for the 1950's decade was 4.2 mb compared with 1.2 mb for 1900–39" (Lamb, 1965). The main effect is that the westerly winds in winter across the Denmark Strait in the early part of the century were replaced by northerly winds during the late 1960s. The last good year class at West Greenland was that of 1963, and it is possible that during the period of northerly winds, the drift of eggs, larvae and 0-group cod from Iceland was reduced. If this is so then we might expect that a period of northerly winds in the Denmark Strait existed up till the second decade of this century, after which supervened one of westerly winds. Then, we would suppose that the West and East Greenland fishery expanded during the 1920s to 30s supported by drift of eggs, larvae and 0-group fish in the Irminger current from Iceland.

The difference in extent of the spawning ground of cod at Iceland between Schmidt's observations in 1903–06 and those in 1963 in the Norwestlant expedition suggests an expansion of the Icelandic stock during the period of warming. As Greenland migrants which are fully mature obviously joined the Icelandic spawning stock never to return to Greenland, there must have been a considerable augmentation of stock at Iceland due to that source alone. The question arises whether there was a native augmentation of the Iceland stock.

Jonsson (1954) published a long series of observations on the age distribution in numbers caught in the Icelandic spawning fishery from 1928 onwards. The North West Working Group of the International Council for the Exploration of the Sea has published a set of stock estimates in numbers calculated by cohort analysis from 1955 onwards (Anon, 1976). Unfortunately there is a gap in the material in the years 1951–4. Cushing (1982) attempted to bridge the gap assuming common exploitation rates by age for the missing years, i.e.

$$N_1/N_2 = \exp -Z = (C_1E_2)/(C_2E_1)$$

where

N_1 is stock in numbers at age n in a year class;
N_2 is stock in numbers at age $n + 1$ in a year class;
C_1 is the catch in numbers at age n in a year class;
C_2 is the catch in numbers at age $n + 1$ in a year class;
E_1 is the exploitation rate in the year class at age n;
E_2 is the exploitation rate in the year class at age $n + 1$.

If $E_1 = E_2$ then $C_1/C_2 = \exp -Z$. Further,

$$\text{in year 1, } C_1 = E_1/N_1$$
$$\text{in year 2, } C_2 = E_2N_2$$

and

$$(C_1/C_2) = (E_1/N_1)/(E_2N_2) = (E_1/E_2) \exp -Z$$
$$E_1 = (C_1/C_2)E_2 \exp Z.$$

With a table of E and ($\exp Z$), E_1 is calculated from E_2 and Z. Then using the estimates of E in 1955 from the North West Working Group material, it is possible to estimate the exploitation rates in the spawning fishery back to 1928 by working backwards. The gap between the two sets of material was bridged using estimates of Z as shown above. Jónsson (1966) calculated the total catches by year classes and such figures were used to correct the estimates of numbers caught at each age group. From material published on length at age, stock numbers in the spawning fishery were converted into weight. Ratios of ages 2/8, 2/9, 2/10 in numbers were averaged for the whole data set and so estimates of numbers in age groups 2–7 were obtained by raising the numbers of the older fish by such averaged ratios; in other words, differences in year classes were established in the younger age groups from differences amongst in older fish. The whole array of stock numbers at age was averaged by age, and then those year classes significantly different from the mean in the spawning fishery were classified as of Greenland origin.

Figure 48 shows the Greenland year classes at Iceland estimated with this somewhat cumbersome method between 1920 and 1967. They appeared

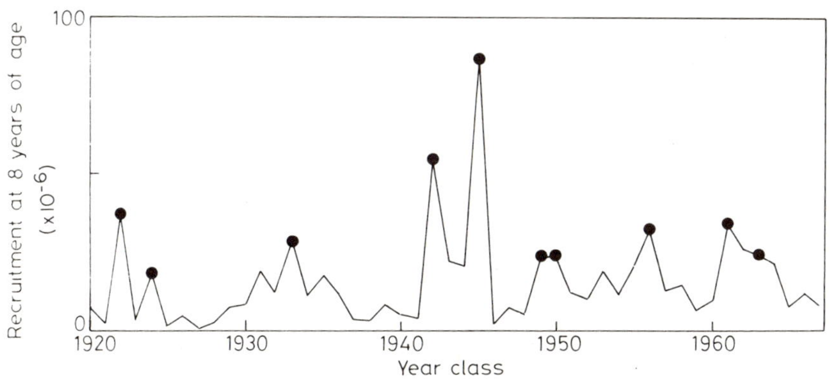

Fig. 48 Greenland year classes (●) in the Icelandic spawning fishery as estimated by Cushing (1982); see text for method.

in 1922, 1924, 1934, 1942, 1945, 1950, 1956, 1961 and 1963. At West Greenland, the notable year classes were recorded by more direct methods in 1917, 1922, 1924, 1926, 1934, 1942, 1945, 1950, 1956 and 1963; those of 1926, and 1961 were not detected by Cushing's cumbersome method. It will be seen that the year classes of 1942 and 1945 were the largest of those "exported" to Iceland. At West Greenland they formed the basis of the much increased catches during the 1950s. Figure 49 shows the time series of Icelandic year classes, i.e. those recorded between 2 and 7 years of age in the pre-spawning fishery. It will be seen that some strong ones are common to Iceland and West Greenland, 1924, 1926 and 1949, but they are only a minority. The most important point, however, is the appearance of five strong year classes in succession, 1931–5, each larger than any other except that of 1924. These strong year classes, hatched towards the end of the dramatic decade, laid the foundations of high stock during the 1940s which persisted into the 1950s because fishing was reduced during the Second World War. The largest Greenland year classes of 1942 and 1945 may have been a direct consequence of the high stock generated by the Icelandic year classes of 1931–5.

The West Greenland stock collapsed in the late 1960s, although a minor recovery with the 1974 year class might have occurred. Hence the input of Greenland migrants to the Iceland spawning fishery was reduced. Cushing (1982) showed that such a reduction in stock might be associated with

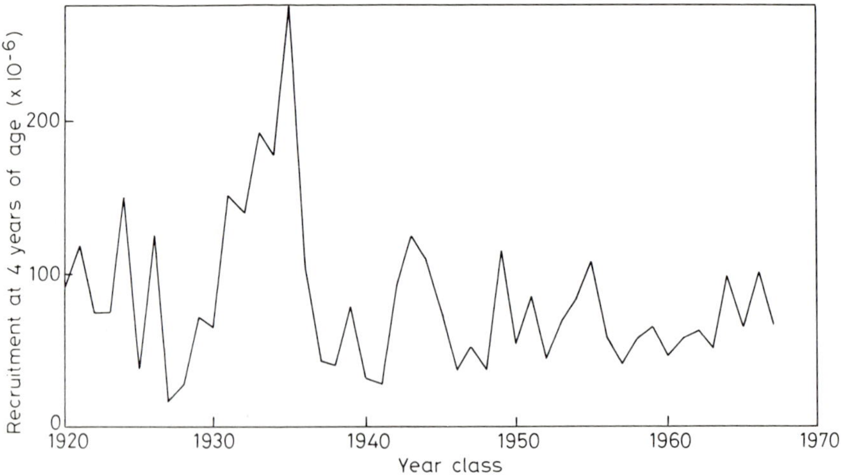

Fig. 49 Icelandic year classes in the Icelandic pre-spawning fishery as estimated by Cushing (1982); see text for method.

reduction in recruitment of about one third. Hence the decline of the West Greenland stock possibly due to climatic factors might well have generated decline in the Icelandic stock. Above it was suggested that the Icelandic stock in 1903–06 was smaller than that in 1963. The rise of the West Greenland cod stock and its subsequent decline was accompanied by a similar, but smaller, rise and fall of the Icelandic cod stock.

The rise of the West Greenland stock cod started in the 1920s, probably with the year classes 1917, 1922 and 1924 and ended perhaps with that of 1963, although minor year classes have been recorded since. The recent period of warming reached a peak in 1945 or 1950, and subsequently one of cooling has supervened. If it is a true period, the stock appeared and disappeared at certain positions in it and we would expect different positions for different stocks and hence appearances and disappearances at different times. If cooling were to continue, we might expect the cod stocks at Iceland to become reduced to some degree and the cod to withdraw from the Barents Sea—a return to the conditions between 1880 and 1920.

THE RUSSELL CYCLE

From 1924 to 1939, Sir Frederick Russell collected weekly samples of macroplankton with a stramin trawl at International Station E1, in the English Channel close to Plymouth. Phosphate observations at the surface and near the sea bed were collected at the same time. After the Second World War samples were collected much less frequently. In October 1965, Sir Frederick noticed the presence of the scyphomedusan *Pelagia noctiluca* (Forskål) with ephyra larvae, and in the following summer the animal was widespread as a result of breeding (Russell, 1967); he regarded this event as an indication of return to the conditions of the 1920s. Sampling was intensified once more, and changes in the ecosystem which had occurred between 1925 and 1935 were reversed between 1965 and 1979. The profound events were named the Russell cycle (Cushing and Dickson, 1976).

There are five prominent components in the cycle: the magnitude of the winter phosphorus maximum; the quantity of fish larvae; the quantity of macroplankton; the presence or absence of the arrow worms, *Sagitta elegans* Verrill and *Sagitta setosa* J. Müller; and the appearance and disappearance of a pilchard population.

In the autumn of 1931, the quantity of macroplankton fell by a factor of four, and *Sagitta elegans* was replaced by *S. setosa,* which itself decreased in numbers between 1931 and 1935. The quantity of winter phosphorus declined from about 0.7 μg-at/l to less than 0.5 μg-at/l between 1929 and 1931; in subsequent years the quantity ranged between 0.4 and 0.5 μg-at/l. The herring population started to decline with the 1926 yearclass, and in 1935 large numbers of pilchard eggs were caught in the stramin trawl (although they had been recorded in 1926). After 1935 the numbers of non-clupeid fish larvae were much reduced; it was originally thought that the numbers of summer spawned larvae declined after 1931 and that the spring spawners diminished after 1935, but only the summer-spawned larvae were well sampled before 1930.

In 1965 the total numbers of fish larvae increased by about an order of magnitude and they (including numbers of summer spawned larvae) increased again in 1966. In the winter of 1970-1, the winter phosphorus increased again to 0.62 μg-at/l, but it had risen to 0.58 μg-at/l in 1969–70. In the autumn of 1970 the quantity of macroplankton increased again by a factor of four. *Sagitta elegans* replaced *S. setosa* in the autumn of 1970, but the abundance did not reach the level of the 1920s until 1978 (Southward, 1980): Southward also showed that in 1970, the numbers of copepods returned to their previous level and that the numbers of pilchard eggs declined in 1971 and in 1973. The cycle is illustrated in six figures (Fig. 50a–e). Figure 50a (Russell, 1973) shows the changes in numbers of some groups of fish larvae. Figures 50b, c, and d are, respectively, Southward's (1980) description of the cycle, the winter phosphorus, the annual numbers of spring- and summer-spawned fish larvae, the annual numbers of pilchard eggs between April and July, and the annual mean numbers of *Sagitta elegans* between May and August. Figure 50e is a summary. Figure 50 displays the periodic nature of the Russell Cycle in so far as a period can be described in half a wavelength.

There were a number of additional events. *Laminaria ochroleuca* de la Pylae, a seaweed recorded in the Plymouth area in the late 1940s, had spread all over the south west by the early 1960s; it declined sharply between 1965 and 1975 (pers. comm. Dr G.T. Boalch). Some species of shell gravel molluscs disappeared between the 1920s and the 1950s: *Myrtea spinifera* (Montagu), *Gouldia minima* (Montagu), *Tellina pygmaea* Loven and *Thracia villosiuscula* (Macgillivray) (Holme, 1961). The stock density of spurdogs (*Squalus acanthias* L.) decreased after 1931, but those of skates and rays increased (Cooper, 1948). Southward (1963) noted that northwesterly plankton forms such as the medusa *Aglantha digitalis* (O.F. Müller), the polychaete worm *Tomopteris helgolandica* Greef, the pelagic mollusc *Spiratella retroversa* (Fleming), the euphausids *Meganyctiphanes norvegica* (M. Sars)

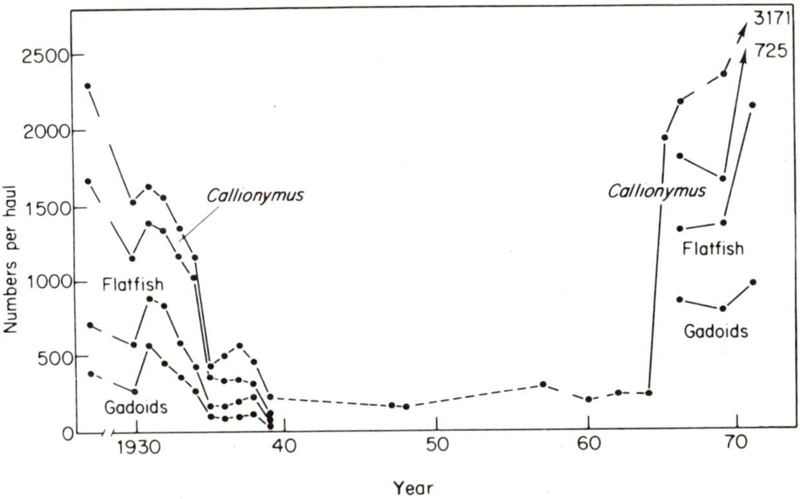

Fig. 50 The Russell cycle:

(*a*) Changes in time series in numbers of some groups or species of fish larvae (Russell, 1973).

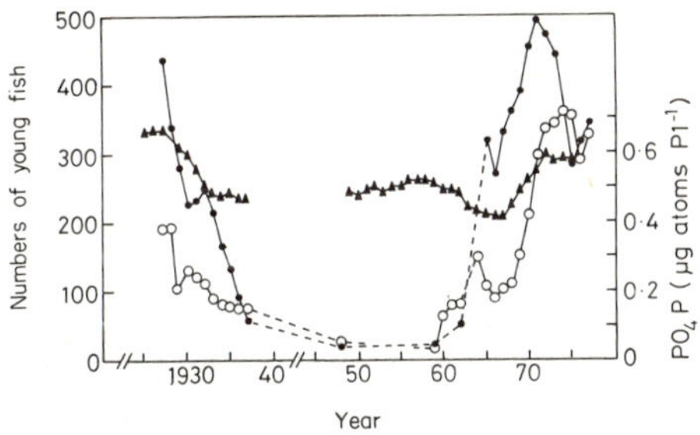

(*b*) Changes in winter phosphorus (▲), in numbers of spring spawned fish larvae (●) and in numbers of summer spawned fish larvae (○) (Southward, 1980).

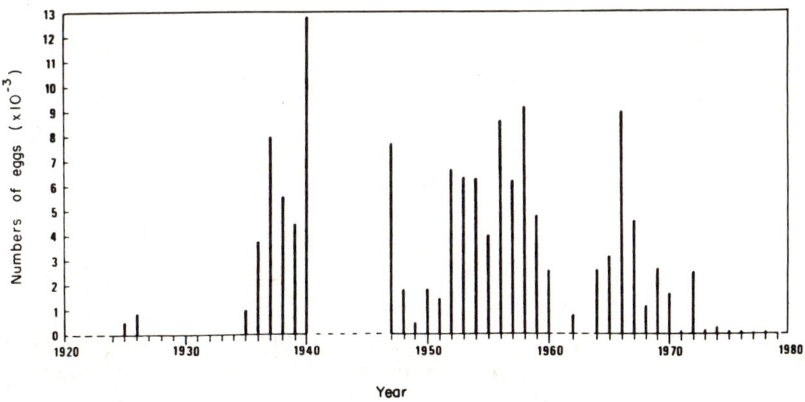

(*c*) Changes in numbers of pilchard eggs, sums of monthly means (Southward, 1980).

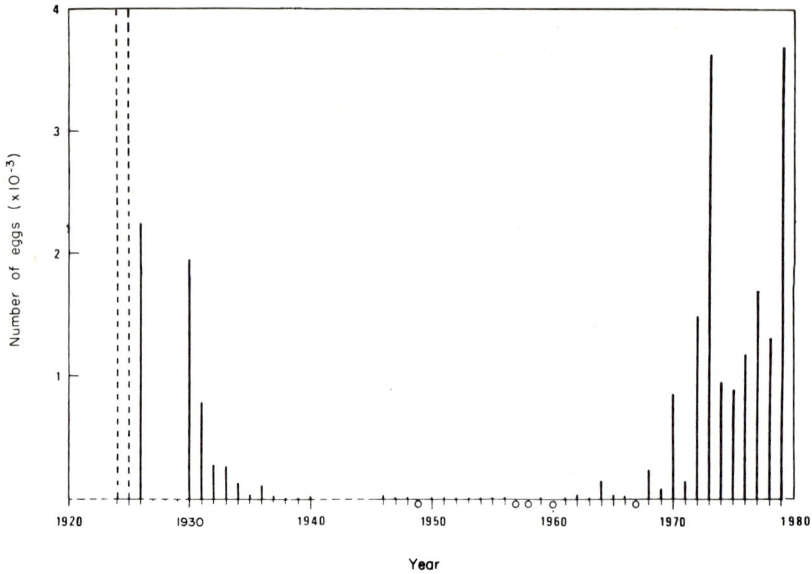

(*d*) Annual mean numbers of arrow worm *Sagitta elegans* between May and August (Southward, 1980).

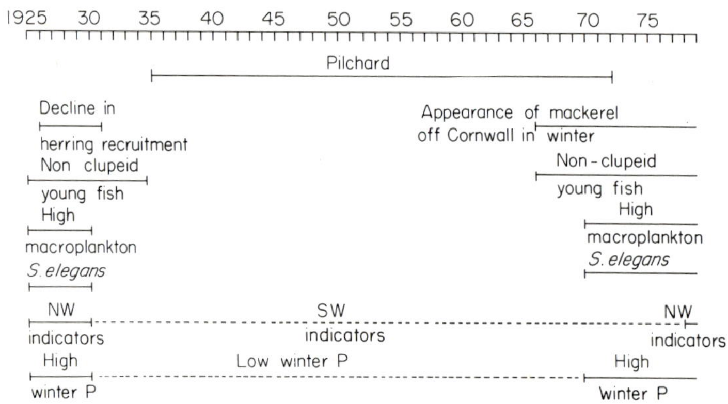

(*e*) Diagram summarizing the main changes described as the Russell cycle.

and *Thysanoessa inermis* (Krøyer) disappeared after 1930 and were replaced by southwesterly forms such as the copepod *Euchaeta hebes* Giesbrecht, *Liriope tetraphylla* (Chamisso and Eysenhardt), and the euphausid *Nyctiphanes couchi* (Bell). Southward (1963) recorded the changes in stock densities of demersal fishes as sampled by the research vessels of the Marine Biological Association in Plymouth (Table 11).

Table 11

Increases and decreases of demersal fish species between 1919–22 and 1949–52 as sampled by the research vessels of the Marine Biological Association at Plymouth (Southward, 1963)

	Increase			Decrease	
× 10	(*Cepiola rubescens* L.)	S	× 100	(*Squalus acanthias*)	G
	(*Merluccius merluccius* L.)	S			
	(*Mullus surmuletus* L.)	S	× 10	(*Capros aper* L.)	S
	(*Trachurus trachurus* L.)	S		(*Trigla gurnardus* L.)	N
× 3	(*Scophthalmus rhombus* L.)	S	× 3	(*Gadus merlangus*)	G
				(*Limanda limanda* L.)	N
Some increase	(*Conger conger* L.)	S		(*Trigla lineata* Gmelin)	S
	(*Lophius piscatorius* L.)	S		(*Zeus faber* L.)	S
	(*Platichthys flesus* L.)				
	(*Scyliorhinus stellaris* L.)	S	Slight increase	(*Gadus morhua*)	N
				(*Gadus luscus* L.)	S
Not recorded earlier	(*Pagellus centrodontus* de la Roche)	S		(*Pleuronectes platessa*)	N
				(*Raia* spp	–
				(*Scyliorhinus canicula* L.)	G
				(*Solea solea*)	S
				(*Microstomus kitt* Walbaum)	N
				Molva molva disappeared	

Distribution S Southerly
N Northern
G General

The species that increased in abundance were all southerly, but those that decreased were of all three distributional categories, five northerly, five southerly, and three general. Perhaps some southern species were replaced by others; if so, southern species increased and northern species decreased. Southward *et al.* (1975) analysed sea surface temperatures at E1 between 1924 and 1972 and detected a secular trend of warming which peaked in 1945. Catches of hake and numbers of summer pilchard eggs were correlated positively with the trends in temperature, whereas cod catches and numbers of autumn pilchard eggs were correlated negatively with them. Southward did not mean that such effects were a necessary direct effect of temperature, but perhaps that changes in temperature were convenient indices of more complex ones. Hence in the Western Channel, hake and cod alternate in warm and cool periods respectively, as do summer and autumn pilchard spawners (distinguished by average egg sizes, Southward, 1974b). Thus there were considerable changes in other parts of the ecosystem which implies that most of its structure was altered.

Cushing (1961) interpreted some of the observed changes in terms of competition between herring and pilchard. Stock densities of recruiting herring were positively correlated with the winter phosphorus *one year after they were hatched*. Numbers of pilchard eggs were negatively correlated with the winter phosphorus *six months after they were hatched* and it is the same quantity of winter phosphorus positively correlated with herring recruitment. Linked by the quantity of winter phosphorus, herring hatched in winter appeared to be inversely related to pilchards hatched in the following summer. The two correlations are illustrated in Fig. 51 (which excludes information now available from the 1950s), and the important point is made that the two correlations succeed each other in time; in other words, competition between the two populations can only be inferred indirectly. Recruitment to the herring population started to decline with the year class of 1925 and continued until it was extinguished with that of 1930. Although pilchard eggs were recorded in June, July and September 1926, they were not observed in abundance until 1935. The sharp increase in eggs in that year probably indicated a strong year class (perhaps 1930) which recruited and spawned in 1935. The pilchards that flourished during the Russell cycle were summer spawners, not autumn spawners.

It was assumed that the decrement of winter phosphorus from 1926 onwards represented an increment of six-month-old pilchards. The numbers of clupeid larvae taken between May and August (probably pilchards) remained low during the decade 1925–35, so any change would have been

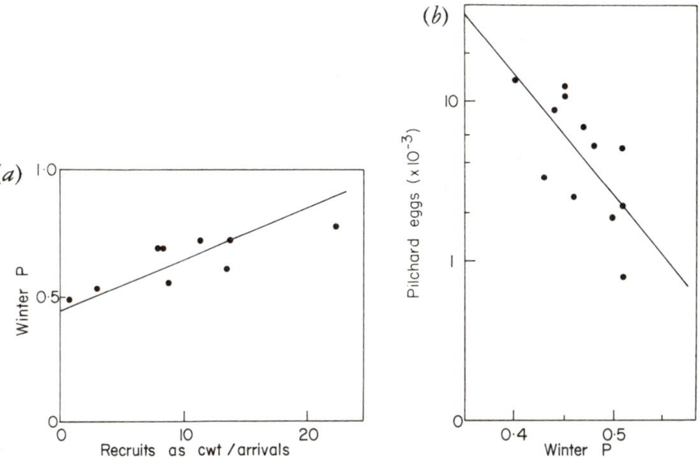

Fig. 51 (*a*) Relationship between the magnitude of recruitment to the herring stock and the winter phosphorus one year after the fish were hatched between 1921 and 1930 (Cushing, 1961);
Fig. 51 (*b*) Relationship between numbers of pilchard eggs as sampled at International Station E1 and the winter phosphorus six months after they were hatched (Cushing, 1961).

in the survival of larvae older than those caught by the stramin net. The major decrement of winter phosphorus occurred between 1929 and 1931, which may have indicated the first major pilchard year class.

Southward (1980) confirmed the original correlation shown in Fig. 51b and found a correlation between the density of pilchard eggs and winter phosphorus in *the winter before,* which may deny the thesis of competition. There are three possible explanations:

(a) Differences in winter phosphorus represent differences in the pilchard population;
(b) Winter phosphorus and pilchards are independent but follow similar long term trends, but then the relationship shown in Fig. 51a remains unexplained;
(c) When the system reaches a steady state, there is a serial correlation between phosphorus and pilchard from year to year; or the correlation between the density of pilchard eggs with phosphorus in the winter before need not surprise us.

The second possibility seems unlikely and as the correlation in Fig. 51b is confirmed, it is possible that the thesis of competition stands.

A further correlation was found between winter phosphorus and the number of pilchard eggs 4½ years later in the pre-war material and 2½ years later in the post-war material (Cushing (1961) suggested that pilchards recruited earlier in their lives after 1950). Hence the postulated increments in six-month-old pilchards were inversely related to the quantity of stock so many years later, and any compensatory mortality must have occurred between six months and the age of the recruitment. An inverse relationship established between numbers of pilchard eggs and pilchard eggs three years later suggests that this interpretation is possibly right. If compensatory mortality occurs in any case, the hypothesis required only increased survival of young fish between two and six months of age. Perhaps the increased survival was due to decreased predation by the herring. When the stock of pilchards declined in the early 1970s, herring were found by research vessels; no fishery developed although Southward (1979) has suggested that in fact mackerel replaced pilchards rather than the herring.

The first event in the Russell Cycle was the decline in herring recruitment between 1926 and 1930. If the decrement of winter phosphorus represented an increment in six-month-old pilchards, a strong year class was hatched in the summer of 1930 which recruited to the adult stock and spawned in 1935. Numbers of fish larvae were reduced in that year, perhaps because they were eaten by the increased number of adult pilchards. The changes were reversed in an extended and variable way in time. The numbers of fish larvae increased by nearly an order of magnitude in 1965 and that increment was sustained and augmented by 1971; at the same time the numbers of pilchard eggs declined and had disappeared by 1973. The winter phosphorus reached 0.58 μg-at/l in 1969 and 0.62 μg-at/l in 1971, which on this hypothesis would represent a decline in six-month-old pilchards. The decrement of macroplankton in 1930 and the increment in 1970 would have been the consequence of increased and then decreased competition with the six-month-old pilchards. The 1930 year class of pilchards may have been a very strong one (like the famous 1904 year class of Norwegian herring), and the changes were consequently very clear. At the end of the cycle, the recruiting year classes had failed and the stock faded as the older fish died away; the failure of recruitment must have started before 1965, but we do not know when. Similarly we do not know why the herring recruitment declined in 1926.

Southward (1979) has also noticed that the reversal of the Russell Cycle might have lasted longer than might have been rigidly expected. Not until 1978 did the numbers of *Sagitta elegans* and *Aglantha digitalis* reach the levels of the 1920s. The Norway pout, *Trisopterus esmarki* Nilsson, and the blue whiting, *Micromesistius poutassou* Risso, appeared on the local trawling grounds in the spring of 1979. Coombs and Pike (1978) have shown from the plankton recorder survey that numbers of blue whiting decreased on the Stornoway–Iceland line from 1965 onward and that they increased on lines west and south west of Ireland numbers increased from 1962 onwards. Hence the appearance off Plymouth of blue whiting in 1979 may have started in the southerly movement of spawners off the continental shelf in 1962–5 (the period when the first signs of reversal of the Russell Cycle appeared). He noted that *Zostera* may be returning to the shores of south west England and Brittany (Glemarec, 1979) and that *Meganyctiphanes norvegica, Thysanoessa inermis* in addition to *Aglantha digitalis,* the northwesterly forms became very abundant in April 1979. Southward suggests that a distributional shift has occurred, a return to the geography of the 1920s, an earlier stage of the Russell Cycle, which are driven by changes in advection, changes in the oceanic circulation. The prime event in this interpretation is the change from one indicator species to another, the arrow worms. As will emerge, the Russell Cycle is linked to changes in the oceanic circulation in a general sense, and it is even possible that the indicators did change in response to some long advective link. It remains mysterious, however, how the same changes in circulation caused the herring recruitment to decline and to set in motion the competitive changes within the ecosystem in the western English Channel. The most remarkable point about the changes was their pervasiveness; the stage was set with one group of actors with understudies waiting in the wings.

Southward (1974) believes that an alternation between herring and pilchard occurred earlier. A summer pilchard period probably took place during the first half of the seventeenth century, the early years of the eighteenth century and in the first thirty years of the nineteenth century. The cycle has been part of the recent period of warming which peaked between 1945 and 1950 and must be linked to the other changes that we have described. It is not then surprising that the Russell Cycle occurred in earlier years, rather like the Norwegian herring period, which may be linked to earlier periods of warming.

CONCLUSION

During the recent period of warming, *Octopus* bred in the south western part of the English Channel in the warmest years, *Physalia* was blown from the central waters of the North Atlantic and Atlantic benthic species spread northwards towards Svalbard and on the Svalbard shelf. Subtropical invaders appeared from southerly waters in the seas round the British Isles, particularly in 1932, although there were records in 1931, 1933, 1934 and 1935. Subtropical invaders were found off the Pacific North West in 1926 and off California in 1931. Arctic species appeared in the Bay of Fundy between 1932 and 1936 and the southern animals were found during the same period in the Gulf of Maine. A more widespread and diffuse migration occurred of boreal animals to the Faroe Islands, Iceland, Greenland, Jan Mayen and the Barents Sea. The records occurred between 1924 and 1949, but most appeared between 1930 and 1935. The West Greenland cod stock increased with the year classes 1917, 1922, 1924, 1926, 1934, 1942, 1945, 1950, 1956 and 1963 and subsequently collapsed. The Iceland cod stock benefited greatly from very strong year classes between 1931 and 1935. The Russell Cycle started in the decade between 1925 and 1935 and reversed between 1966 and 1979. The overwhelming impression is of changes both profound and widespread in that dramatic decade between 1925 and 1935.

But some changes were less linked to that decade but to the more general period of warming between the twenties and 1950. Such were the strengths of West Greenland year classes of cod, the appearance of boreal animals at Iceland between 1924 and 1949 and perhaps some of the outer events associated with the Russell Cycle such as the disappearance of some shell gravel animals. As indicated in Chapter 4, the recent period of warming was probably accompanied by an intensification of atmospheric and oceanic circulation. Figure 118 (p. 329) shows that the strength of the upper westerlies started to diminish in the mid 1920s. Then the influx of "indicator" organisms into the North-east Atlantic during the dramatic decade, 1925–35, occurred as the intensification of circulation may have relaxed. Iselin (1940) suggested that the gyre tightens under intenser circulation and spreads when it slackens. An analogue of such a thesis might explain how the swordfish and its counterparts reached Iceland, Greenland and the Murman coast during the period after 1925. However, it is as yet impossible to sustain or deny such a thesis in physical terms. It is interesting to note that the general westerly circulation over the British Isles peaked in about 1925 and then declined. The immediate result of slackening winds in the north-west Atlantic was to retain more heat in the sea because there was less evaporation, and peak temperatures were recorded between 1945 and 1950.

The biological consequences of the events of the dramatic decade were as follows: subtropical invaders appeared off California, the Gulf of Maine, and the Bay of Fundy and the British Isles; the year classes of the Icelandic cod stock were the strongest recorded; and the Russell Cycle started. The more general consequences of the recent period of warming were: the appearance of boreal invaders in high latitudes in the North-east Atlantic; the spread of barnacles on British coasts; the colonization of the banks off West Greenland by cod, probably from Iceland; the spread of Atlantic benthic animals on the Svalbard shelf; and the appearance of *Octopus* in the English Channel and of *Physalia* off the coasts of Cornwall. The events of the dramatic decade are possibly associated directly with the intensification of the North Atlantic subtropical gyre most pronounced between 1926 and 1931 whereas those associated more generally with the recent period of warming between 1920 and 1945–50, mainly after the dramatic decade, were consequential. The colonization of West Greenland by the cod from Iceland is perhaps associated with local meteorological events, which of course might be steered by changes in the more general wind distribution of the North Atlantic.

The economic consequences of these changes were considerable. The fishery for cod at West Greenland flourished only during the recent period of warming; annual catches reached as much as 450 000 tonnes, which no longer exist. The Iceland cod stock was sustained by the great year classes from the last half of the dramatic decade and by immigrant fish from Greenland; the yields are today perhaps a third less than in the 1930s and 40s. The fishery for cod in the Barents Sea started in the late 1920s (as recruitment to the stock recorded in the Vestfjord), and the year classes of 1976–8 are well below average. It is possible that as cooling supervened since about 1950, between 500 000 and 1 000 000 tonnes of cod have been lost purely as a result of the climatic change.

Chapter Six

The Short-Term Changes in Fish Populations

As shown in earlier chapters, fish populations respond to climatic changes in one way or another. The death of the tilefish was possibly a direct effect of cool water from the north on the whole population. The changes, however, shown in the herring, pilchard and West Greenland cod populations in the last chapter are changes in recruitment, or year-class strength, from time to time of several orders of magnitude: the 1904 year class of the Atlantoscandian stock of herring was probably such an example as compared with its predecessors. Such are major changes over long time periods. This chapter will describe the effects of climatic change from year to year upon recruitment, i.e. the short-term changes in fish populations.

OTTESTAD'S CORRELATIONS WITH UNSPECIFIED CLIMATIC FACTORS

Ottestad (1942) introduced a method of linking records of fish catches (or catches per unit of effort) with periods extracted from the differences in the widths of annual rings on sections of pine and spruce trees in long series. Erlandsson (1936) had examined trees from four localities in northern Sweden and northern Finland. The measurements from each tree were corrected for age because, as in fish, the increments of growth decrease with age. The corrected measurements were shown to be normally distributed and correlated with temperature. The material from Karesuando was smoothed by moving means of threes and analysed by Fourier analysis: the following periods above noise level were extracted:

2.75, *8.50*, *11.33*, 18–19, *23*, *35*, 38 and *57*

Ording (1941) extended the observations to include four forests in Nordland near the southern edge of the Vestfjord in northern Norway; in this group there were three samples of 20 trees and one of 5. He constructed a compound curve from the underlined periods above and correlated it with the observations; for the period 1887–1936, $r^2 = 0.58$. Ottestad constructed a similar curve, with four periods, 11, 17.5, 23, and 57 years. Annual numbers of cod caught in the Vestfjord were used between 1885 and 1929, and since 1929 catches in weight were converted to catches in number. They were set back by seven years because differences in catch from year to year were assumed to be due to differences in the magnitudes of recruit year classes hatched seven years before. Ottestad's original work showed a clear correlation between the compound curve and the lagged annual catches. In later analysis, Ottestad (1969) added two periods and fitted data on catches to his compound curve between the years 1885 and 1956 (Fig. 52a). Figure 52b shows another relationship based on the catches per unit effort of the cod. The 11-year cycle is the sunspot cycle. The 17.5-year cycle appears in Mitchell's analyses of Sirén's tree ring data from Lapland (Lamb, 1977). The 23-year cycle is the double sunspot one, when the sun's magnetic field changes, which was detected by Mason (1976) in a power spectrum of the Manley record of Central England air temperatures (with the long-term trend removed); the sunspot cycle was hardly detected in this series. The 57-year cycle appears in the power spectrum of $\delta^{18}O_2$ samples from the Greenland ice cap (56 years with a wide bandwidth) (Dansgaard *et al.*, 1971). Thus there is some independent evidence that Ottestad's periods are of climatic importance. Zupanovitch (1968) made a similar analysis of catches of the Adriatic sardine between 1930 and 1960 and was able to make successful forecasts of catches. Ishevskii (1961) related catches of cod from the Arctonorwegian cod stock to a compound curve made of three periods (4.5, 9.0–10.0, 18.5 years) extracted from material on sea temperature and sea level.

The series of tree ring material from Ording (1941) was examined with spectral analysis with digital filtering (with a program of John Shepherd's): Ottestad's first three periods were detected significantly above red noise. All four periods used by Ottestad originally have been detected in other data series, all four of which might be relevant to a study of fish populations.

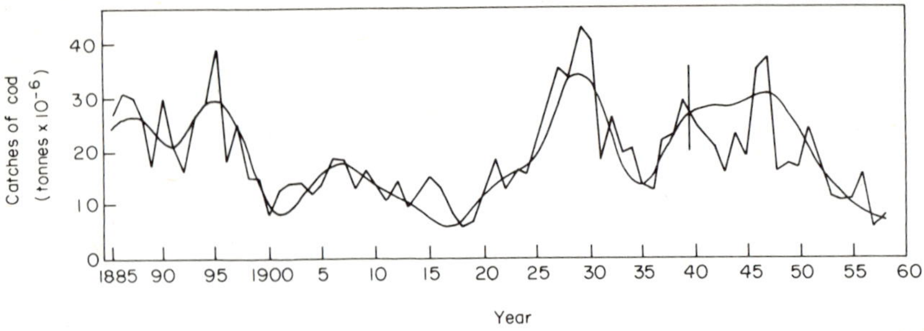

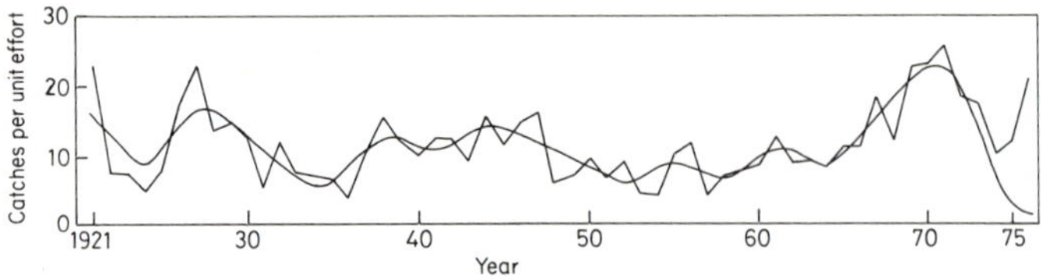

Fig. 52 Ottestad's correlations with fish catches or stock densities with a form of harmonic analysis: (*a*) the correlation between catches of the Arcto-norwegian cod stock from 1885–1956 and a curve derived from five periods extracted from differences in widths of annual rings on pine trees; (*b*) the correlation between stock density (catch per unit of effort) of the same stock of cod and a similar curve.

Yet no causal connection between a particular period and meteorological events has been proposed, let alone between that and the generation of recruitment.

If, however, differences in catches or stock densities as indices of differences in year class strength are really correlated with differences in tree ring widths, the link is necessarily unspecified but the choice of factors is restricted. The physical factors common to the two environments are sunlight and temperature, which are themselves related to wind strength and cloudiness. Such are the factors that govern the primary production in the sea, in which case a component of recruitment may be associated with annual differences in primary production.

THE COMMON YEAR CLASSES IN THE NORTH ATLANTIC

It is a common experience for fisheries biologists to find strong year classes in stocks of the same species that live far apart from each other. For example the 1976 year class of cod in the North Sea was strong, as it was in the Irish Sea, and the 1976 year class of sole was also strong in both regions (Anon., 1979). The phenomenon was analysed by Templeman (1965) who examined ten stocks of cod, haddock and herring between 1902 and 1962. Year classes were ranked in five groups, but in the earlier years of the series the age distributions were not expressed in numbers or in catches per unit of effort and there is an element of subjectivity in this classification. Templeman's results are summarized in Table 12, in which some of the outstanding year classes are listed.

In some years (e.g. 1949, 1950, 1956) quite good year classes are common to six stocks or more; further, in 1904 and in 1950 outstanding year-classes were common to gadoids and clupeids. Of 60 year classes a small proportion would be expected to be strong in any case, but 9 out of 10 in the one year (1950) appears improbable.

Templeman (1972) approached the problem again with the period 1941–70 for cod and haddock stocks only, and all estimates of year classes were expressed in stock density. He classed the recruitments in five groups (one to five stars) and they may be added by years in each region, provided that none of the stocks suffers from recruitment overfishing. Figure 53 shows such indices for the whole period in the North Atlantic, separately

Table 12
Strong year classes common to cod, haddock and herring stocks in the North Atlantic between 1902 and 1962 (Templeman, 1965)

1904	Atlantoscandian herring★★★★★; Arctonorwegian cod★★; Icelandic haddock★
1922	West Greenland cod★★★; Iceland cod★★★★; Arctonorwegian cod★; Icelandic haddock★★
1934	West Greenland cod★★★; Iceland cod★★; Arctonorwegian cod★; Atlantoscandian herring
1943	All cod stocks and Grand Bank haddock
1945	Six stocks★ and ★★
1949	Seven stocks★ and ★★
1950	All ten stocks except Grand Bank haddock; Arctonorwegian haddock★★★★; Arctonorwegian cod★★★★; Atlantoscandian herring★★★★★
1955	Iceland cod; Grand Bank haddock
1956	All (except Atlantoscandian herring)★ and ★★

The stars indicate individual strengths of year-classes.

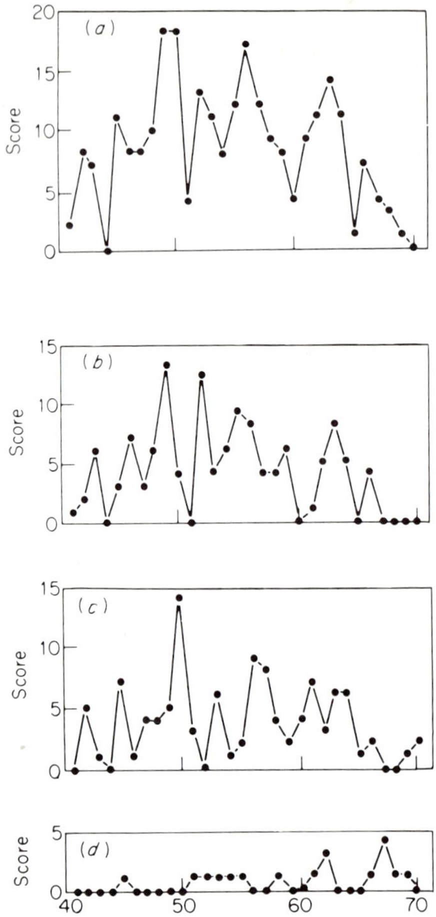

Fig. 53 Templeman's indices of gadoid recruitment from 1941 to 1970: (*a*) in the north-west Atlantic, the ICNAF area; (*b*) in the north-east Atlantic; (*c*) in the North Sea.

in the north-west Atlantic (the then ICNAF area), in the north-east Atlantic (including West Greenland) and in the North Sea. The estimates from 1967 onwards include year classes which are not fully represented because the time series ends in 1970. In both the north-east and north-west Atlantic the year classes rose towards a peak in 1949 (in the west) and 1950 (in the east) after which they declined; when we recall that the recent period of warming reached a peak between 1945 and 1950 and was succeeded by a period of cooling, the distributions suggest that the gadoid stocks responded directly to the climatic changes expressed in the long-term changes in air or sea temperature. Stronger year classes in 1949, 1952 and 1955 in the north-west Atlantic were succeeded by analogues in 1950, 1953 and 1956 in the north-east, which is an additional point of evidence that the determining processes are common to broad areas. Towards the end of the period in the 1960s in the North Sea the cod, haddock, whiting and saithe stocks increased by a factor of three (see Chapter 7). The gadoid outburst, in the North Sea succeeded the period of high recruitment in more northern waters. In earlier times, such changes might have been interpreted as a migration from high latitude waters to the North Sea, but since the gadoid stocks in the North Sea are genetically distinct from those further north (Jamieson and Turner, 1979) the changes are due to increments and decrements of recruitment *in situ*.

Garrod and Colebrook (1978) extended Templeman's approach considerably with 18 stocks (including gadoids, flatfish and herring) for 22 years, 1950–72; the recruitments to all these stocks were estimated by cohort analysis. Their first result is shown in a correlation matrix of recruitment to all 18 stocks (Table 13); the coefficients were contoured at $r = 0.20$ and $r = 0.40$. Recruitment to the stocks of Iceland haddock, West Greenland cod, East Greenland cod and Faroe haddock are all highly correlated with each other, reasonably because all four stocks inhabit the same region. But the recruitment to the stocks of North Sea sole are well correlated with those of North Sea plaice, Georges Bank haddock and Gulf of St Lawrence cod, stocks spread right across the North Atlantic. In 160 correlations we would expect 8 to arise randomly (at $P = 0.05$) and there are 20 in Table 11; hence the correlation matrix reveals real associations across the North Atlantic as Templeman had originally suggested.

If year class strengths of disparate stocks are related to some degree across the North Atlantic they must be determined by causes which are pervasive on an oceanic scale. The number of physical factors common to such an array of regions on an oceanic scale must be small.

THE VARIABILITY OF RECRUITMENT TO FISH STOCKS

The recruitment to fish stocks varies considerably from year to year by about an order of magnitude, sometimes by a factor as low as three, sometimes by as much as two orders of magnitude. If adult natural mortality is not highly density dependent, there may be 16 or 25 age groups in an

TABLE 13

Matrix of correlation coefficients between recruitments to North Atlantic fish stocks

		1	2	3	4	5	6	7	8	9	10	11	12	13	14	15	16	17	18
Arctonorwegian haddock	1	•	61	9	22	12	22	18	18	20	−10	−1	−24	−33	22	−15	−14	−28	8
Arctonorwegian cod	2		•	34	40	53	39	15	4	36	−28	22	−3	1	29	1	−3	−28	−3
North Sea sole	3			•	45	42	44	21	12	8	8	13	23	19	−7	−31	−32	−31	−25
North Sea plaice	4				•	46	24	0	8	41	−8	29	0	16	12	16	−23	−48	−41
ICNAF haddock	5					•	36	33	46	52	25	18	−8	22	0	27	7	−25	−52
Gulf of St. Lawrence cod	6						•	61	30	37	11	14	25	18	9	−10	8	−34	−11
Iceland haddock	7							•	60	55	48	0	27	15	11	23	26	−30	−15
West Greenland cod	8								•	59	50	4	15	14	8	27	12	−25	−48
East Greenland cod	9									•	25	30	13	21	35	43	12	−16	−23
Faroe haddock	10										•	−16	12	15	2	26	22	2	−38
South Newfoundland cod	11											•	26	66	52	10	13	−10	14
Nova Scotia cod A	12												•	41	12	−14	−1	−18	1
Nova Scotia cod B	13													•	40	16	37	−7	−9
Iceland cod	14														•	35	21	7	30
ICNAF herring	15															•	39	5	−12
Faroe cod	16																•	14	0
North Sea haddock + whiting	17																	•	24
West of Scotland herring	18																		•

unexploited and long-lived stock and then that stock is four to five times less variable than the annual recruitment. This comparison alone suggests that the function of the stock with many age groups is to reduce the variability in the recruitment from year to year. In other words the population fecundity with which each new year class starts represents the average of many year classes.

Figure 54 shows the variability of year class strength for some stocks of cod, herring and sockeye salmon for periods of decades. Each set of estimates was calculated by cohort analysis by my former colleague David Garrod. The year classes of the three cod stocks, Arctonorwegian, Iceland and West Greenland vary steadily about a geometric mean (the scale in recruitment is logarithmic), by an order of magnitude. In the 1950s and 60s, the recruitment to the Icelandic cod stock varied much less, by about a factor of three. The variability of the West Greenland cod stock is a little higher, particularly during the 1920s. That of the Arctonorwegian cod stock appears to vary in rather a different way, in almost a *quasi*-cyclic manner (see Fig. 88); in the late 1960s, there were four poor year classes in succession which must have threatened the existence of the stock at the extant level.

The year classes of North Sea herring remained steady, with rather a low variability until the late 1950s when the stock started to decline; when split into the three component groups of North Sea herring—Downs, Buchan and Dogger (Cushing and Bridger, 1966)—the decline in the 1960s was shown more sharply, due to recruitment overfishing (see below). In contrast, the year classes of the Norwegian or Atlantoscandian herring gradually increased through the decades until the stock finally collapsed due to recruitment overfishing. Recruitment to the North Sea plaice stock has remained roughly steady although since the 1950s the year class strength may have increased. In the Karluk salmon (estimated as escapement, not by cohort analysis) there is a long and continued decline from the first decades of the present century.

Figure 54 shows another group of recruitment estimates. The stock of Georges Bank haddock (like that of the North Sea haddock) is very variable, for year classes differ by nearly two orders of magnitude; the collapse in 1965 and 1967 was due to recruitment overfishing, probably the only gadoid stock to so fail. Very recently, strong new year classes have entered the stock of Georges Bank haddock. The recruitment to the North Sea sole stock varies by nearly two orders of magnitude. There are two herring stocks, the Downs in the southern North Sea and the British Columbia

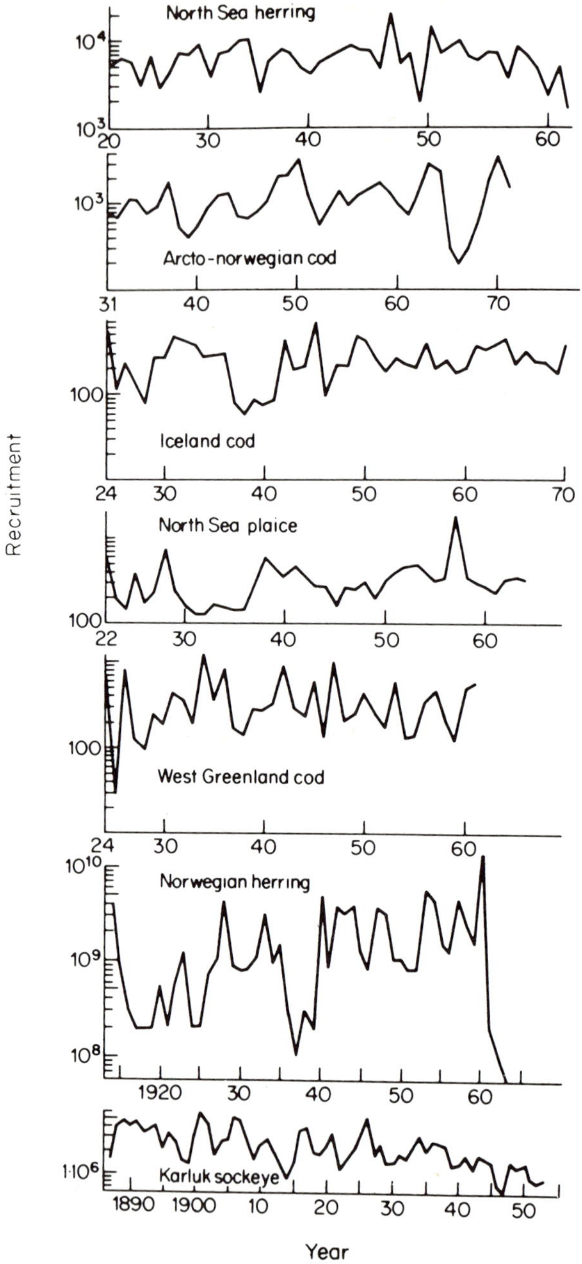

Fig. 54 Variability of recruitment on a logarithmic scale to cod, herring, plaice and sockeye salmon through decades.

Lower East stock which spawns in the Strait of Georgia inside Vancouver Island. The figure shows the slow decline of the Downs stock during the 1950s due to recruitment overfishing. From 1915 onwards the year class strength of the British Columbian Lower East stock rose steadily to 1945 by nearly an order of magnitude.

If we disregard the effects of recruitment overfishing, the recruitment to the gadoid stocks appears to vary steadily about its geometric mean. In contrast, the year classes of the herring and salmon stocks rise or decline with time and vary about a trend. Such groups of fishes tend to be vulnerable to recruitment overfishing whereas the cod-like fishes are less so; such are described as being resilient, being able to resist environmental change to some degree whereas the stocks of herring-like fishes respond to such changes. For example the herring stocks may have increased or decreased by several orders of magnitude as shown in Chapter 3, whereas the Arctonorwegian stock of cod may have been reduced only by a factor of two during the last cool period.

Kasahara (1962) examined three series of recruitment indices for the mackerel (*Pneumatophorus diego* Ayres), the Californian sardine (*Sardinops caerulea* Girard), and the Hokkaido herring. The year class strengths of mackerel varied by a factor of 70; that of the Californian sardine changed by a factor of 8. The recruitment to the Hokkaido herring between 1907 and 1948 varied by a factor of 55 and declined by about an order of magnitude in each decade. All are pelagic stocks and display the responsive character elicited from Fig. 54.

A different form of variability is demonstrated by the Malabar oil sardine (*Sardinella longiceps* Valenciennes). At the end of the south-west Monsoon in July and August in south-west India, the sardine spawns in the upwelling system off the Malabar coast of India. Within twelve months the offspring are caught in a canoe fishery; they spawn twice in their lives before they die at the age of two and a half. Catches which have been recorded for some decades are very variable and they have declined sharply since 1968. Let us denote catches in a four year period by C_1, C_2, C_3 and C_4. Then the stock may be represented by

$$(C_3/E) + (C_2/E) \exp -Z + (C_1/E) \exp -2Z$$

and recruitment by (C_4/E), where E is exploitation rate and Z is the instantaneous coefficient of total mortality. Let $F = 0.5$, where F is the instantaneous coefficient of fishing mortality and let $Z = 1.0$; with a series of catches going back to 1940 a curve of recruitment upon parent stock may be constructed. A Ricker curve was fitted with errors to that curve (Cushing and Harris, 1973) which was a near linear one, but there was a remarkable number of anomalies outside the expected range of error. Such anomalies are plotted in time series in Fig. 55. The rise in catches in the 1950s depended on the three anomalously high recruitments, but the decline after 1968 may have depended upon lack of rainfall as Raja (1969) suggested; when the June rainfall is less than 10 mm, the sardine ovaries suffer from

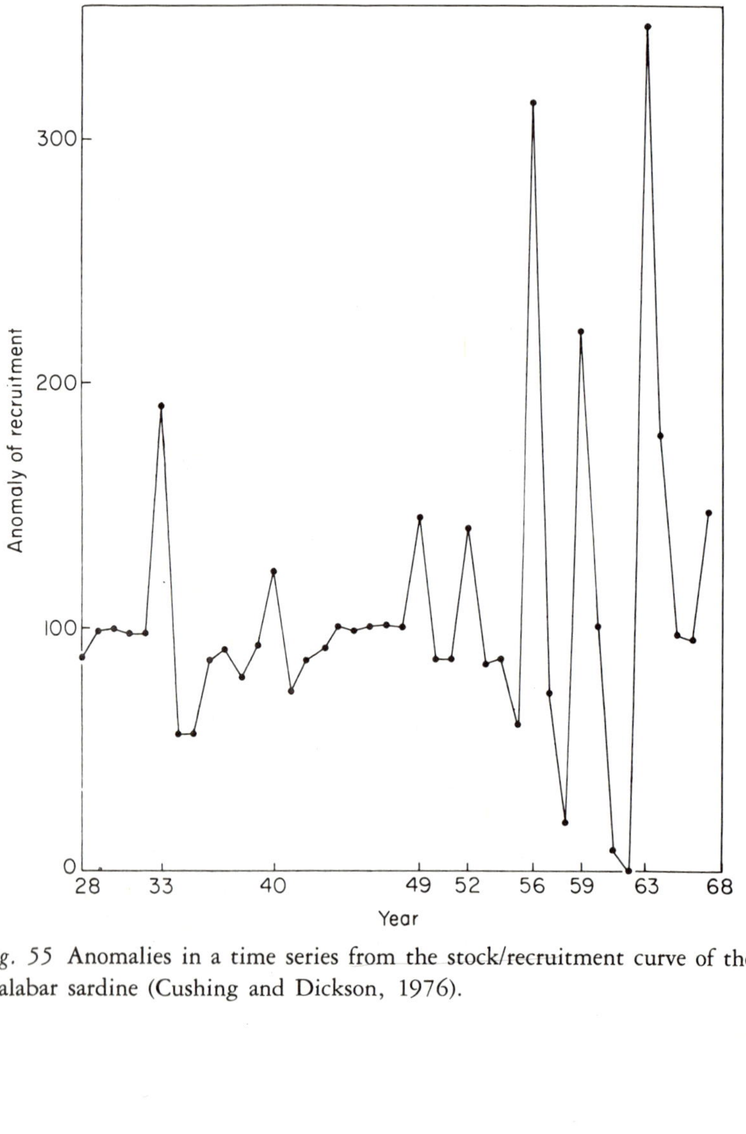

Fig. 55 Anomalies in a time series from the stock/recruitment curve of the Malabar sardine (Cushing and Dickson, 1976).

egg breakdown or atresia. But the causes of the three high year classes remain unknown; one of the interesting points is that the high anomalies are followed by low ones two years later. As the fish live for only two years, it is possible that the adults of the abundant year classes eat their juvenile successors in large numbers. The Malabar sardine stock is perhaps a special case which combines an upward trend in recruitment with sharply increased variability; it is obviously a stock which would repay further attention.

Garrod and Colebrook (1978) examined 34 stocks of herring and gadoids throughout the North Atlantic. They classed them in four groups: the north-west Atlantic coastal shelf, the central oceanic Atlantic (Iceland and Faroe Islands), the north-west European seas, and the Arctic Rim (Arctic, West Greenland, Labrador, and East Greenland). The variability was least at Iceland and the Faroe Islands and high on the Arctic Rim; in the north-west European seas, variability was low in the Irish Sea, but high in the northern North Sea. This is the first time that the variability of the recruitment to fish stocks has been broken down on a regional basis.

They extended their analysis with principal component analysis and extracted two components which between them accounted for 35% of the variability of the data. They plotted one component on the other and the stocks arranged themselves in groups in the following way:

A. Arctonorwegian haddock
Arctonorwegian cod
North Sea sole
North Sea plaice
Georges Bank haddock
Gulf of St. Lawrence cod

B. Iceland haddock
West Greenland cod
East Greenland cod

C. Faroe cod
Faroe haddock
Iceland cod
South Newfoundland cod
Nova Scotia cod
Georges Bank herring

D. North Sea haddock and whiting
West of Scotland herring

Groups B, C and D are regional, but group A comprises elements from all regions.

There are two conclusions from this survey of recruitment to the fish stocks (*a*) that year class strength varies about an order of magnitude and that some stocks are resilient whereas others respond to climatic changes, (*b*) that variability may be grouped by regions, and is sometimes common to some stocks across the ocean.

THE DEPENDENCE OF RECRUITMENT UPON ENVIRONMENTAL FACTORS

For many years fisheries biologists have tried to relate year class strength to environmental variables—often temperature and salinity which are easy to measure. For example (as noted in Chapter 3) Johansen (1925) related the stock densities of mackerel to salinity and temperature in the Kattegat for the period 1909–23. The first correlations of year class strength on environmental factors were made by Johansen (1928) who related the stock densities of small plaice in the southern Kattegat and Belt Seas to salinity differences between 1904 and 1926. Jensen (1952) related the stock densities of small plaice in the transition area (between the Baltic and the Skagerrak) to surface temperature and salinity at Schultz's Ground LV. All three studies were made in an area of sharp differences in temperature and salinity, which themselves probably indicate difference in Atlantic input into the Kattegat (see Chapter 9).

In some stocks recruitment is positively correlated with temperature. Dickie (1955) correlated the abundance of the giant scallop (*Placopecten magellanicus* Gmelin) with temperature in the Bay of Fundy between 1930 and 1946. Kislayakov (1961) made a similar analysis of the Arctonorweigian cod stock, as did Hermann *et al.* (1965) for the West Greenland cod stock (Fig. 56). Zupanovitch (1958) found an analogous relationship for the Adriatic sardine. Dow (1969) showed that the abundance of Maine lobsters was directly dependent upon temperature. Favorite and Maclean (1973) found that year classes of the south-east Alaskan herring were very high when the temperature anomalies off Queen Charlotte Island, off British Columbia, were high; indeed, one occurred in 1958 the extraordinary year of warming on the west coast of North America, the year when the sockeye salmon returned to the Strait of Georgia *north* of Vancouver Island instead through the Strait of San Juan de Fuca.

An inverse relation between year class strength and temperature was shown by Uda (1952) for the Hokkaido herring, by Marr (1960) for the Californian sardine, by Ketchen (1956) for the English sole (*Parophrys vetulus* Girard) in Hecate Strait, and by Martin and Kohler (1965) for the New Brunswick cod. Dickson *et al.* (1973) related the recruitment of North Sea cod inversely to temperature for the years between 1956 and 1973, including those years in the 1960s when the gadoid outburst took place (see Chapter

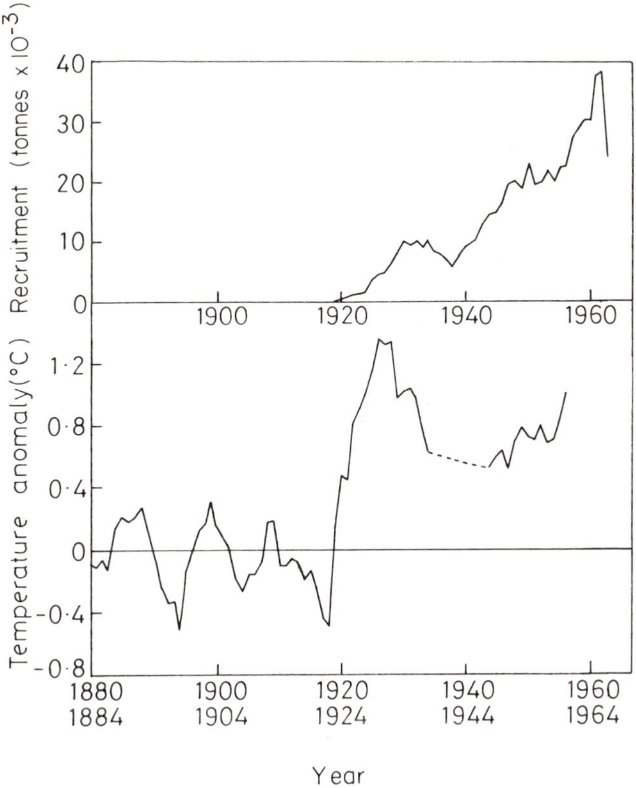

Fig. 56 Dependence of recruitment to the West Greenland cod stock upon temperature during the early years (Hermann *et al.*, 1965).

7). They showed that in space the best correlations were found on the Flamborough Off Ground, the major spawning ground of cod today, between the Dogger and the English coast; the spatial network was based on the tracks of the *bateaux routiers* which collect surface samples of temperature and salinity on merchant shipping routes across the North Sea. The ships take samples about once a month and the highest correlations were found in March/April and in January. Cod spawn in March/April and the study of correlations suggest that the critical events on formation of year classes occur on the spawning ground or near it. The high correlations in January are unexplained.

Yamanaka and Yamanaka (1970) related the year class strength of tuna species to sea surface temperature in the ocean:

	Yellowfin (W. Pacific)		Yellowfin (W. Indian)		Bigeye (E. Pacific)		Albacore (Coral Sea)	
	Warm	Cold	Warm	Cold	Warm	Cold	Warm	Cold
Strong	1	3	0	6	0	8	1	4
Weak	1	1	8	2	10	3	6	1

In each case the strong year classes appeared when the sea surface temperatures were cool and vice versa. The tuna-like fish spawn across extensive areas of ocean for as much as nine months of the year. The cool water probably indicates a distribution of divergences under local wind stresses. In other words strong year classes of four tuna-like species depend upon distributions of oceanic but local and transient updwellings.

It is unlikely that all correlations on temperature can be explained in the same way. The appearance of the cod at West Greenland may be associated with increased mixture of the Irminger and East Greenland Currents, and the cod may have survived more readily in the resulting somewhat warmer water. The inverse relation between temperature and the abundance of the Californian sardine is probably an effect of increased upwelling; we shall return to this point in the next section of this chapter. As will be shown in Chapter 11, it is possible that the delayed development of fish eggs in cooler water plays a part in the generation of year classes, and hence we might expect inverse correlations of recruitment upon temperature.

In Garrod and Colebrook's principal component analysis referred to in the last section, the first two components of variance were estimated from the year class strengths (1950–72) and six hydrographic sets of observations of temperature, temperature anomalies and salinity anomalies at six stations spread across the North Atlantic between the Sambro LV off Nova Scotia to the Kola Meridian in the Barents Sea. Figure 57 shows the first two components of variance for each year between 1950 and 1972; the figure also shows the periods of predominantly high and low atmospheric pressure between 20°W and 30°W. The first components tended to be positive when the pressure is high and vice versa. Garrod and Colebrook show that the magnitude of recruitment reflect the annual differences in the first component. The simplest explanation of this relationship is that differences in

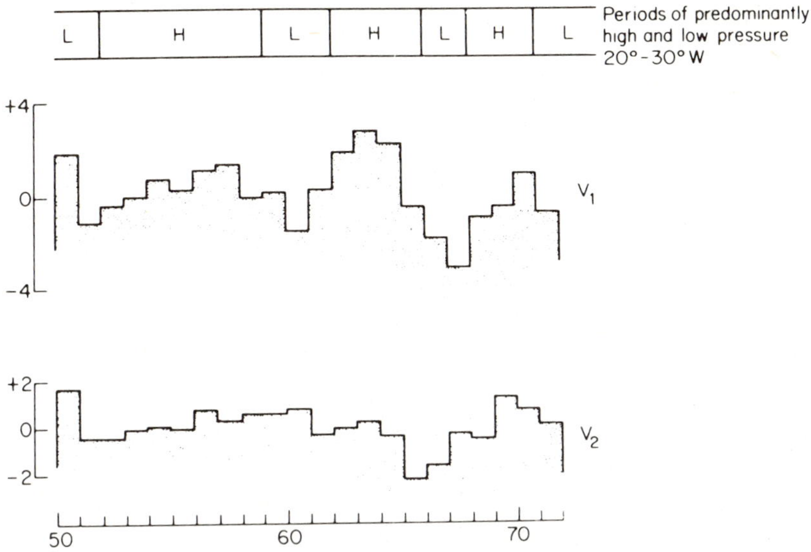

Fig. 57 The first principal component of variance of year class strength and six environmental factors compared with the average pressure distributions in the North Atlantic (Garrod and Colebrook, 1978).

recruitment from year to year are affected by differences in wind strength and direction. In Chapter 11 the importance of this observation will be stressed when we examine the possible biological mechanisms involved.

Garrod and Colebrook also show that during the period 1950–72 the number of days of westerly winds over the British Isles decreased steadily (see Fig. 42). We shall return to this phenomenon in Chapter 11, and all we need note here is that significant changes in the wind distribution have occurred since 1950. This is the period of cooling after the recent period of warming described in Chapter 5. Such events suggest that changes in recruitment such as the decline in many of the North Atlantic gadoid stocks, the gadoid upsurge since 1962 in the North Sea and the reversal of the Russell cycle were all linked events, in a broad sense.

There are a number of correlations between catches (or year class strength) and the outflow of freshwater to the sea; Gran (1923) suggested that freshwater favoured primary production. Indeed the reproductive rate of the algae is accelerated with a certain dilution of sea water. Bakken (1961) showed a remarkably good correlation between sprat catches in western Norway with the depth of the 34‰ isohaline at Utsira (in western Norway);

high catches were associated with deeper isohalines, that is, more fresh water. Analogous arguments are pursued by Sutcliffe *et al.* (1973, 1977) for fish stocks affected by the outflow from the St Lawrence. Such relationships are possibly linked to other climatic factors.

THE LINK BETWEEN UPWELLING, PRODUCTION AND YEAR CLASS STRENGTH

In the long-term *Calcofi* cruises, zooplankton has been sampled by month and quarter between 1951 and 1966 on a grid of stations up to 300 miles offshore between Point Concepcíon and Punta San Eugenio, California. The quantities were estimated as ml displacement volume per 1000 m hauled in the top 300 m of the water column. Smith (1971) has summarized the material by quarters for the months January, April, July and October. Bakun (1973) estimated upwelling indices for the same region as Ekman transports calculated from the atmospheric pressure fields in the North-east Pacific. With annual estimates, upwelling indices are poorly correlated with zooplankton volumes ($r^2 = 0.16$, $n = 16$).

The material is shown in time series in Figure 58a. The correlation fails in the years 1958 and 1959, the outstandingly warm years off the west coast of North America. The lower insert shows the degree of appearance of the pelagic crabs (*Pleuroncodes planipes* Stimpson) between Punta San Eugenio and Point Concepcíon. The northerly invasion was associated with the flooding of the eastern tropical Pacific at the same time as the trans-equatorial flow towards the coast of Peru which is El Niño. Walsh (1978) has linked El Niño with the appearance of the pelagic crabs off southern California. Hence when the pelagic crabs appear towards Point Concepcíon the waters off southern California may well be flooded with water from the eastern tropical Pacific, as may have happened in 1926 and 1932 (Chapter 5). In 1958–59 we believe that the water off southern California contained planktonic populations different from those of other years which originate from the north in the California current. Figure 58b shows the dependence of annual average quantities of zooplankton (in displacement volume) upon the upwelling indices of $m^3/s/100$ m of coast line, excluding the observations in the years 1958 and 1959 ($r^2 = 0.47$; $n = 13$). Bakun and Parrish (1980) have published a correlation between zooplankton off Central California in spring with upwelling indices at 39°N averaged for the previous twelve months. But we recall that Bernal and McGowan (1981) show with

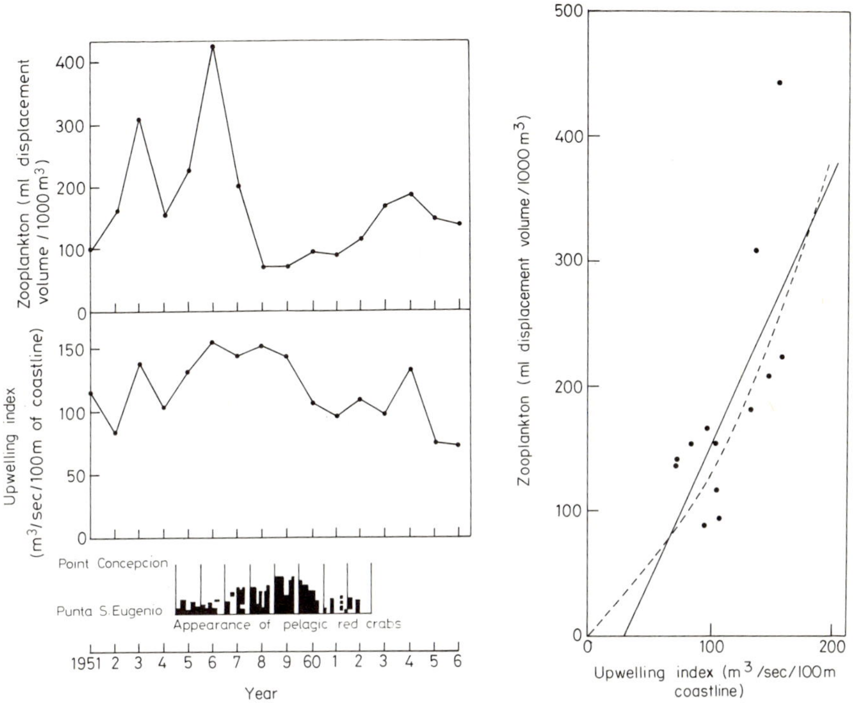

Fig. 58 Zooplankton and upwelling off California: (*a*) the full series from 1951–66; (*b*) the correlation between zooplankton and upwelling excluding the El Niño years 1958 and 1959.

a much larger data array that the production of zooplankton is correlated with southerly transport (as indicated by differences in position of the 34‰ isohaline) rather than with upwelling; in other words, production is supplied by nutrients from the north, rather than from the coast. However, the dominant physical process in annual production may well be offshore divergence, differences in which may be linked to differences in southerly transport. The higher production of zooplankton may depend upon the degree of upwelling or off shore divergence. Hence it is likely that the primary production also depends upon upwelling or offshore divergence. But upwelling which varies by a factor of two generates difference between years in zooplankton production by a factor of four. It is assumed that differences in zooplankton production depend on differences in primary production. The true dynamic relationship may well be more complicated than a simple two-stage link between upwelling and zooplankton production.

The same upwelling indices have been related to other parts of the ecosystem. Peterson and Miller (1975) showed that more plankton was produced in years of high upwelling off Oregon. The annual catches of the Dungeness crab (*Cancer magister* Dana) between northern California and Washington State were related by Peterson (1973) to the upwelling indices with a lag of about 1½ years which suggests the good survival of a year class. Botsford and Wickham (1975) have detected a further lag of 9 or 12 years by autocorrelation which may be an effect of cannibalism by adults upon young stages. The recruitment in the Pacific mackerel stock (*Scomber japonicus* Houttuyn) was related by Parrish (1976) to Bakun's upwelling indices. Recently Bakun and Parrish (1981) have related log recruitment of the Pacific mackerel with eight variables in two periods (April–June or May–July) and found three high correlations in the first and five in the second; each of the variables was an index of upwelling, of which the most valuable was the wind stress curl index. They also showed that ln (recruits/stock) of the California sardine depended upon upwelling indices and wind curl stress together with stock. In addition they refer to detailed work on other species; year class strength of hake (*Merluccius productus* Ayres) is negatively dependent upon January offshore transport (Bailey, 1981) which may imply that the larvae sink into the water to be upwelled; the relative recruitment of Pacific bonito (*Sarda chiliensis* (Cuvier)) is correlated with upwelling indices off northern Baja California during the spawning season from March to July. In the years of high upwelling there is high production of zooplankton, pelagic carnivores such as the sardine and the mackerel, predators like the bonito, and benthic animals like the Dungeness crab and the hake. In microcosm the whole ecosystem responds to differences in upwelling from year to year almost certainly directly through differences in primary production. The distributions of fish eggs and larvae off California are spread far offshore, far beyond the dynamic boundary to the region of coastal upwelling at about 50 km from the shore. The correlations may originate in both upwelling and divergence driven by the same wind system, if recruitment is determined in the larval stage, or in the strict coastal upwelling zone if it is determined at or around metamorphosis. The large mass of material available in the Calcofi data files might yield some correlations by chance, but the links with some of the more important components of the ecosystem may not so arise. Hence the link between recruitment and primary production may well be real. We recall that strong tuna year classes appear in the ocean when divergences at the surface were pronounced.

The argument in this section states that differences in upwelling generate differences in zooplankton (presumably through differences in primary production) with consequences which pervade the ecosystem governing the recruitment to stocks of sardine, mackerel, bonito, hake and crab. Of course the dynamic relationship between primary production and year class strength cannot be the simple direct one implied in a correlation. The only outcome of such correlations is to indicate that such dynamics are worth investigating. The question arises whether analogous effects are found in the production systems of temperate waters. Figure 59 shows the variability of primary production during a decade in the western English Channel (Boalch *et al.*), and differences are shown from year to year by a factor of three. Smith and Eppley (1982) have correlated primary production in the region off La Jolla with the sea temperatures off Scripps pier for a decade or so; the long time series of temperatures was converted to estimate primary production for a long period, in which trends may be detected. If the argument from the upwelling system may be extended, we would expect such differences in primary production to be equally effective throughout the ecosystem. Differences in recruitment to fish stocks from year to year might well be quite closely related to the differences in primary production in the simpler systems of subtropical seas. The dynamic relationship in its annual development must be complex even if the differences from year to year remain.

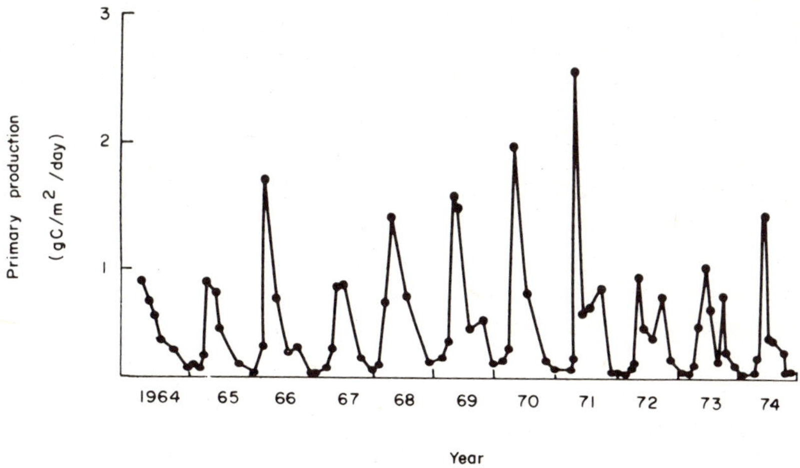

Fig. 59 Variability of primary production in the western English Channel for the decade 1965–1974 (Boalch *et al.*, 1977).

THE STAGES OF THE LIFE HISTORY AFFECTED BY CLIMATIC CHANGE

Environmental factors might be expected to affect any stage of the life history to some degree, but fisheries biologists have long thought that year class strength was determined in young or juvenile stages. An early form of explanation suggested that differences in recruitment might be determined by the degree to which eggs and larvae were blown by the wind in shallow seas towards the expected nursery ground. This was the basis of correlations attempted by Carruthers (1938) between recruitment of Downs herring (and of North Sea haddock) and the wind strength from particular directions. For example, it was suggested that haddock larvae were blown by the wind over the Norwegian deep water and were lost to the stock; however, Saville (1959) has shown that the larvae never reached the deep water, so that explanation no longer stood.

Hjort (1914) suggested that there was a critical phase in the life history of fishes when the yolk sac larvae started to learn to feed and further that the major differences in year class strength were attributable to the availability of food during this period. A difference in survival has never been properly established then, but there is little doubt that Hjort was right in broad terms. As noted above, Hunter (1972) showed that it takes nearly a month for larvae of the northern anchovy (*Engraulis mordax* Girard) to learn to feed fully. Once the little fish can feed, they grow avidly and their death rates are high (see Chapter 2); one would expect year class strength to be established during such periods of high growth and mortality. Fishes are extremely fecund, and many die to ensure sufficient food for the survivors. As noted above, Rauck and Zijlstra (1978) have demonstrated that the recruitment to the Southern North Sea plaice is established before the metamorphosed fish reach the beaches as 0-group fish. In other words the processes are essentially complete in about 140 days. In this somewhat developed sense, Hjort's thesis that year class strength depends upon the availability of food in larval (and post larval) stages remains true.

In Chapter 2 it was noted that larval fish in general feed first for a day or so on algae and then on the nauplii of copepods, which themselves feed on algae. The production system in any sea is primarily that of the phytoplankton to which the production of herbivorous zooplankton is closely linked. Thus the larval fish feed at the very root of any production cycle, and if recruitment is determined at an early stage in the life history there should be a link between primary production and year class strength, as was suggested above for the ecosystems in upwelling areas. There is no reason why this should not be true of all three forms of production cycle—temperate, upwelling and tropical oceanic. In the first, production starts from very low levels in spring and builds up to a peak after which it dies away, the whole cycle lasting for some weeks. In an upwelling area the cycle resembles the temperate one in that it starts from very low levels at the bottom of the euphotic layer, augments in the upwelling water, peaks, and dies away as the upwelling plume is drifted offshore. In the deep subtropical and tropical ocean the production is probably in a quasi-steady state which is perturbed from place to place and from time to time by surface divergences. The temperate cycles vary in time of onset at the fixed positions where fish spawn. In an upwelling area, the plumes vary in time and in position along the coast, and the fish may well link their spawning to the plumes by feeding on the algae at a late stage in their reproductive cycle: then the larvae would grow in a patch of zooplankton (Cushing, 1978). In the steady state system in the deep ocean, spawning extends for nine months of the year across thousands of miles of ocean and the chances of a good year class perhaps depend upon the number of divergences within the frame of time and space within which the spawning fish are found.

Colebrook (1965) wrote that the temperate production cycle varies in amplitude, spread and time of onset, so there is considerable potential for variability in the system. Robinson (1970), following Williamson (1961), used the long-term results of the plankton recorder network within waters around the British Isles to show that the time of onset of the production of phytoplankton depended upon Craig's "temperature difference", which is the mean surface temperature for May, June and July less that for March, an index of vertical stratification. Colebrook (1979) extended the observations in time from 1948–76 and in space across the whole North Atlantic. The seasonal cycle of phytoplankton (indicated by "greenness") and of copepods was portrayed for 33 areas from off Portugal to the Norwegian Sea and from West Greenland to the Gulf of Maine. With this material Colebrook used the standard deviation of surface temperature in season, which is highly correlated with Craig's temperature difference; he showed that the time of onset of the temperate production cycle throughout the North Atlantic is significantly correlated with that standard deviation. When the time of onset is late, the standard deviation of surface temperature is low, and vice versa; throughout the ocean the time of onset differs by nearly two months. The evidence presented by Colebrook establishes that a relationship exists between time of onset of production and depth of mixing in general manner. What is required to gain further understanding of the processes is a model based on the biological and physical processes which can be used to develop forms of prediction.

Analogous processes occur in other forms of production cycle. The position of an upwelling plume depends upon the wind direction with respect to the coastline. Its time of onset depends on when the wind starts to blow, and the subsequent production cycle depends on the rate of upwelling and on its intermittency. In the tropical and subtropical ocean, the best production for fishes occurs there and in the divergences, seaward of the coastal upwellings—the products of local wind stresses.

Production in temperate waters starts when the critical depth exceeds the depth of mixing. Sverdrup (1953) defined the critical depth as that depth to which the integrated production exceeds the integrated respiration. If the depth of mixing is greater than the critical depth, respiration exceeds production because the algae are swept below the critical depth for it is assumed that algal cells become randomly mixed in depth within the depth of mixing. The time of onset of the production cycle depends upon the irradiance and those factors which determine the depth of mixing, convection and wind stress. As spring advances in latitudes >40°, the sun rises higher in the sky and the winds slacken. The compensation depth is defined as that at which the rate of primary production equals the rate of respiration, and the effective reproductive rate of the algae is that determined by the ratio of compensation depth to depth of mixing. As the sun rises higher in the sky (and days become longer), the average compensation depth becomes greater and the depth of mixing less. The rate of increase of algal reproductive rate depends upon the sunlight, day length, cloudiness, heat input and upon the wind stress.

The essential relationships may be put formally:

$$D_{cr}/[1 - \exp(-KD_{cr})] = (1/K)(I_0/I_c)$$

where

D_{cr} is the critical depth in m;
I_0 is the average subsurface irradiance (i.e. just below the surface) of the same spectral distribution at the compensation depth;
I_c is the irradiance of the average compensation depth during the day;
K is the extinction coefficient.

Then

$$D_{cr} = \exp(K)(D_c/K)$$

where

D_c is the compensation depth in m (Sverdrup, 1953). Steele (1965) wrote.

$$D_{cr} = (p_m/r^1)(e/K)\{1\text{-exp-}(I_0/I_m)\}$$

where

p_m is the maximum rate of photosynthesis in gC/gC/h;
r^1 is the rate of respiration in the same units;
I_0 is the photosynthetic energy (between 380 and 720 nm) reaching just below the surface in ly/min;
I_m is the photosynthetic energy which generates p_m in ly/min.

Because the rate of respiration amounts to about 0.1 p_m, on a daily basis $(p_m/r^1) = 5$. With this method the dependence of photosynthesis on irradiance is well described (whereas it is merely linear in Sverdrup's formulation). Further there is no need to estimate the compensation depth where the spectral distribution of irradiance is much narrower than that near the surface.

The real problem is how to estimate the depth of mixing. There are various algorithms to express the depth of mixing due to wind stress (for example, see Tully and Giovando, 1963), but in midwinter in the North Atlantic and the North Pacific the depth of convective mixing may reach several hundred of metres across most of the oceans north of 45°N (for example, see Ellett and Martin, 1973). Woods (1980) has examined the diurnal and seasonal variation of convection. He defines a thermal compensation depth, D, at which the rate of solar radiation absorption equals the rate of surface heat loss. When $D < H$ (depth of the wind mixed layer), it is the lower boundary of a layer (the depth of which is C) in which the turbulent kinetic energy is generated by buoyant convection. Buoyant convection in the wind mixed layer provides a source of turbulent kinetic energy restricted to a layer the depth of which, $C < H$. At night $C = H$, and convection contributes significantly to deepening the mixed layer by turbulent entrainment. During most of the day, when the rate of heat input from the sun is greater than the rate of surface loss, $C < H$ and convection is less. Woods estimates D_{min} and p_{10}, the proportion of the day when $D > 10$ m, by latitude, season at the equinox and solstice and degrees of cloudiness. Hence the depth of mixing should be averaged diurnally to take into account the turbulent mixture due both to convection and to wind stress, which is not incorporated into present estimates of critical depth.

There are two problems for the biologist. The first is to determine the critical depth in different regions poleward of 40° latitude, and for this purpose the depth of mixing estimated by Woods' method is the best available. The second problem is to establish the modification of the algal reproductive rate by the depth of mixing after the critical depth has been reached. Cushing (1962) called the ratio of compensation depth to the depth of mixing, the production ratio; the actual reproductive rate of the algae is the potential rate multiplied by the production ratio. This approach assumes that production starts before the thermocline is established. Radach (1980), reporting on the results of the FLEX experiment in the North Sea, established that once the summer thermocline is established, the production cycle accelerates. However, the critical depth is usually determined considerably earlier, and a period of transients occurs before the thermocline is finally established. But Woods's method based on the diurnal input of heat is in fact designed to estimate and average the depth of mixing under such conditions.

An interesting special case is found in the Southern Bight of the North Sea, where the depth of mixing is effectively the depth of water (because of the strong tidal streams) so the production ratio is (D_c/Z) where D_c is compensation depth and Z the depth of water (Cushing, 1972). Lee and Folkard (1969) had shown that under easterly winds the water was relatively clear (60–80%) transmission but under north-westerly winds it was much less clear (20–60% transmission). At 20 m or 40 m water depth, production would be expected to start at any time between February and May, depending on the wind direction. With easterly winds (82% transmission) production would start in early February in shallow water (20 m) and in early March in deeper water (40 m) in the Southern Bight. With northwesterly winds production would be much delayed because of the increase in turbidity.

The importance of the time of onset of the production cycle is seen in Fig. 60. In the North Sea and the Northeast Atlantic from 1948 to 1969, material from the Continuous Plankton Recorder network shows a decline in the biomass of zooplankton, a reduction in the zooplankton growth season and a delay of about a month in the time of onset of the production cycle in the North Sea. It is reasonable to suggest that the reduction in biomass, numbers and growing season of the zooplankton was a consequence of the delay in time of onset of the production cycle. This may be only part of the story in the North Sea because the smaller copepods may have suffered in competition with the larger *Calanus* (see Chapter 7). The decline

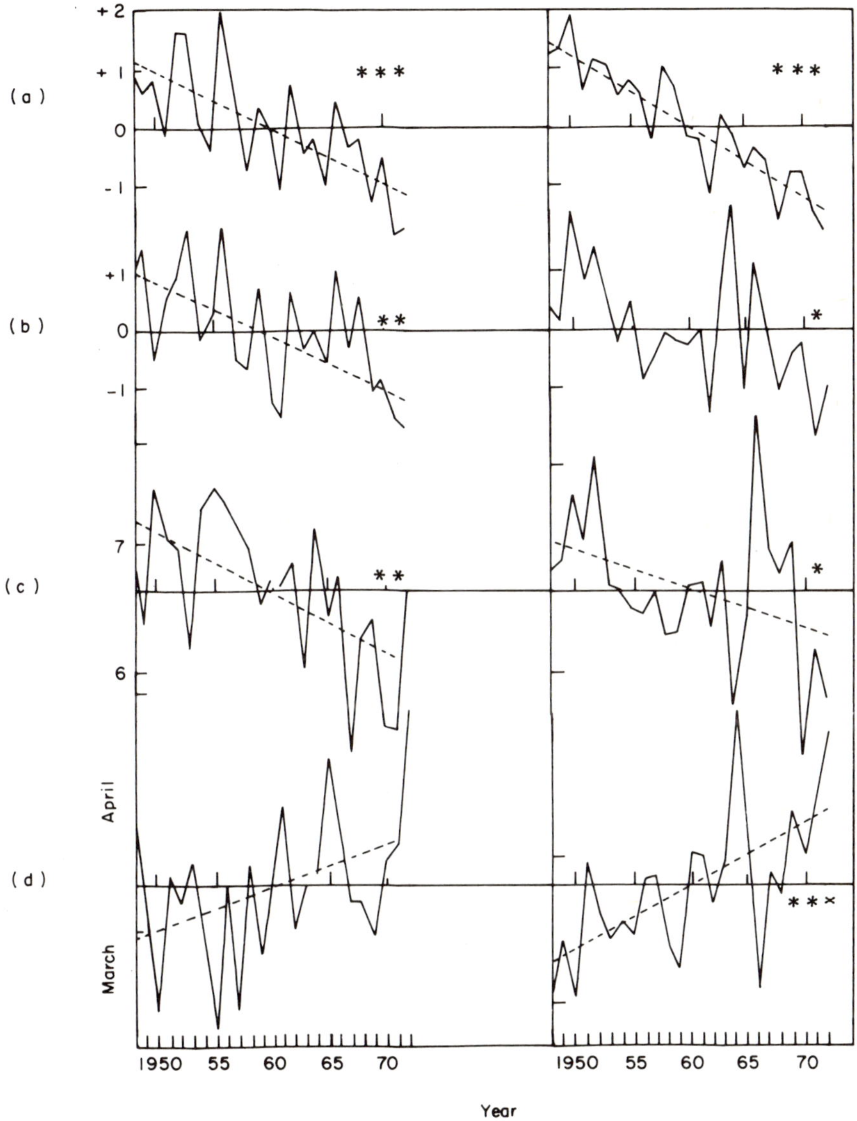

Fig. 60 Decline in zooplankton biomass, growing season, and the delay in time of onset of the production cycle in the North Sea and North Atlantic (Glover *et al.*, 1972).

in zooplankton biomass and numbers is as pronounced in the north-east Atlantic as in the North Sea, but the delay in the time of onset of the production cycle is less marked.

If the time of onset of the production cycle can vary by as much as a month or six weeks in temperate or high-latitude waters—a period often as long as the production cycle itself—the consequences on the fish populations may be considerable (see Chapter 6). Such a trend is probably the consequence of a shift in physical factors such as wind strength, wind direction, irradiance and heat.

Fish larvae live in the plankton and usually feed upon larval or juvenile herbivores, so it is not surprising that their growth and mortality may well be determined by the quantities of food available. We know now that the magnitude of recruitment of one stock of fish at least is established before the little fish reach the nursery ground and that the critical processes must have been completed mainly during the larval drift or just after metamorphosis. Because the production of herbivores is linked quite closely to that of the phytoplankton, we should not be surprised to find that the year class strength depends upon the variability of the production cycle to some degree.

CONCLUSION

In this chapter an argument is conducted which leads to the conclusion that the magnitude of year class strength is determined during the larval drift or during the very early juvenile stages. The interesting point is that the conclusion was reached by analysing the possible links with climatic change. Although the link is a potentially open one, the fish stocks are sufficiently remote from climatic events to restrict the number of physical factors common to both. Such factors as wind or sunlight predominate and may well determine the rates of natural increase of fish stocks.

Throughout the North Atlantic strong year classes are often common to disparate stocks on both sides of the Atlantic. The fishes themselves live thousands of miles apart yet are affected by common and pervasive factors. Templeman's earlier work introduced the problem, but it was Garrod and Colebrook's fuller analysis that established it statistically and from which they concluded that there were subdivisions within the North Atlantic; the subdivisions were not geographically, but biologically defined. Hence the physical factors that determine the magnitude of year class strength must be spread across the ocean at about the same time. Such a consideration limits the number of factors which are possible. From the discussion in Chapter 4 we should not be surprised by links across the ocean if the Rossby wave pattern in the upper atmosphere can generate events related to each other 5000 km apart.

Now that recruitment may be estimated in a number of stocks for a few decades, it is well established that it may remain steady or trend up or downward. In general, the year class strength of the cod-like fishes does not rise or fall with time; hence it is resilient in that it tends to resist or be insulated from environmental change. The herring-like fishes are responsive because their year classes increase or decline with time as one or other environmental factor prevails. We distinguish between stocks that are resilient and those that are responsive.

Variability in recruitment in the North Atlantic falls into geographical groups. Year class strengths of herring stocks in temperate waters are more variable in deeper water, probably because the time of onset of the production cycle is more variable. Yet the geographical grouping is independent of depth of water, which implies that differences in recruitment are governed by common factors within regions. If year class strength is determined at an early stage in life history and is linked to factors within regions and across oceans, it may well depend upon the processes within the production cycle.

There are three links between primary production and recruitment: that developed within the upwelling ecosystem off California; that between the year classes of tuna-like fishes and cooler water at the surface of the Pacific; and that shown by the collapse of the 1972 and 1973 year classes in the Peruvian anchoveta stock when primary production was reduced by a factor of three. It is difficult to model the production cycle in any predictive way because the physical and biological processes cannot yet be properly separated. Yet the links are sustained.

Chapter Seven

The North Sea

The North Sea is one of the most studied regions in the world ocean. Hydrography, plankton and fish stocks were sampled with varying intensity from the first decade of the century until the Second World War, but since then sampling has been intense. It is physically a somewhat complex region, stratified in summer north of the Dogger Bank and dominated by the tidal streams to the south. The circulation in the sea remains a little obscure, but water enters in the North and in minor part from the English Channel. The Baltic outflow streams north at and near the surface over the Norwegian Rinne and below it an Atlantic stream enters the North Sea flowing south. The conflux of the two currents generates a jet stream which flows westward and mixes the waters. This is a qualitative description of the dynamic structure of the North Sea, which is perhaps not yet fully established.

The North Sea sustained catches of herring, sprat and mackerel for centuries, but the historical records are not quantitative. Since the first decade of the present century, catches of herring amounted to as much as 1 500 000 tonnes before all North Sea stocks collapsed through recruitment overfishing. During the 1960s, catches of North Sea mackerel reached 700 000 tonnes before the stock collapsed through recruitment overfishing. One of the most remarkable events in the last three decades is the growth of the industrial fisheries for juvenile herring, sprat, sandeels and Norway pout (a small gadoid). By 1974, nearly 2 000 000 tonnes of such fish were caught for industrial purposes. Amongst the demersal fishes in the North Sea the most remarkable event of the century was the gadoid outburst which started in 1962 and which persists to the present day, the rise in the stocks of cod, haddock, whiting, saithe and Norway pout.

It has been suggested that all three events are linked—the decline the two pelagic stocks, the gadoid upsurge, and the considerable increases in catches of industrial species. Andersen and Ursin (1977) have proposed that the decline of the herring and mackerel stocks released food which fuelled the increase either of the gadoids or the industrial species; their names are associated with multispecies models by which some of the vital parameters of individual stocks are linked. Because animals eat each other at all stages of their life histories the links are pervasive, but we are a little uncertain which are the important ones. This chapter is concerned only with the influence of climatic changes upon the fish stocks and the plankton.

CHANGES IN THE NORTH SEA PLANKTON

The first information on the North Sea plankton was obtained from the international cruises executed in the first decade of the century under the auspices of the International Council for the Exploration of the Sea. Material also exists for particular cruises, but a regular series of stations was worked each month on a line between Flamborough Head, on the north-east coast of England, and the Dogger Bank between 1933 and 1962 (except for the Second World War); the time series was called the Flamborough Line. However, the main source of information on the plankton of the North Sea is the plankton recorder network; continuous plankton recorders are towed behind merchant vessels at 10 m on fixed routes each month. There are records dating back to 1931, but the present series, all processed in the same way, extends back to 1948.

Figure 61a shows the changes in numbers per cubic metre of *Calanus* on the Flamborough Line between 1933 and 1962; the animals were identified as *C. finmarchicus* Gunner, but a small proportion of *C. helgolandicus* Claus may have been included in the counts. In 1950 there was a sharp increase by a factor of about three, which persisted subsequently with some variability. The same change was detected in the plankton recorder material from certain squares in the southeastern (Lw) and southwestern North Sea (Nu), the area where the young herring grow. In the Northern and Central North Sea an increase in numbers of *Calanus* occurred but it was not so marked (Colebrook, 1963). Glover (1957) detected a more widespread change, one from a "neritic" type (*Pseudocalanus, Paracalanus, Acartia, Centropages hamatus* Lilljeborg) to an "oceanic" type (*Clione limacina* Phipps, *Spiratella* spp., *Calanus, Centropages typicus* Kroyer and euphausiids) (Fig. 61b). Williamson (1961) used principal component analysis to establish the common forms of variance between 1948 and 1958 for the array of plankton entities in the North Sea. He linked the change from "neritic" to "oceanic" plankton to one in the degree of vertical mixing and hence perhaps to the time of onset of the spring outburst of plankton in the North Sea.

Williamson showed that the first principal component for the decade 1948–58 was correlated with differences in stratification, i.e. Craig's difference, that in surface temperature between March and July. Hence Williamson concluded that the change from a "neritic" plankton to an "oceanic" one was associated with changes in wind strength (and direction) which might affect change in stratification and hence the time of onset of the production cycle. Tait (1957) and Martin (1976) have noticed that the water in the Faroe–Shetland Channel sometimes comprises a part which originates in the southern Norwegian Sea and in the East Icelandic Current, particularly between 1949 and 1952. Tait and Martin (1965) refer to 1949–1956 as a period in which the "arctic influence" was strong and to 1957–1959 as one in which that influence was moderate. This little time series corresponds quite closely to the changes in *Calanus* recorded on the Flamborough Line and in squares Nv and Lw shown in Figure 61a (Colebrook, 1963; Cushing, 1966). The abundance of *Calanus* increased sharply from 1950 to 1956 after which it decreased somewhat. Figure 62 shows the average distribution of *Calanus finmarchicus finmarchicus* Gunnerus in the North Atlantic from the Plankton Atlas (Anon., 1973): the animal is most common in the North-East Atlantic in the Norwegian Sea and in the East Icelandic Currents. Hence when the "arctic influence" in the Faroe–Shetland Channel increased, it is not surprising that *Calanus* entered the North Sea in greater quantitites. There appear to be conflicting explanations of the changes in the North Sea as being due to advection through the Faroe–Shetland Channel or due to differences in stratification as suggested by Williamson. Presumably the advective change was associated with changes in wind distribution over the North-East Atlantic and the two apparently distinct phenomena might have a common origin. The increment in *Calanus* in the North Sea in the early 1950s was probably due to increased advection from the Faroe–Shetland Channel. At the same time the degree of stratification in the North Sea changed which may have affected the communities, whether "neritic" or "oceanic".

Colebrook (1963) correlated the annual abundances of *Calanus finmarchicus* Stages V and VI (i.e. maturing and spawning animals) between statistical squares around the British Isles as sampled by the plankton recorder net-

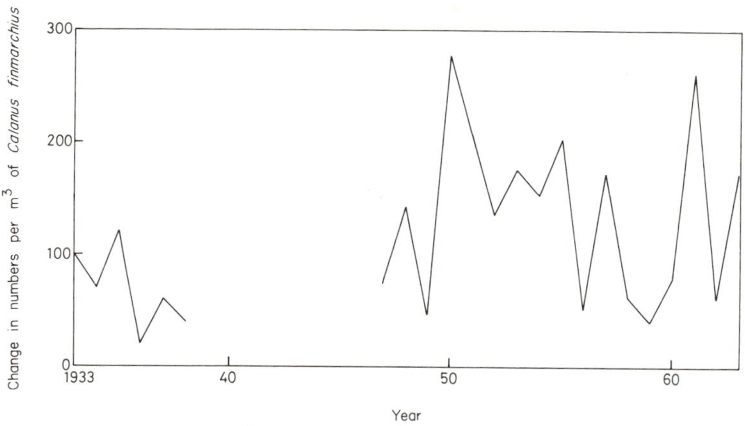

Fig. 61 (*a*) Changes in numbers per cubic metre of *Calanus finmarchicus* on a line of monthly stations between Flamborough Head and the Dogger Bank between 1933 and 1962 (Cushing, 1966)

work. He found that samples from the north western North Sea and from the waters west of Scotland were related; similarly, those from the southern North Sea and eastern North Sea were related but were distinct from the northern group. The latter region in the southern and eastern North Sea includes the area sampled by the Flamborough Line, with results as shown

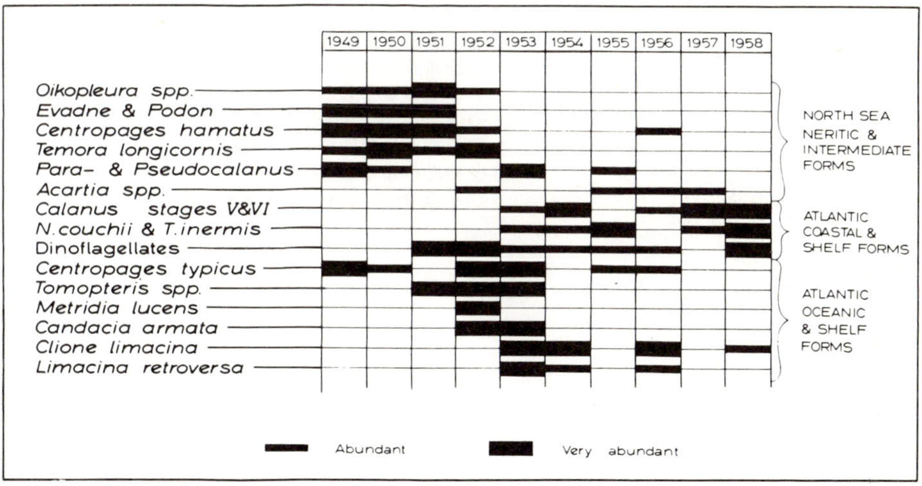

Fig. 61 (*b*) Change from a neritic to an oceanic type of plankton in the North Sea during the 1950s (Glover, 1957)

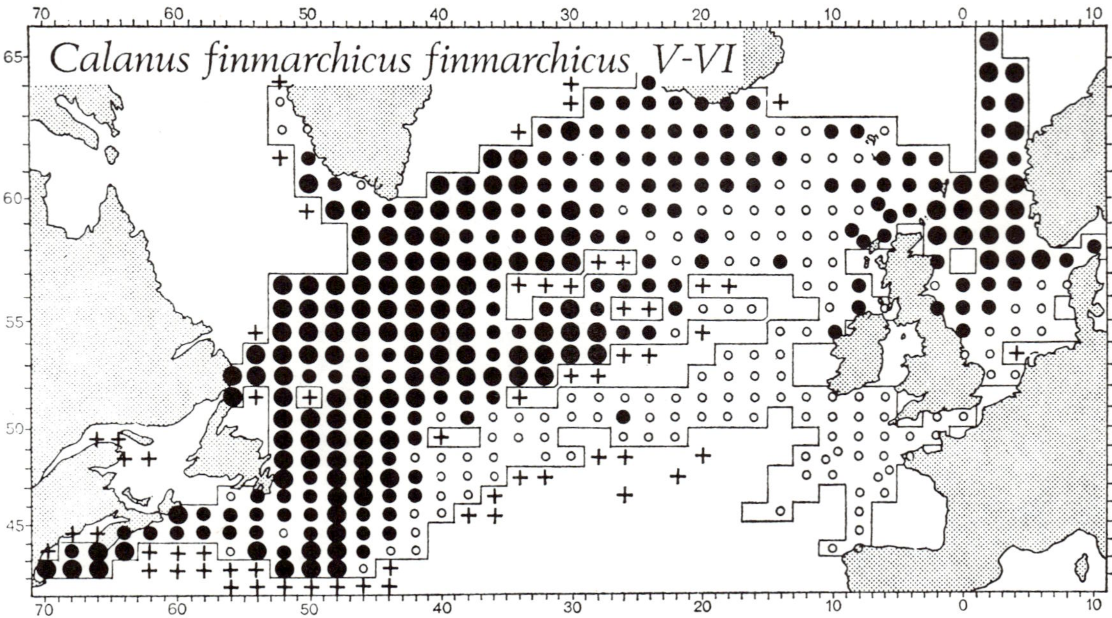

Fig. 62 Distribution of *Calanus finmarchicus* in the North Atlantic.

in Figure 61. Between the two major groups, the west central North Sea samples occupied an intermediate position.

There are three currents which bring water and, presumably, *Calanus* into the northern North Sea. The first is the "Caesium" Current which enters the North Sea through the Pentland Firth. It was discovered by following the output of ^{137}Cs from the Windscale plant in the Irish Sea, and Jefferies *et al.* (1973) showed that the course of the stream lies in Scottish coastal waters. The second input is the Dooley Current which passes between the Orkney and Shetland Islands and which comprises mixed coastal and Atlantic water in a swift and narrow south-going stream which turns eastward towards the Skagerrak in about 57°N (Dooley, 1974). The third input is the Atlantic stream not mixed with coastal water which flows in 200 m from the Faroe–Shetland Channel which turns south round Tampen Bank into the Norwegian Rinne (Svansson, 1965; Dooley, 1974) and in the course of the recent FLEX experiment it was shown to be a prominent feature, reaching even to the surface. This current was recognized by the great Swedish oceanographers who worked in the Skagerrak in the first two decades of the century. The three inputs to the North Sea are not independent, but differences in flow between them must occur from time to time. It would seem reasonable for example that the main *Calanus* inflow since 1949–52 has occurred in the Tampen Bank current rather than in the Dooley current, because the Tampen Bank current originates in about 200 m from the Faroe–Shetland Channel where one might expect to find the arctic intermediate water when it exists there. Riepma (1980) analysed current measurements during the FLEX experiment under different wind conditions; the south easterly flow of the Tampen Bank current was present under all wind directions, and indeed one might conclude that the major flow in the North Sea is inward through the Tampen Bank Current and outward through the Baltic outflow. The other inflows noted are not excluded, but are probably of less importance. Hence the changes described in Fig. 61 may indicate true changes in the input of *Calanus* into the north-eastern North Sea.

The most remarkable event in the North Sea and in parts of the north-east Atlantic is the decline in abundance of certain entities during the last three decades (Fig. 60). Colebrook (1978) has studied the zooplankton records from the plankton recorder network and found a decline of about two standard deviations in standard measure in "Total Copepods", Chaetognaths, zooplankton biomass, *Temora, Candacia, Spiratella, Centropages typicus* Krøyer and *Pseudocalanus*. The decline lasted from 1948 to 1977,

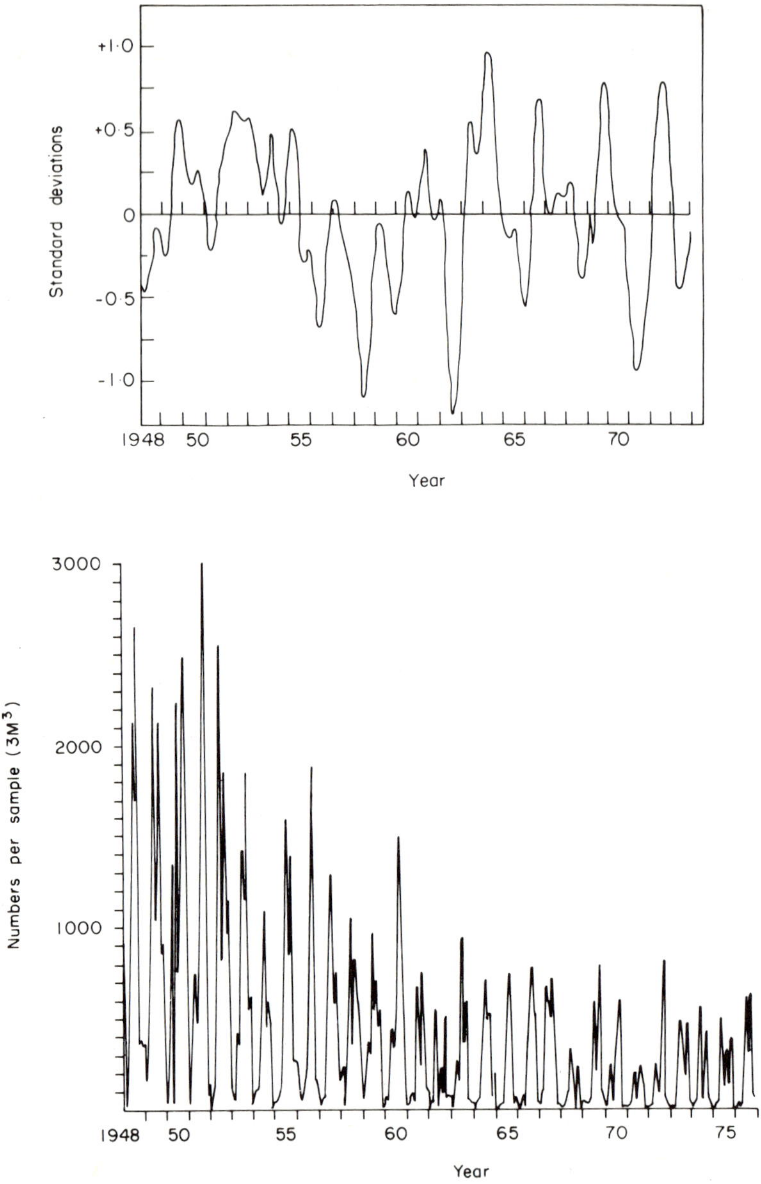

Fig. 63 The stock density (*a*) in standard measure of *Calanus finmarchicus* in the North Sea between 1948 and 1977 (Colebrook, 1979) and (*b*) in numbers per 3 m^3 of *Pseudocalanus elongatus* (Glover, 1979).

and the trend line passed through the mean between 1960 and 1962. This group includes parts of Glover's neritic group and parts of his oceanic group. Numbers of *Pseudocalanus* decreased by a factor of five during the three decades, but the stock density of *Calanus* did not decline in the North Sea at all, although it became more variable (Fig. 63a). One of the most interesting points about Fig. 62a is that the peak quantities of *Calanus* were higher in the 1960s and 70s in the eastern North Sea than in the 1950s. In other words, it is likely that whatever events augmented the *Calanus* populations in the North Sea during the 1950s, they were more intense during the 1960s and 70s. If we combine the results shown in Fig. 61 and 63a, it is likely that the stock of *Calanus* in the North Sea increased in 1950. Subsequently the smaller and more abundant copepods decreased during three decades. Figure 63b shows the decline in stock density of *Pseudocalanus elongatus* Boeck. From observations made by Cushing (1955) and Cushing and Vucetic (1963) off the Northeast coast of England in 1948, 1949 and 1954, *Calanus* stocks before 1950 comprised less than half the copepod population in the Central North Sea by weight, whereas in 1954 they comprised more than 80% by weight.

The changes in stock density of *Calanus* in April, May and June are shown in Fig. 64. There are six areas in the North Sea described in the plankton recorder network: B1 and B2 in the north, C1 and C2 in the centre and D1 and D2 in the south: the B1, C1 and D1 squares lie in the east and the B2, C2 and D2 ones in the west. In B1 and C1, the stock densities of *Calanus* between 1948 and 1978 varied about a mean. In B2, there was a peak in 1952–7 after which numbers declined and in C2 numbers peaked in 1950–1, after which they diminished. In D1 and D2, numbers were very low, but after 1970 in D2 they were exceedingly low. Summarizing, in the east central and north eastern North Sea, stock densities remained high, if variable. In the western North Sea, north and central, there was a peak in the early 1950s, after which stock densities declined and in the southern North Sea, stock densities were very low in the west after 1970. The changes recorded on the Flamborough Line and in the statistical squares Nv and Lw are really eastern North Sea changes and are perhaps linked to the input of *Calanus* from the Norwegian Sea and the East Icelandic Current via the Tampen Bank Current into the Norwegian Rinne.

It is interesting to examine the ratio of *Calanus* to *Pseudocalanus* in numbers per 3 m^3. In the North Sea *Calanus* is abundant in April, May and June whereas *Pseudocalanus* predominates between May and September.

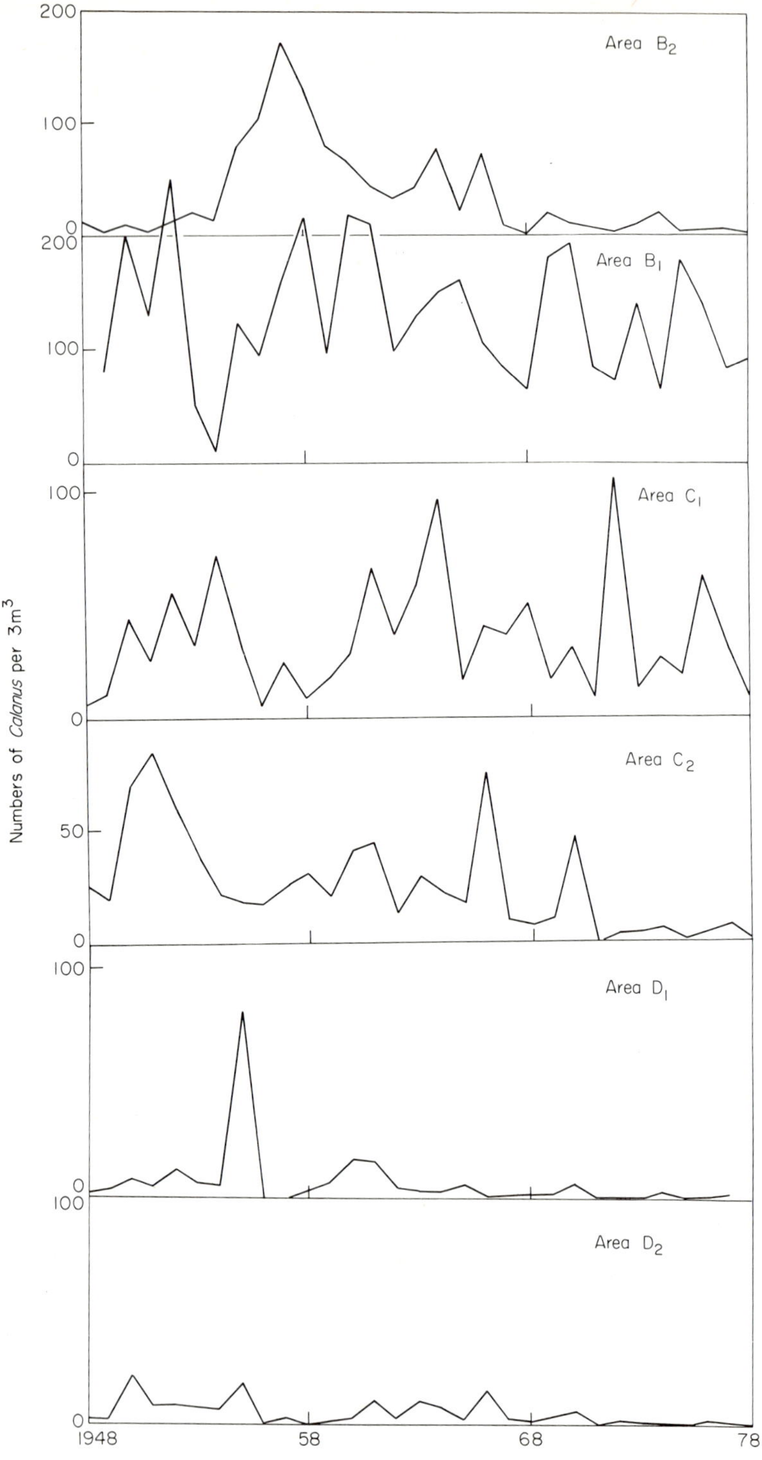

Area B_2
Area B_1
Area C_1
Area C_2
Area D_1
Area D_2
Numbers of *Calanus* per $3m^3$
200
100
0
50
1948
58
68
78

Stock densities were averaged for the periods of abundance and ratios calculated for each year and each of the six areas (B1, B2, C1, C2, D1 and D2). Then the ratios were averaged by five-year periods and the results are shown in Fig. 65. In the north eastern North Sea (B1) the ratio of *Calanus* to *Pseudocalanus* in stock density increased throughout the three decades and in area C1 it increased until 1968 and then declined a little. In the other four areas of the North Sea, west and south, the ratios remained at about the same level; in other words, the increase of *Calanus* with respect to *Pseudocalanus* was confined to the north-eastern and east central North Sea.

The dramatic rise in *Calanus* stock densities in the early 1950s was probably associated with the increase in "arctic influence" in the Faroe–Shetland Channel which was perhaps passed into the North Sea in the Tampen Bank Current. Hence stock densities remained high in the north-eastern and east Central North Sea (if more variable) throughout the three decades under discussion. From the structure of the Faroe–Shetland Channel, some quantities of *Calanus* must enter the northern North Sea in the Dooley Current, and we cannot expect the two populations (Tampen Bank Current and Dooley Current) to be distinct either in origin or in destination within the North Sea. Hence within the North Sea there are four components of variation—the two currents and two *Calanus* populations—and we must expect them to become mixed in the course of time. But in general terms we would expect the north-eastern North Sea to be fed by the Tampen Bank Current and the north-western North Sea by the Dooley Current. Further south we might expect the two components to become mixed, particularly as both streams tend to end their courses in the Skagerrak.

Colebrook (1979) has analysed the decline in smaller copepods in some detail. The array of some twenty or so plankton entities in three decades was analysed by principal component analysis and the first component displayed the same sort of decline as shown by *Pseudocalanus*. During the same period, the frequency of days per year of westerly weather over the British Isles declined. Figure 66 shows this decline and that of total zooplankton. Each trend in standard measure is fitted by polynomial curves and in a rough way both decrease with time in the same sort of way. Table 14 shows that the decline in zooplankton was correlated with surface tem-

---

Fig. 64 Changes in numbers per m^3 of *Calanus finmarchicus* in the six areas of the North Sea, B_1, B_2 (north-east, north-west), C_1, C_2 (east and west central) and D_1 and D_2 (south-east and south-west).

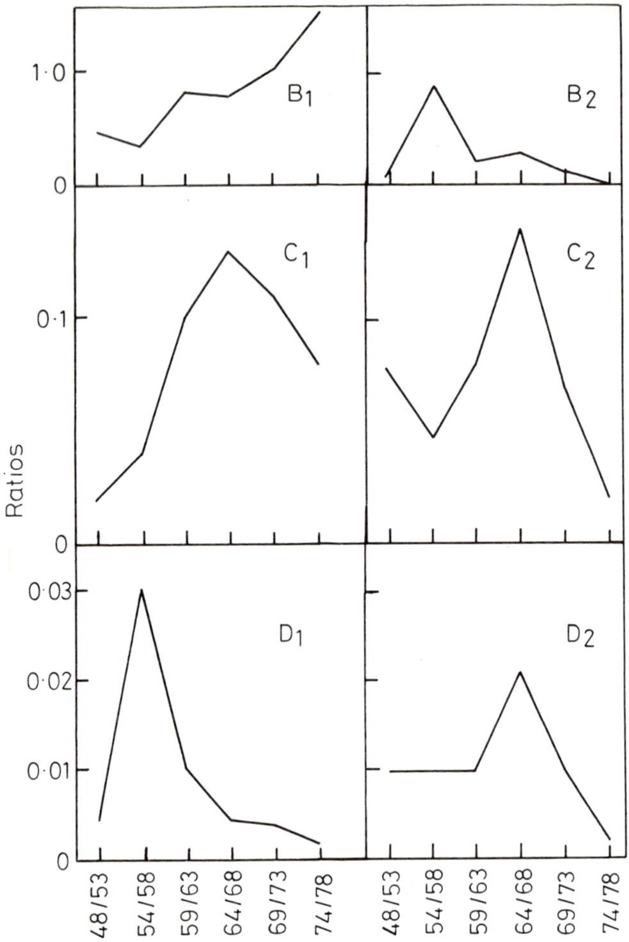

Fig. 65 The ratio of stock densities of *Calanus finmarchicus* averaged in April, May and June to those of *Pseudocalanus elongatus* averaged between May and September in five year periods for the six areas in the North Sea.

perature at the Ocean Weather Stations and with the number of days of westerly weather (with a two-year-lag, Lag 2; with no lag at all, Lag 0).

Thus there is no doubt that the first principal component of zooplankton abundance in the north-east Atlantic is correlated with the decline of surface temperature at the Ocean Weather Ships. It is unlikely that the decrease is a direct function of temperature, if only for the reason that the sea surface

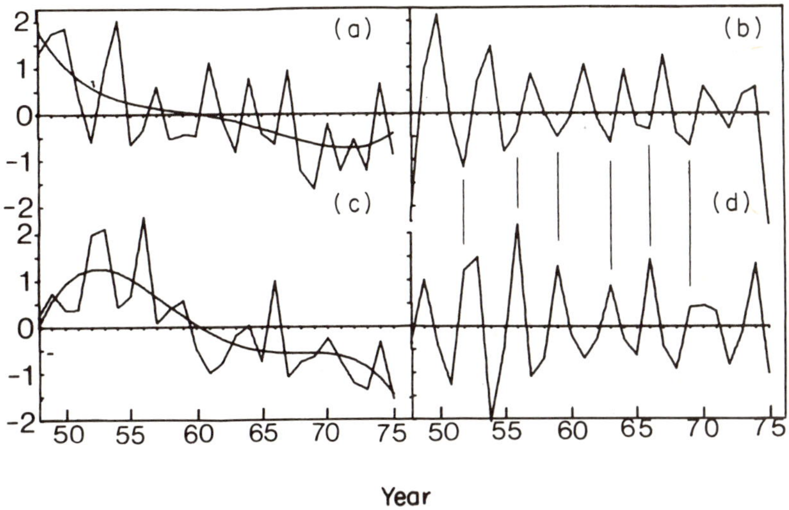

Fig. 66 (a) Decline in numbers of days of westerly wind across the British Isles between 1948 and 1977 (Colebrook, 1979).

(*b*) Decline in the principal component of changes in total zooplankton biomass between 1948 and 1977 (Colebrook, 1977).

(*c*) Variability of westerly wind indices over the British Isles with the downward trend filtered out.

(*d*) Variability of total zooplankton biomass with the downward trend filtered out.

temperature is a convenient, if indirect, index of a whole complex of climatic factors. The correlations with days of westerly winds in the same year (Lag 0) are not very convincing, but those with westerly winds two years earlier are, particularly in the Central North Sea. At first sight such a relationship appears to be absurd because zooplankton animals do not live very long, with three to five generations each year.

Colebrook's next step was to analyse the data about the trends shown in Fig. 66a. Figure 66b shows the variability of the indices of westerly winds and the first principal component of zooplankton abundance with the high-frequency elements remaining after the trend had been filtered out. There is an inverse relationship between the two variables, with a periodicity of about 3 1/2 years. Hence there is an inverse relationship between the abundance of the smaller zooplankton between 1948 and 1977 and the indices of westerly wind. From our discussion in Chapter 6 of the

Table 14
Correlations between the first principal component of zooplankton abundance and two environmental factors

	Area	Surface temperature at Ocean Weather Stations, first and second principal components	Days of westerly weather	
			Lag 0	Lag 2
Oceanic	B5	0.53+[a]	0.42+	0.52++
	B4	0.44+	−0.04	0.57++
	C5	0.62+++	0.23	0.51++
	D5	0.70+++	0.19	0.54++
Shelf	B1	0.60+++	0.32	0.47+
	C4	0.66+++	0.25	0.69+++
	D4	0.46	0.42+	0.25
Northern N. Sea	B2	0.35	0.29	0.36
Central N. Sea	C2	0.44+	0.46+	0.60+++
	C1	0.43+	0.41	0.61+++
Southern N. Sea	D2	0.52++	0.50++	0.43+
	D1	0.32	0.19	0.47

[a] Correlation significant at: $P < 0.05$, +; $P < 0.02$, ++; $P < 0.001$, +++

dependence of the spring outburst of plankton in temperate seas upon differences in wind strength and direction we might expect a relationship. From Fig. 66b we might suggest that the periodicity was in the original data. But the real puzzle is why the decline in the smaller zooplankton might be related to westerly winds two years earlier and inversely year by year with westerly winds in a given year. Perhaps there are two distinct mechanisms: (1) the dependence of the production cycle upon a wind direction other than westerly; and (2) upon the advection of water into the North Sea driven by a more general wind stress two years earlier.

The most remarkable event in the North Sea (and in the north-east Atlantic) was the delayed time of onset of the spring outburst described in Chapter 6. During the period of three decades the production cycle was delayed by about three or four weeks. At the same time the seasonal production of zooplankton biomass has gradually been reduced from seven or eight months to about five or six months. Perhaps one or two generations were lost (Fig. 67a, areas C1, C2, D1 and D2 in the North Sea; Glover, 1979). At the same time the total zooplankton in the North Sea was reduced to about half, although there might have been a recovery since 1975 (Fig. 67b; Glover, 1979). The two events—the delayed spring outburst and reduction of zooplankton biomass—are probably linked. Such secular changes are probably linked to analogous changes in wind strength and direction and, as suggested above, to advection. The maintenance of *Calanus* stocks in the eastern North Sea is less important quantitatively than the general decline in zooplankton biomass. However, the *Calanus* population may play a particular part in the recent biology of the North Sea (see below).

Reid (1978) examined the phytoplankton records in a little more detail between 1958 and 1973 in each of the six areas of the North Sea (B1 and B2 in the northern North Sea, C1 and C2 in the central North Sea, and D1 and D2 in the southern North Sea). The year-to-year variation in seasonal abundance of "greenness" (the measure of phytoplankton density as estimated from the plankton recorder network) is considerable, and the variability increases in the north in the deeper water. The abundance of diatoms remained high in all sectors, except B1 in the north-eastern North Sea, but after 1968 the autumn outburst of diatoms ceased. In the central North Sea (C1 and C2) there has always been a pronounced autumn outburst of the same amplitude as the spring production cycle (Colebrook and Robinson, 1965). Cushing (1967) associated the presence of early autumn-spawned herring larvae and post larvae from the Buchan and Dogger spawners with the occurrence of that autumn outburst in the west central North Sea where the larvae are hatched and in the east central North Sea, the area to which the larvae drift. Between 1977 and 1981 all capture of herring in the North Sea was banned as the stocks had suffered from recruitment overfishing. The failure of a component of the autumn production cycle may discourage the recovery of those stocklets that spawn on the Scottish coasts and around the edges of the Dogger Bank.

As the westerly winds over the British Isles have diminished, so has the zooplankton biomass, but the variations about both downward trends are inversely related to differences in the westerly wind. During the same period, the time of onset of the spring outburst has been delayed, and presumably as a consequence the productive season of zooplankton has been shortened. These events appear to depend upon differences in westerly wind two years earlier which might affect advective changes and hence the overwintering populations, as Colebrook (1979) has suggested, within the same system, and independent changes in the *Calanus* populations may have been sustained by input from the Norwegian Sea and the East Icelandic Current. However, there may be distinct and variable components which enter the North Sea by the Dooley and Tampen Bank Currents.

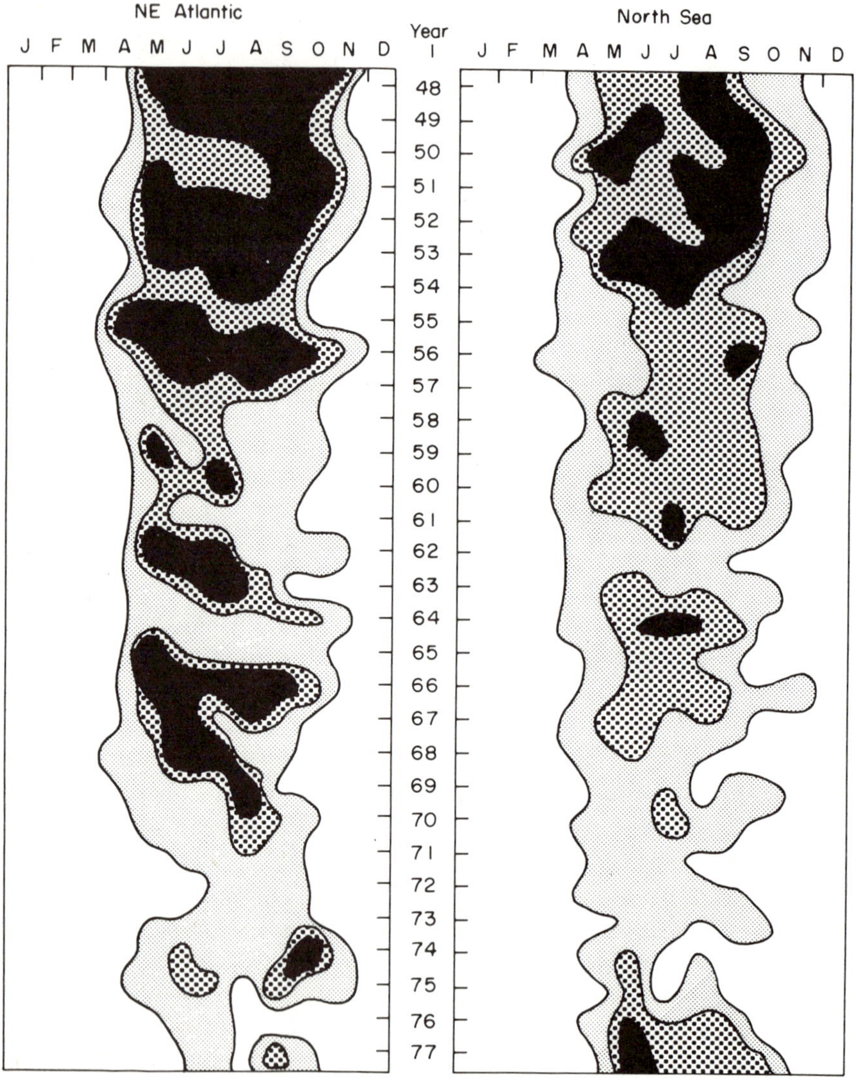

Fig. 67 (*a*) Seasonal production of zooplankton biomass between 1948 and 1974 (Glover, 1979)

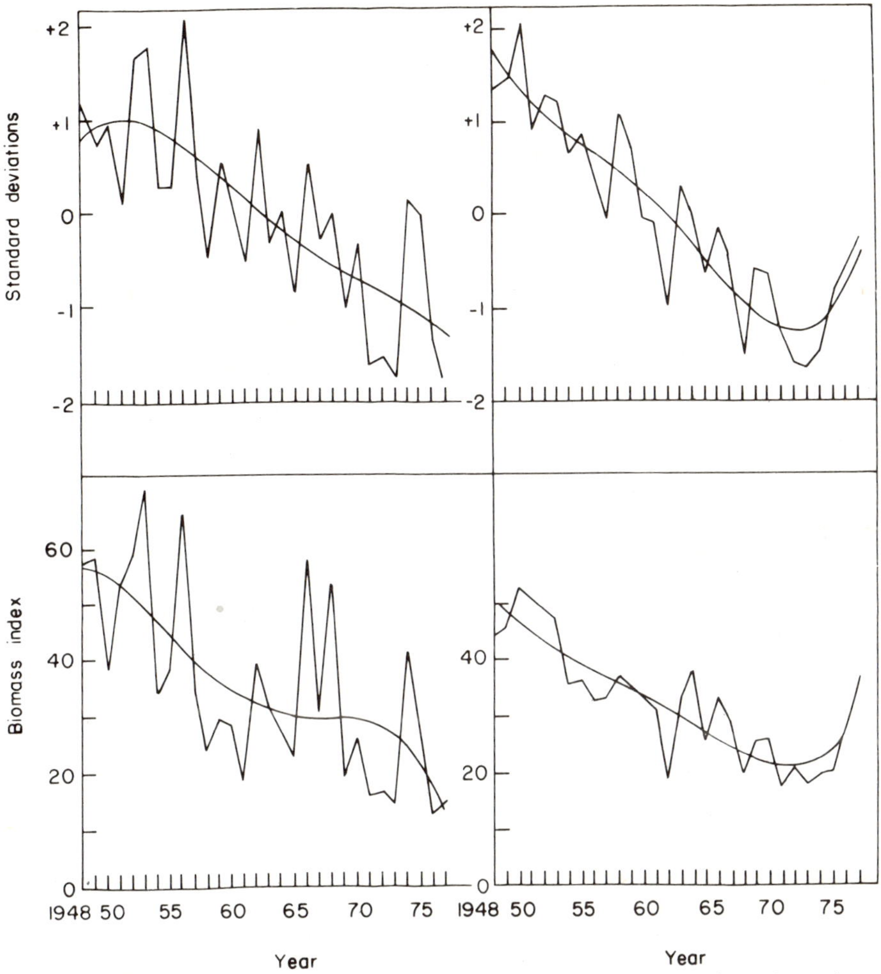

Fig. 67 (*b*) Reduction in total zooplankton biomass between 1948 and 1974 (*left*) North Atlantic, (*right*) North Sea.

CHANGES IN THE NORTH SEA HERRING POPULATIONS

In the North Sea, herring spawn in spring and autumn but since the end of the 1930s spring spawners such as those that spawned in the Firth of Forth have tended to die out; however, they continued to flourish in the Norwegian Sea at least until the Atlantoscandian stock was much reduced by fishing in the late 1960s. There are three groups of autumn spawning herring in the North Sea: the Downs group, which spawns in November and December between the *Galloper* L.V. off Harwich and the Baie de la

Seine in the eastern English Channel; the Dogger group, which spawns in September and October on the edge of the Dogger Bank and on the English coast near the Bank; and the Buchan group, which spawns in August and September off Shetland, Orkney and the Scottish north-east coast. Saville (1979) has found that part of the herring stock exploited in the northern North Sea comprised recruits from the stock which spawns off the West coast of Scotland. The three autumn-spawning groups in the North Sea differ in spawning season and region, meristic characters, growth, year class strength, and total mortality (Cushing, 1955; Cushing and Burd, 1957; Cushing and Bridger, 1966; Saville, 1979). There are also minor groups which spawn at the mouths of rivers and in estuaries such as the Blackwater, Milford Haven or the Sussex rivers: Yarrell (1836) called such groups Leach's herring.

There have been considerable changes in the North Sea herring populations since the early 1950s. Figure 68a shows the trend in l_1 (the first summer's growth recorded on the scales of herring in the Downs stock) between 1937 and 1961; between 1937 and 1949, the growth in the first summer on the nursery ground remained steady and it increased sharply between 1950 and 1952, the years when the abundance of *Calanus* increased in the eastern North Sea as shown in the last section (the l_1 material is taken from Iles, 1967). In Fig. 68b, l_1 is plotted on the *Calanus* anomalies from the small statistical squares Nv and Lw in the eastern North Sea. There are two sets of observations, 1948–56 and 1957–9, distinct but with similar slopes. Recruitment to the Downs stock of herring started to decline in the second half of the 1950s and that in the three years, 1957–9, was only 60% of that in the earlier period which is the reason why the observations were split between 1956 and 1957. The time series is closely related to that already noticed in "arctic influence" in the Faroe–Shetland Channel.

If the first year's growth depended upon *Calanus* anomalies as shown above, we might expect that of the three-year-old herring, the first adult age group, to be similarly related and the quantity of food available during the three years of growth (Fig. 69). One consequence of recruitment overfishing is that as recruitment declines, so does stock, until it comprises only a single year class. Hence the mean length of four-year-old Downs fish is inversely dependent upon the stock ($r^2 = 0.39$) and also upon the stock of Downs fish to which was added one-third of the Dogger stock ($r^2 = 0.75$); the Downs fish migrate south across the Dogger area, where they were exploited, and Cushing and Bridger (1966) suggested that about

(*a*)

(*b*)

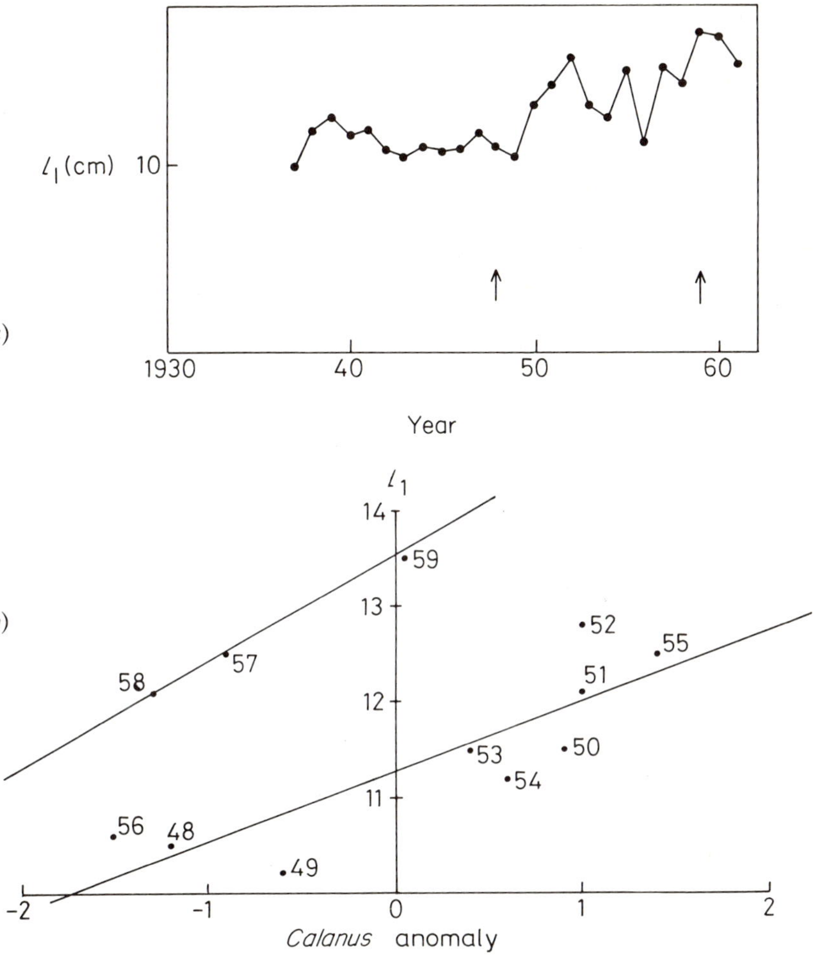

Fig. 68 (*a*) Trend in l_1 of Downs herring between 1937 and 1961. (*b*) Dependence of l_1 (the first summer's growth as recorded on the scales of herring) on *Calanus* anomalies in the eastern North Sea: the relationship between 1948 and 1956 and between 1957 and 1959 when recruitment had been reduced to 60% of that in the earlier period.

one-third of the Dogger catch was taken from the Downs stock. Deviations from the last regression are positively correlated with *Calanus* anomalies in the East Central North Sea ($r^2 = 0.33$), which means that the growth of herring in the North Sea would be affected by differences in *Calanus* abundance if the stocks were not suffering from recruitment overfishing.

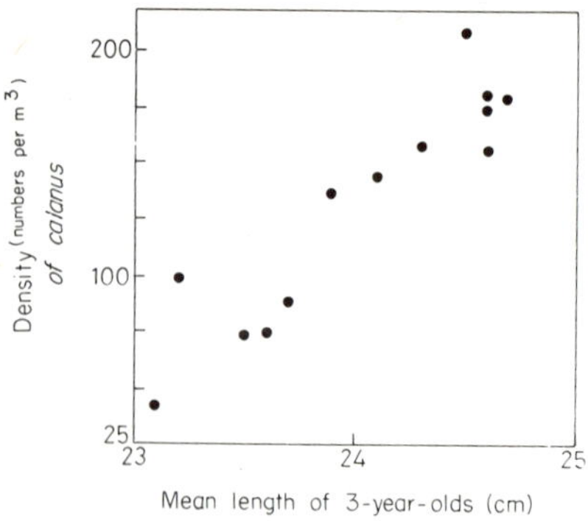

Fig. 69 Dependence of the mean length of three-year-old herring of the Downs stock upon the density of *Calanus* averaged during the years of growth.

We conclude that the length of four-year-old Downs herring depends inversely upon recruitment (and hence stock) and positively upon *Calanus*, much as does the length attained in the first summer.

An interesting point is that the lengths of four-year-old Buchan herring are positively correlated with those of the Downs stock ($r^2 = 0.58$), which is not surprising because the two groups grow up on the same nursery ground in the eastern North Sea where *Calanus* has remained abundant since the early 1950s. Because the recruitment at least during the late 1950s and the 60s may have comprised fish from the West of Scotland, the length of four-year-old herring should not be correlated with it. Similarly if stock is dominated by recruitment, a correlation between the lengths of four-year-old fish and stock would not be expected.

The three groups of North Sea herring differ in size at four years of age; the Buchan herring are about 2 cm longer than those of the Dogger group, which are 2 cm longer than those of the Downs group. During the 1930s a typical four-year-old Downs herring measured 24.0 cm in length and by the 1950s it had increased to 26.0 cm and by the late 1960s to more than 27.0 cm (Cushing and Bridger, 1966). Figure 70 shows the remarkable increase in length for the three spawning groups.. We have shown that from 1950 onwards, the increase in length was associated with the increase in *Calanus* in the North Sea and with the decline of stock. However, before that date, there was an increase in length of the herring of the Downs stock during the 1930s when stock levels were about constant and when levels of *Calanus* were low. From Fig. 68, l_1 the first summer's growth recorded

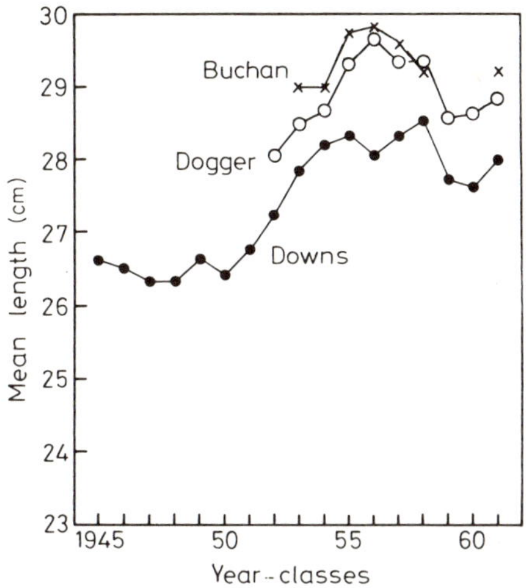

Fig. 70 Increase in length of North Sea herring for each of the three spawning groups—Downs, Dogger and Buchan (Cushing and Bridger, 1966).

on the scales, did not change very much between 1933 and 1949, a period during which the herring stock had increased during the Second World War. Hence there must have been a change independent of stock, independent of growth in the first year and perhaps independent of *Calanus* stock density. This earlier growth change may have been related to an unrecorded planktonic component, as the herring is planktivorous.

Figure 70 also shows that between 1950 and 1952 the lengths of four-year-olds of both Downs and Buchan stocks both increased by about 1 cm—a very sharp increment indeed. Burd and Cushing (1962) atrributed it to a rise in the abundance of *Calanus*. Cushing (1966) examined samples of *Calanus* from the plankton recorder network in four statistical squares in the North Sea expressed as anomalies from the mean between 1948 and 1962. Table 15 shows correlations between *Calanus* anomalies and l_1, l_2 and l_3 (growth in the first year, in the first two years, and in the first three years) of the Downs group of fish for the western (W) and eastern (E) squares in the central North Sea. *Calanus* anomalies were averaged for two and three years for the correlations with l_2 and l_3.

The growth of the Downs herring was positively correlated with the *Calanus* anomalies in April, May and June, the growing season in both the western and eastern parts of the Central North Sea, which is the area where the young fish grow. In the first summer of their lives, herring grow

Table 15
Correlation between *Calanus* anomalies and growth characteristics of the Downs herring (l_1, l_2 and l_3) (Cushing, 1966)

	April, May, June		July, August, September	
	W	E	W	E
l_1	−0.20	0.10	−0.39	−0.14
l_1	0.69^{++}	0.58^{++}	0.18	−0.01
l_3	0.60^{+}	0.64^{++}	−0.11	−0.25

in shallow water on the Danish, German and Dutch coasts (and also in the Thames estuary). The grounds on which the juveniles grow in summer lie in cold water patches east and north west of the Dogger Bank (Burd and Cushing, 1962). Hence the correlations shown in Table 15 and the relationships in Fig. 70 are not unexpected.

Burd (1962) suggested that herring leave the nursery ground at a critical length (21 cm for the Downs herring), irrespective of age and start to mature in the subsequent feeding season; consequently, a fish must mature at a particular age—three, four or five. Figure 71 shows the percentage of recruits to the Downs, Dogger and Buchan stocks at ages three, four and five. In 1925, about 10% of the Downs fish recruited at three years

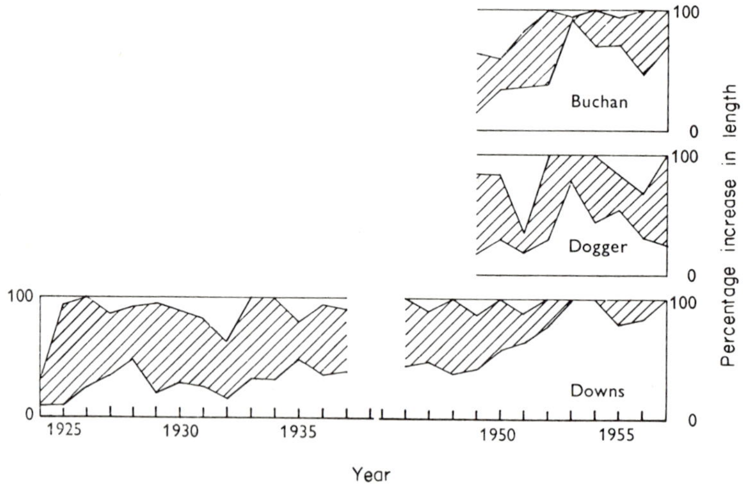

Fig. 71 Percentages of recruits at ages three, four and five to the three spawning groups between 1924 and 1957 (Cushing, 1966).

of age, but by 1953 all had recruited at that age. In general, from the mid 1950s onwards in all three stocks, herring have fully recruited at three years of age. It is a consequence of the major increase in length in all three autumn spawning stocks, due probably first to an unspecified change in the plankton in the thirties affecting herring of the Downs stock, secondly to the increase in *Calanus* in the 1950s and thirdly to the decline in all three stocks from the mid 1950s onwards.

The effect of these somewhat complex changes was to increase the weight of individual herring and to lengthen the period during which they were exposed to fishing as adults. Hence the index of stock, catch in weight per drifter shot did not decrease as sharply as it might have done had there been a constant weight of individual fishes. One of the most important factors in the decline in the North Sea herring stock was the development of an international industrial fishery for juvenile herring in the eastern North Sea: 21 200 tonnes were landed in 1950, but by 1955 catches had reached 115 900 tonnes (Madsen, 1978). The North Sea herring stock collapsed because of overexploitation of both adult and immature fish (Shepherd, in press). The development of the industrial fishery in the eastern North Sea was perhaps a consequence of the increased growth rate of the herring in the early 1950s. If herring recruited to the adult stock earlier in their lives, they probably passed through the nursery ground at an earlier age. There would probably have been no increase in *Calanus* on the beaches where the little herring spend their first summer. On the nursery ground, however, in the eastern North Sea, the increment of their preferred food might have made the herring leave much earlier. There were three distinct effects upon the stocks in the North Sea: (1) an increment in the weight of individuals of 10–20%; (2) a reduction in the age of recruitment; (3) a concentration on the nursery ground which may have stimulated the industrial fishery. All three are consequences of increased growth because of an increase in *Calanus* in the eastern North Sea, but each had different effects upon the fisheries. More generally, the increase in *Calanus* made the North Sea herring stock more vulnerable to the effect of fishing.

As noted above, Saville (1979) has discovered that young fish migrate from the West of Scotland into the northern North Sea as larvae and back again as juveniles. Hence the northern North Sea stock must have comprised a mixture of Buchan spawners and West of Scotland juveniles; young fish of two and three years of age were probably returning to the West of Scotland whereas older fish were Buchan spawners. Figure 72 shows the numbers of recruits to the three areas of the North Sea (IVa, b and c)

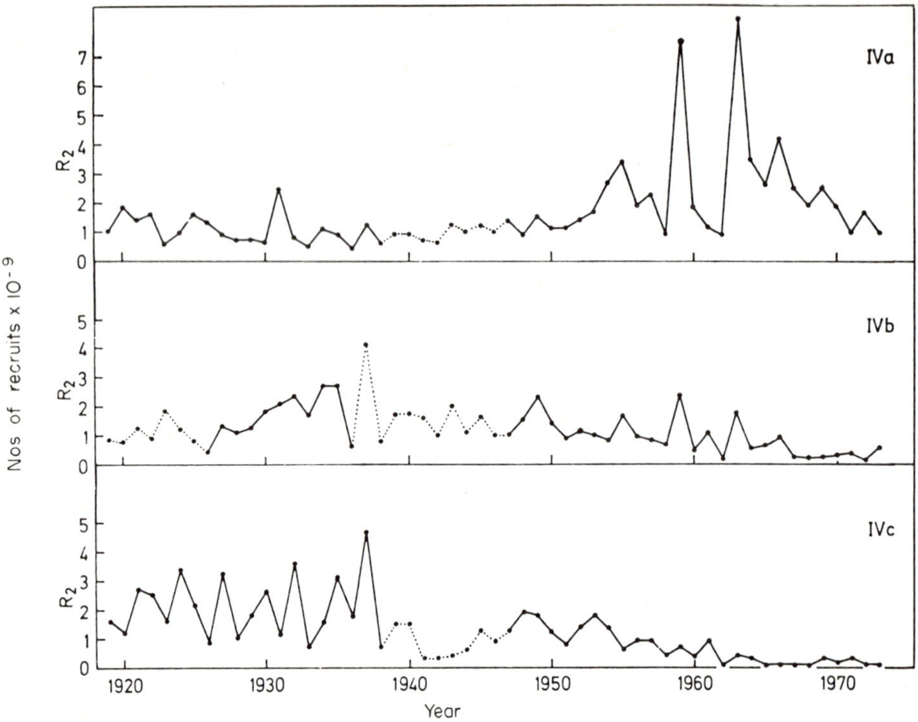

Fig. 72 Changes in recruitment in subareas IVa, IVb and IVc in the North Sea, indicating the recruitment to each of the three spawning groups (Burd, 1978).

corresponding very roughly to Buchan, Dogger and Downs stocks (Burd, 1978). The estimates were made by cohort analysis, i.e. they were derived from estimates of abundance of older fish. Hence the estimates of recruitment in IVa must comprise components of both the Buchan stock and that off the West coast of Scotland. Similarly the estimates of Dogger stock probably include a third of the Downs spawners as they migrate south to spawn. The Dogger recruitment increased during the 1930s when the herring started to grow. This may have been a genuine increase in the Dogger stock or an increase in the vulnerability of the Downs stock on the Dogger grounds; the apparent decline in Downs recruitment during the 1940s may well be an artefact due to increased capture in IVb. However, by far the most important effect occurred in IVa from the early 1950s onwards. It is not important whether the recruits were of Buchan or West of Scotland origin: whatever their source many of them grew up in the eastern North Sea. As noted in the previous section, the ratio of *Calanus* to *Pseudocalanus* had increased steadily in the north-eastern North Sea, which effectively means that the *Calanus* stocks increased there in the early 1950s and maintained their abundance subsequently. Is it possible that the high level of recruitment between the early 1950s and the middle 1960s was a consequence of increased *Calanus* abundance in the eastern North Sea? There is no correlation between recruitment and abundance of *Calanus,* but in view of the mixed stock sources in the area we might not expect one. The decline of recruitment in all three areas, in IVc from the late 1950s onward, in IVb from the early 1960s onwards and in IVa in the late 1960s was due to recruitment overfishing.

In the earlier section on changes in the North Sea plankton it was found that whereas the smaller copepods had declined in general since 1948, there had been a pronounced increase in *Calanus,* particularly in the north-eastern North Sea which was probably maintained, if with considerable variation, at least till 1977. It is not known what caused this change, but *Calanus* invades the North Sea from the Norwegian Sea and could well be maintained by the Atlantic flow into the Norwegian Rinne. Whatever the ultimate cause of the change, there were considerable effects upon the herring populations in the North Sea. They grew more quickly, changed their patterns of recruitment, and became more vulnerable to fishing both as adults and as juveniles. Further it is possible that the sharp increase in recruitment in the northern North Sea was related to the increase in *Calanus* in the eastern sector. Many of these changes are entangled with the effects of fishing, which are not part of our present account. However, as a consequence of heavy fishing there has been a ban on fishing herring in the North Sea (and in the Norwegian Sea, off the West coast of Scotland and the Celtic Sea), and there is considerable concern about the prospects of full recovery. If the *Calanus* stock densities are maintained in the eastern North Sea we might expect growth rates to be maintained and possibly high recruitment again in the North—if the stocks recover. In the previous section we referred to the disappearance of diatoms in the autumn outburst since 1968. We do not know what effect this will have on the structure of the production cycle, but it may make it harder for the herring larvae to survive particularly from the Dogger area and perhaps from the Buchan area. Hence there are good natural omens of recovery, but they depend on how well the *Calanus* populations subsist in autumn in the absence of diatoms.

CHANGES IN THE DEMERSAL STOCKS IN THE NORTH SEA

With the development of cohort analysis (or virtual population analysis) the recruitment to fish stocks in numbers can be estimated well, provided that the numbers caught at each age are sampled efficiently. Today, such estimates extend back for two or three decades for a number of commercial fish stocks, but only for herring do the time series extend back as far as the 1920s and 30s. Beverton (1962) plotted the annual logarithmic ratio of catches to the long-term mean between 1906 and 1957 (Fig. 73). Catches are affected by the amount of fishing effort exerted as shown by the differences before and after both world wars, but the figure shows three phenomena of greater import:

(1) Sole catches increased by a factor of ten and, as will be shown below, recruitment increased during the 1960s and 70s, so the stock has probably increased in the North Sea throughout the present century. During this period southerly species increased in the North Sea.

(2) Catches of plaice and turbot were higher after the Second World War than before it. Catches may have increased somewhat by improved management of the plaice stock, but the magnitude of recruitment increased after the Second World War. There is no evidence that the stock had suffered from recruitment overfishing before that war and the increase took place after it at somewhat higher levels of stock. We shall show below that both the distribution of plaice and of their spawning extended in the German Bight.

(3) Catches of haddock declined during the 1930s and early 1950s, and during the 1890s the stock density was perhaps five times higher than during the subsequent 60 years (Lundbeck, 1962). Yet in 1962 (see below) the recruitment to the North Sea haddock opened the remarkable gadoid outburst of the 1960s and 70s, being 25 times larger than the previously recorded average (back to the early 1920s).

In a paper to the Aarhus Symposium on changes to the North Sea fish stock Holden (1978) tabulated estimates of recruitment to five major stocks and averaged them as follows:

	Cod	Haddock	Plaice	Sole	Whiting
1926–38	47	98	267	53	822
1944–59	74	83	406	107	309
1960–69	123	392	460	116	787

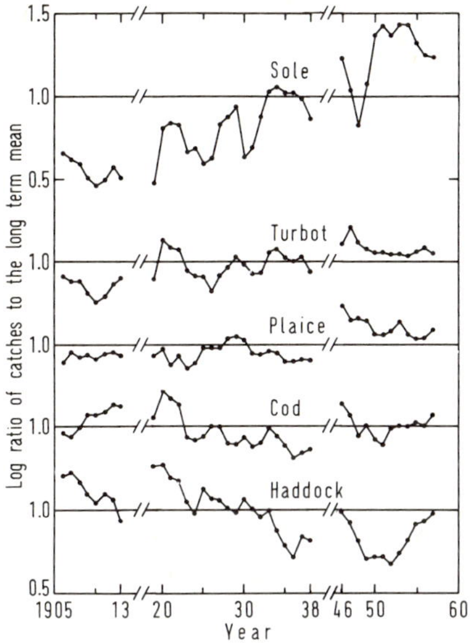

Fig. 73 Annual logarithmic ratio of catches of each of five commercial species to the long-term mean between 1906 and 1957 (Beverton, 1962).

(The methods of estimation given in Holden's paper are basically cohort analyses extended back in years converted from catches per unit of effort; but the numbers cannot be compared between species. Indeed whiting are not superabundant.)

For cod, plaice and sole, recruitment increased steadily during the period examined: indeed the increment in sole stocks continued during the 1960s and 70s. The increase in the stock of plaice is associated with a shift of spawning ground from the Southern Bight to the German Bight. Bückmann (1961) examined a long time series of samples from the German Bight (1895, 1904, 1905, 1906, 1908, 1909, 1920, 1926, 1935, 1937, 1953, 1955, 1960 and 1961) and showed that in 1955 the stock of eggs increased by an order of magnitude in the German Bight but that in the Southern Bight remained constant during the period. The spread of the spawning stock to the north is shown in a chart of the egg distributions published by Harding *et al.* (1978). As will be shown below, this spread is reflected in the changed patterns of distribution of fishing activity.

The increase in sole catches since the beginning of the century is part of an invasion by southerly species. Aurich (1953) recorded the densities of eggs/m^2 of sole, horse mackerel (*Trachurus trachurus* L.), anchovy (*Engrauilis encrasicolus* Linnaeus) and pilchard in two periods, 1904–30 and 1949–52.

	1904–30	1949–52
Sole	10–35	100–276
Horse mackerel	23–77	100–846
Anchovy	<25	1000+
Pilchard	0	588

Postuma (1978) has recorded catches in the North Sea (and other ICES regions) of such southerly species. He found that pilchards were rarely caught in the northern North Sea, that they peaked in the Central North Sea during the 1950s and in the southern North Sea in the late 1950s, after which they declined. On the other hand, sea bream catches peaked in the northern North Sea during the 1960s. The general trend is that southerly species became more abundant in the 1950s, after which they declined somewhat but not very much.

General trends in distribution are rather hard to establish. Comparing charts published by Olsen (1883), Thompson (1912), Masterman (1912) and Anon. (1909a,b), the general impression is that the distribution of demersal species have remained about the same, but it must be recalled that all such observations were made in a period which was climatically somewhat homogeneous. Figure 74 shows charts of distributions of British catches per unit of effort of plaice, sole, cod, haddock and whiting in three periods, 1923–5, 1958–60 and 1977–8. It is well known that catch per unit of effort as index of stock varies with changes in power of fishing vessels and with extremes of abundance. Hence we do not compare magnitudes of stock densities between periods, but the positions only of peak apparent abundance. Of course such distributions may become a little distorted by the biases in catch per unit of effort as indices of stock density, but we assume them to be spatially distributed.

Figure 74a shows the changes in distribution of the plaice stock in the North Sea. Between the early 1920s and the late 1950s the centre of distribution shifted from the Southern Bight to the German Bight, probably as the German Bight spawning increased in 1955. By the late 1970s, the main concentrations had shifted north to the eastern Dogger and the Klondyke grounds west of the Norwegian Rinne. Thus the centre of distribution and the major spawning ground have shifted north as the German Bight spawning group has progressively increased in importance.

Figure 74b shows the changes in distribution of the sole stock. In the early 1920s, the centre of distribution lay off the Thames estuary and by the late 1950s it had shifted into the German Bight, much as has had happened to the plaice stocks. By the late 1970s there were two centres of distribution—in the German Bight and off the Thames estuary. We have no direct evidence of the spawning ground changes except that the fish do spawn in the two places. The causes of change in the centres of distribution are perhaps due to relative changes between spawning groups or stocklets.

The centres of distribution of the gadoids for the three periods are shown in Figs. 74c (cod), 74d (haddock) and 74e (whiting). In the cod and whiting there was a distinct southerly movement between the 1920s and the late 1950s and a slight return to the north by the 1970s. There are three centres of cod spawning—off northern Scotland, off the coast of Yorkshire, and off the island of Texel in the Southern Bight; perhaps the changes observed are again relative changes between the three spawning groups. The centres of haddock distribution shifted northward between the early 1920s and the late 1950s, with perhaps an indication of southerly return in the late 1970s.

Plaice and sole shifted into the German Bight and towards the north eastern North Sea. Cod and whiting moved southerly and perhaps back north again, whereas the haddock might have shifted northerly and then back to the south again. It is reasonable to suppose that such changes are governed by relative changes in spawning distributions, perhaps as conditions for survival changed from decade to decade in the North Sea. The spawning groups or stocklets of plaice are probably not genetically distinct (Purdom and Thompson, 1976), but adults tagged on one spawning ground do not stray to another (de Veen, 1962). If survival improved in the German Bight during the 1950s, so would recruitment, and hence stock and the spawning ground would have become more extensive, as did that of the northern anchovy off California after stock expansion in the 1950s (Ahlstrom, 1966). If this train of argument were to be sustained on the lines suggested in an earlier chapter, we would expect the production cycle in the German Bight to have shifted towards the time of spawning, thus improving the match of spawning distribution to that of the larval food. To continue the argument further, we would expect that from the mid 1950s onwards the wind strength and direction of the German Bight had become more favourable for survival. In such a way, differences in the strength of wind and its direction might have affected the distribution of the more important fish stocks in the North Sea. However, the argument is, of course, speculative.

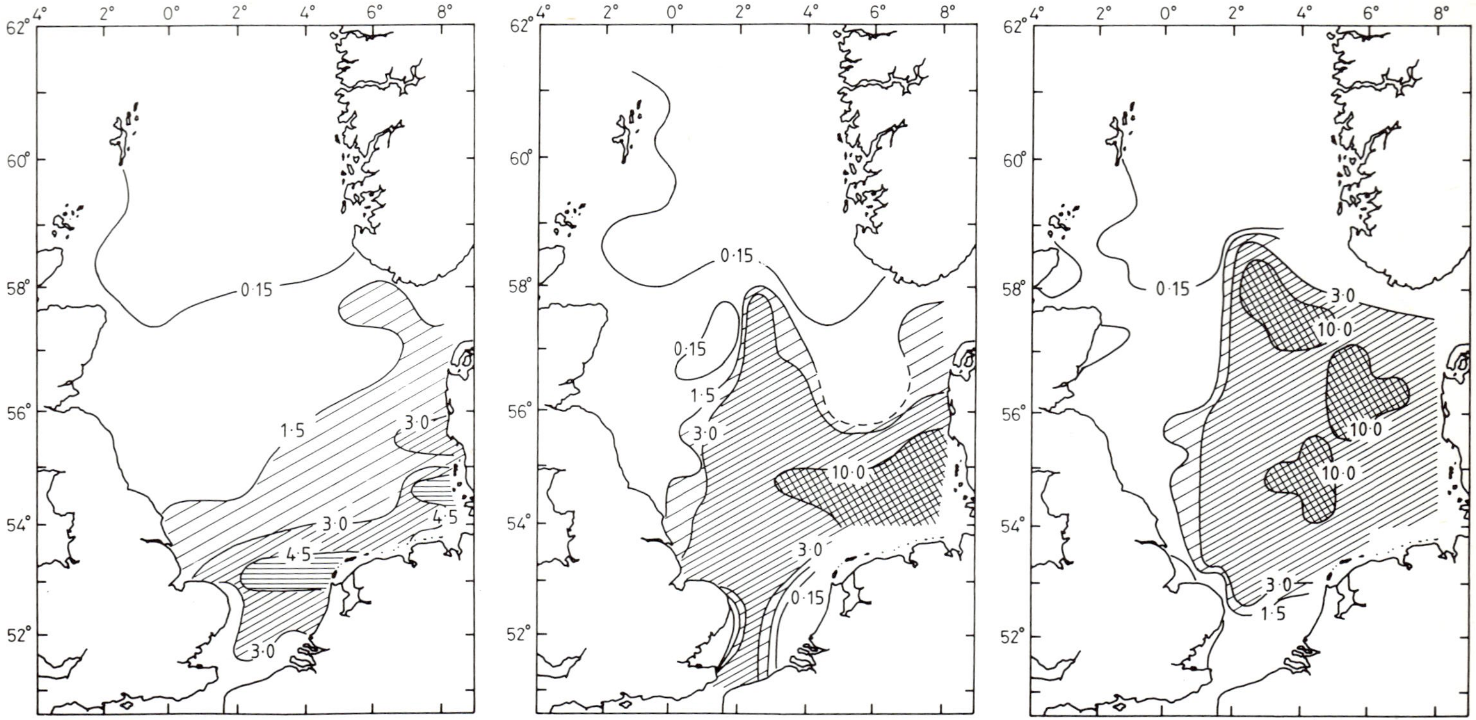

Fig. 74 (*a*) Charts of distribution of catches of plaice per unit of effort (tonnes per 100 hours) of British trawlers in 1923–5 (steam); 1958–60 (motor) and 1977–8 (motor).

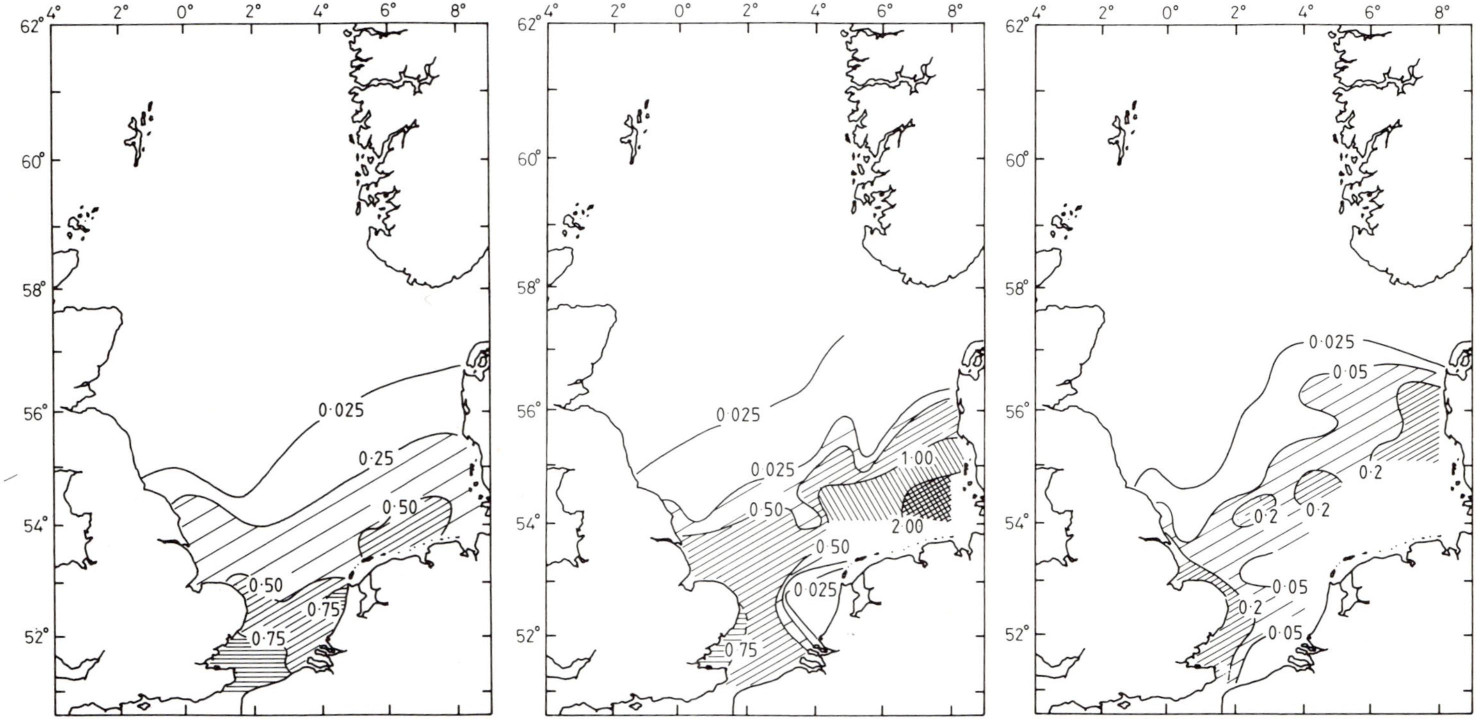

Fig. 74 (*b*) Charts of distribution of catches of sole per unit of effort (tonnes per 100 hours) of British trawlers in 1923–5 (steam), 1958–60 (motor) and 1977–8 (motor).

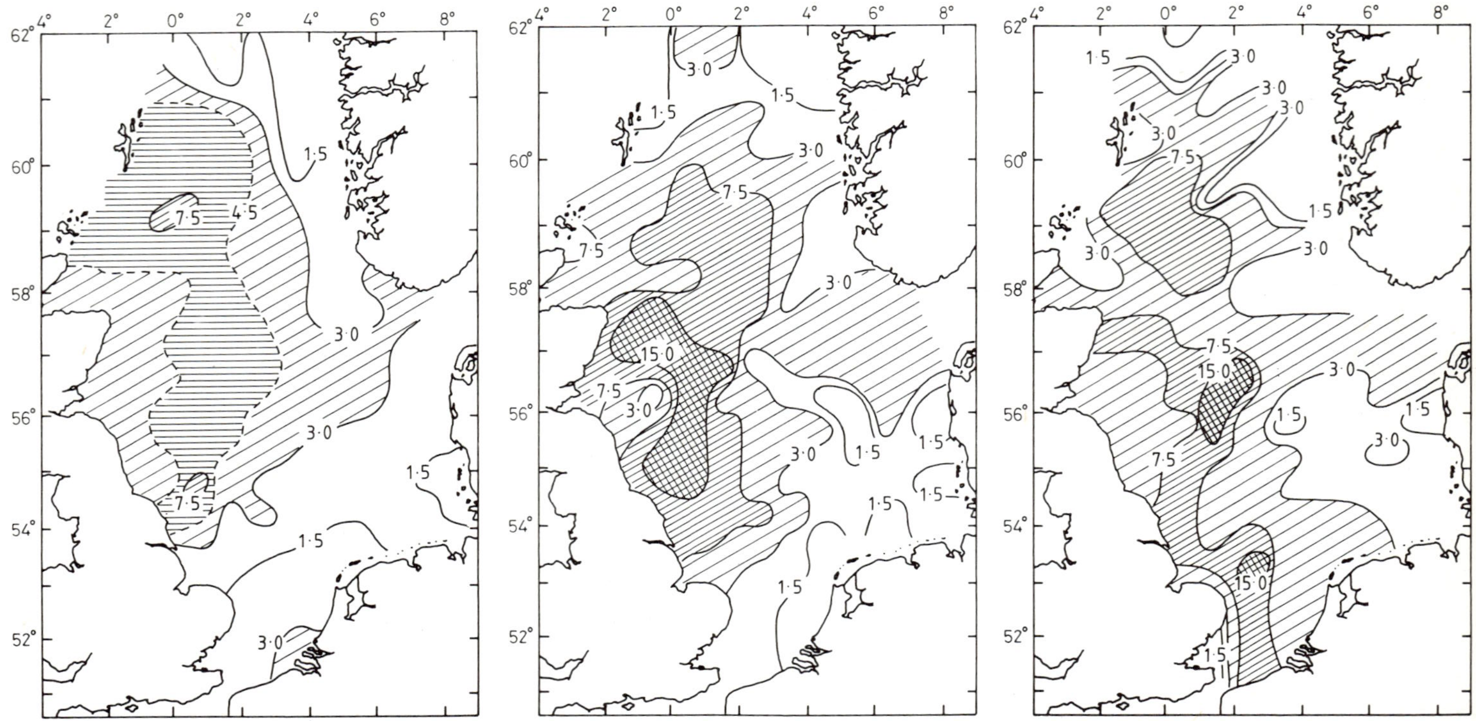

Fig. 74 (*c*) Charts of distribution of catches of cod per unit of effort (tonnes per 100 hours) of British trawlers in 1923–5 (steam), 1958–60 (motor) and 1977–8 (motor).

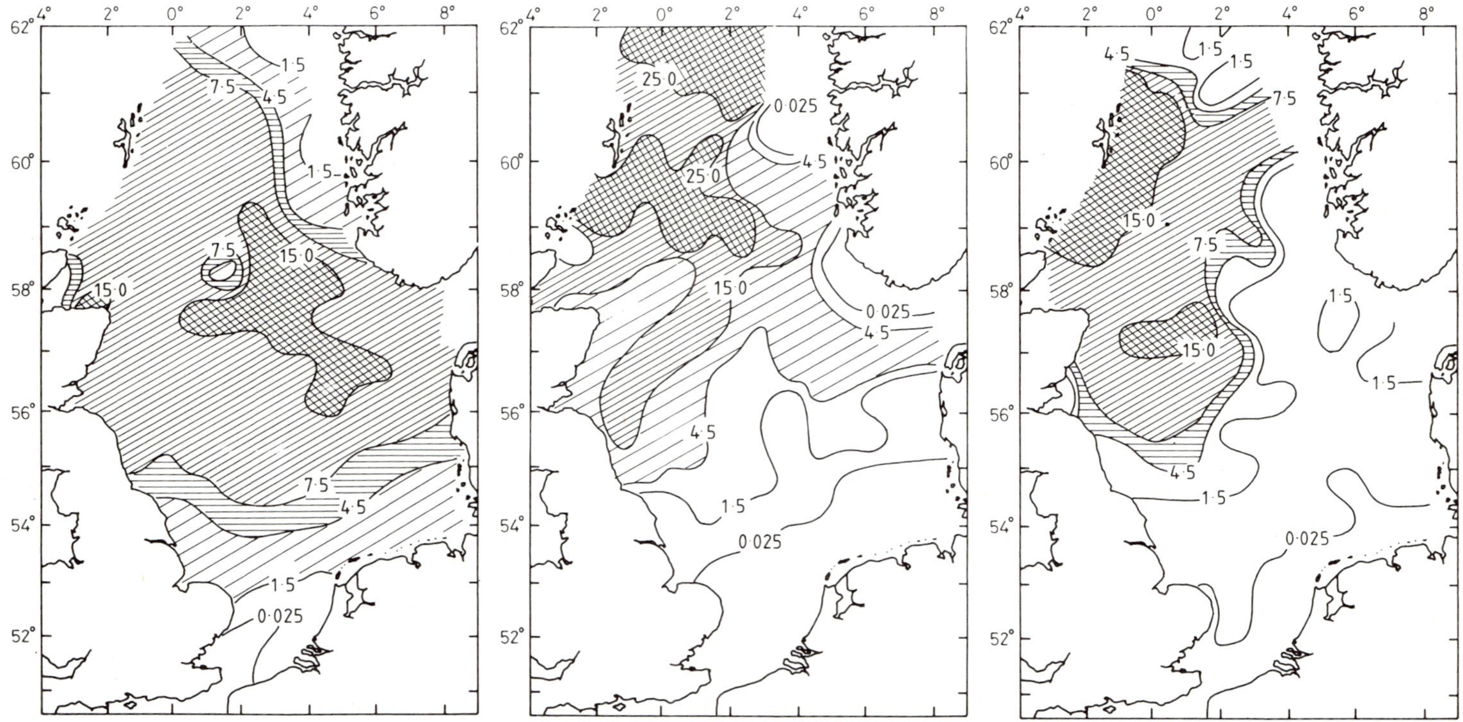

Fig. 74 (*d*) Charts of distribution of catches of haddock per unit of effort (tonnes per 100 hours) of British trawlers in 1923–5 (steam), 1958–60 (motor) and 1977–8 (motor).

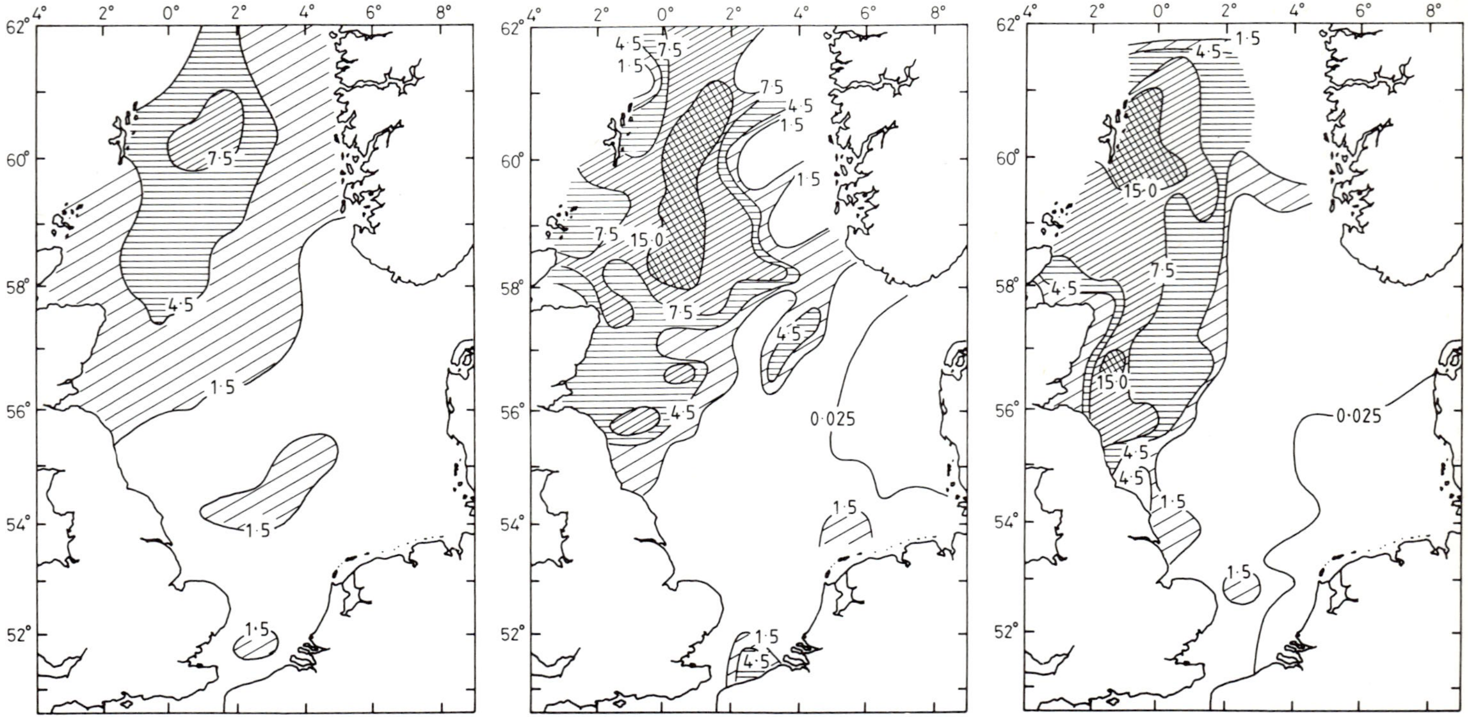

Fig. 74 (*e*) Charts of distribution of catches of whiting per unit of effort (tonnes per 100 hours) of British trawlers in 1923–5 (steam), 1958–60 (motor) and 1977–8 (motor).

The most remarkable event that occurred in the North Sea was the so-called gadoid outburst which started in 1962 with a very large year class of haddock; in the same year there was a strong year class of whiting. Subsequent very large recruitments of haddock and whiting were hatched in 1967, 1974 and 1980. High cod year classes appeared in 1963, 1964, 1965, 1966, 1969, 1974 and 1976. Saithe year classes increased in 1962 compared with the previous four years and were particularly large in 1966, 1967 and 1968. The total yield from cod, haddock, saithe and whiting increased by a factor of three during the 1960s. Significant catches of Norway pout were made in 1959, and they rose above 100 000 tonnes in 1962; in the mid 1970s catches increased to about 400 000 tonnes on average, but it is not known whether the augmented catches represented an increment of recruitment (Holden, 1978).

There are two ways in which the increase in gadoid stocks can be represented. Figure 75a demonstrates the augmentation of the stocks on an arithmetic scale to show the scale of the event. Catches of cod, haddock, whiting and saithe increased from 259 000 tonnes each year in the 1930s to 670 000 tonnes in the decade 1962–73 (Holden, 1978), an increment of 2.6 times or more than 400 000 tonnes. Figure 75b shows the same material on a logarithmic scale which shows that the rate of increase of each stock was about the same during the decade and common to all the gadoid stocks. A correlation matrix linking all five stocks shows them to be related (Cushing, 1980). This implies that the recruitments of the five stocks were affected by a common factor, but it can be no more than an implication, although the correlations between cod, haddock, whiting and coalfish are high (Table 16).

Cushing (1980) tried to analyse these events in terms of the release of food from the reduced herring and mackerel stocks and by the relaxed predation on cod larvae by the two diminished pelagic stocks. He concluded that the most likely recipients of released food are the post-metamorphosis cod, but they would have to share it with the much more abundant planktonic carnivores. Similarly the relaxed predation was hard to identify; cod larvae might be eaten by adult herring (or adolescents on the way from the nursery ground), but during the period of abundance they were scarcer in the herring guts than might have been expected. They might have been eaten on the eastern nursery ground, but the cod were probably already too big when they arrived there. Yet the I group herring as they leave the nursery ground might have eaten the 0 group cod that remained there. This phenomenon is very hard to establish.

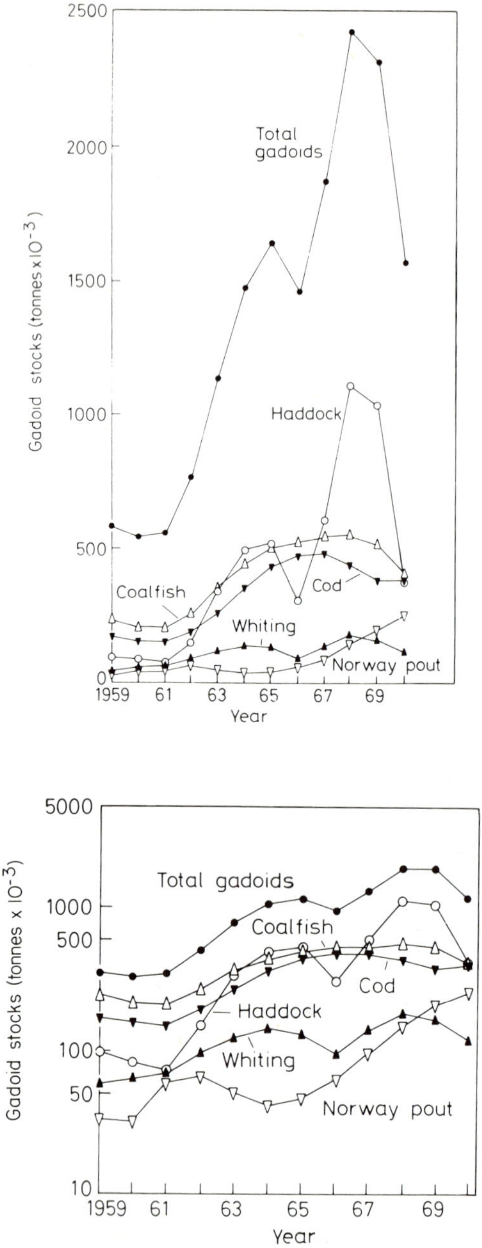

Fig. 75 Increase in five gadoid stocks during the 1960s: (*a*) on an arithmetic scale; (*b*) on a logarithmic scale.

Table 16
Correlations between stock estimates of five gadoid species, 1959–70 (stock estimates from Andersen and Ursin (1977) Table 24)

Coalfish	Cod	Haddock	Whiting	Norway pout	
—	0.975^{++}	0.814^{++}	0.853^{++}	0.414	Coalfish
	—	0.704^{++}	0.765^{++}	0.430	Cod
		—	0.933^{++}	0.583^{+}	Haddock
			—	0.524	Whiting
				—	Norway pout

The real difficulty is that the haddock year class of 1962 was hatched in the northern North Sea at a time when the herring in that region was more abundant than it had been since the 1920s. Hence the cod population which grows up in the central, eastern and southern North Sea would be a better candidate to benefit from released food or relaxed predation.

In an earlier section we referred to an increase in the ratio of *Calanus* to *Pseudocalanus* (and *Paracalanus*) in the north-eastern North Sea over the Norwegian Rinne, which because of the decline of the smaller copepods, implies that the stock densities of *Calanus* increased there in 1950 and remains high, if variable. It will be recalled that overwintering *Calanus* might reach the North Sea from the Norwegian Sea or from the East Icelandic Current. *Calanus* might also enter the North Sea in the Dooley Current, but only in the north-eastern North Sea have the levels of *Calanus* been maintained since the early 1950s.

Cushing (1966) referred to a correlation between *Calanus* stock densities on the Flamborough Line and haddock year classes between 1933 and 1961. The haddock spawns in the Northern North Sea, and the stations on the Flamborough Line probably sample water and animals which have drifted from the north-eastern North Sea. The correlation collapsed with the outstanding haddock year class of 1962 with which the gadoid outburst started (Fig. 76a). The failure of the correlation might not have deserved more notice than any of the others that litter the subject had it not been succeeded by another. Cod year classes in the North Sea between 1962 and 1967 depend significantly upon the stock density of *Calanus* in the north-eastern North Sea. It is an accident that the time series started in 1962 when the

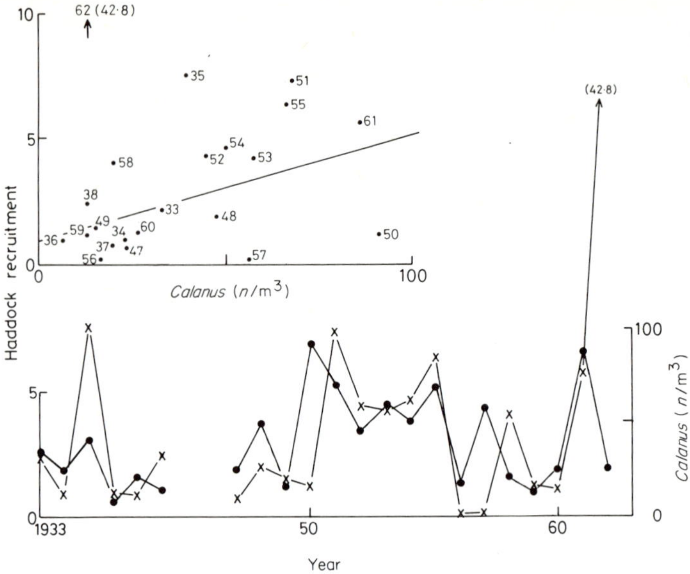
62 (42·8)
(42·8)
Haddock recruitment
Calanus (n/m3)
Calanus (n/m3)
1933
50
60
Year

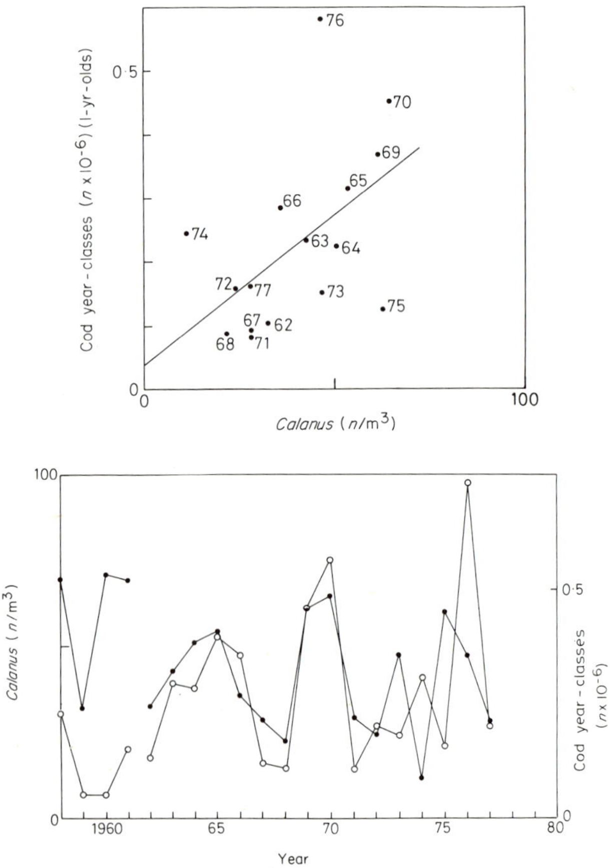
Cod year-classes (n x 10-6) (1-yr-olds)
Calanus (n/m3)
Calanus (n/m3)
Cod year-classes (n x 10-6)
1960
65
70
75
80
Year

cohort analyses were initiated. Holden (1978) extended the time series back to 1954 with the use of catches per unit of effort of two-year-old cod taken by Grimsby trawlers (which might not represent the North Sea cod population between 1961 and 1971 (the base of Holden's estimate). The correlation between cod recruitment estimated in this way and *Calanus* stock density in the north-eastern North Sea between 1954 and 1977 is not significantly different from zero (Fig. 76b). To put it another way, the *Calanus* stock densities in the north eastern North Sea in 1958, 1960 and 1961 were as high as any value later, before the gadoid outburst. We have two correlations that succeed each other in time, one before the gadoid outburst started which failed in 1962 and one which flourishes during the gadoid upsurge, but could not be established before 1962.

Jones (1973) noted that both cod and haddock larvae fed on *Calanus* and indeed both probably grow with the rising cohorts of *Calanus*. From the evidence of the two succeeding correlations the year-to-year variability of both cod and haddock year classes probably depend upon the stock density of *Calanus*. If that were properly so, we would expect a correlation between haddock year classes and *Calanus* from 1962 to 1977. It does not exist, although the 1967 and 1974 year classes were very strong when *Calanus* stock densities were at their highest in the Central North Sea (square Kv). But haddock year classes are *inversely* related to *Calanus* stock densities in B_1 ($P < 0.02$) which might suggest an inverse dependence of haddock on cod year classes (because cod eat haddock); there is no correlation, nor with a lag of one or two years in either direction. Haddock may depend on *Calanus* in the Dooley Current in the north-western North Sea, whereas the cod may depend on the stock density in the Tampen Bank current; but there is no reason to suppose an inverse relationship between them. What is much more likely is that three processes are confounded, year class strength of cod and haddock on two sources of *Calanus* and an inverse dependence of haddock year classes on those of cod which cannot be detected.

The most important point that separates the two correlations, causing one to fail and the other to start, is the increase in magnitude of the year class strengths of five gadoid species from 1962 onwards. I compared the seasonal distribution of *Calanus* in B_1 in the north-eastern North Sea for two periods, 1949–61 and 1962–78. Figure 77 shows the two distributions (transformed by natural logarithms), and it will be seen that in the later

---

Fig. 76 The failing correlations of haddock and cod recruitments upon *Calanus* stock densities: (*a*) haddock, 1933–61; (*b*) cod, 1962–77.

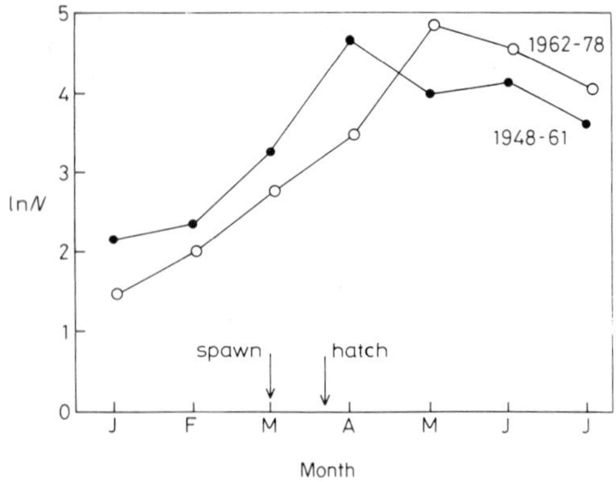

Fig. 77 Seasonal distributions of *Calanus* in the north-eastern North Sea (area B_1) in the plankton recorder network) in the periods 1949–1961 and 1962–1978; the arrows mark the average dates when cod spawn and when their larvae hatch.

period, that of the gadoid outburst, the production of *Calanus* reaches its peak about a month later (with monthly observations). The same phenomenon has been recorded for the smaller copepods (as noted above), but for both *Calanus* and the smaller copepods the trend is a continuous one, not broken neatly in the year 1962. In the main part of the North Sea the cod spawns in March (but in February in the Southern Bight) (Daan, 1978). If the eggs hatch in about three weeks, the larvae may hatch in early April. In the early period, the young larvae would grow after two weeks or so in a period of declining *Calanus* stock density. In the later period, the young larvae would have a period of six weeks after hatching before the stock density of *Calanus* declined. Indeed the *Calanus* were more than twice as abundant in the later period, during the critical weeks when the cod larvae were growing up.

Cushing and Dickson (1976) suggested that the year class strength of a fish population was strong or weak according to whether the production of larvae was matched or not to that of their food. Figure 77 suggests that during the period of the gadoid outburst the chance of the larvae obtaining more food was greater than during the earlier period. Then the increase in magnitude of the gadoids during the 1960s might well have been due to the delayed production of *Calanus,* although of course it cannot be shown precisely.

However, an index of delay was developed in the following way. The seasonal production of *Calanus* peaks one, two, three or four months after the peak date of cod spawning in March. Figure 78a shows the dependence of cod year classes upon the delay in months between 1954 and 1977. Figure 78b shows the lack of dependence of haddock year classes upon the same index of delay, but the largest haddock year classes (1962, 1967 and 1974) all occurred in years when the production of *Calanus* peaked in May. There are two conclusions from this remarkable relationship (1) that the match/mismatch relationship put forward by Cushing (1975) may be plausible, and (2) the gadoid outburst might well have been the result of delayed *Calanus* production in the north-eastern North Sea.

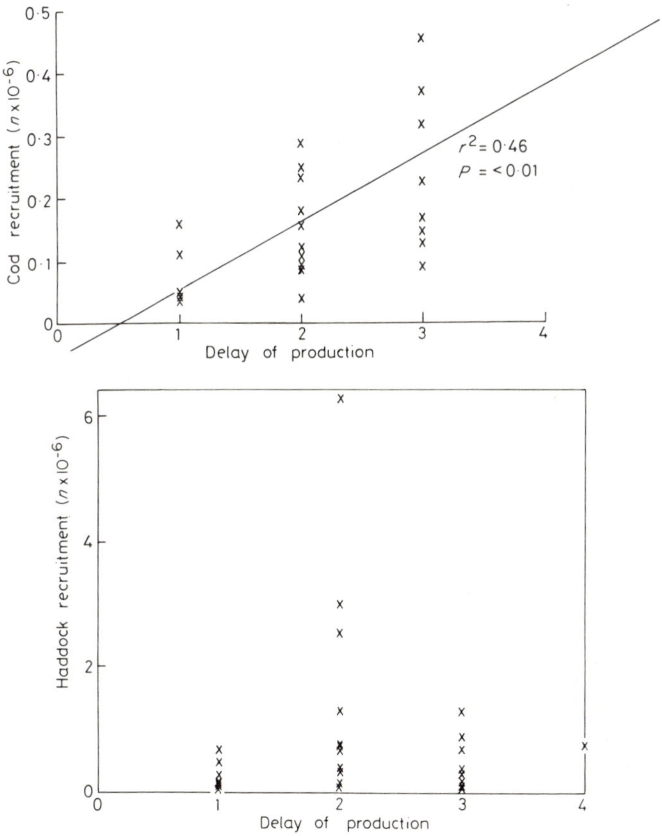

Fig. 78 Dependence of recruitment in the North Sea upon an index of delay (months of delay of *Calanus* peak production after cod spawning): (*a*) cod 1954–1977; (*b*) haddock, 1948–1977.

Another change in the North Sea fish stocks is shown by the changes in length of some commercial fishes (Fig. 79). An increase in the length of the herring has already been mentioned, and another series is reproduced here, that of the high and low maturity stage fish of four years of age in the Buchan fishery. The increase in length was sustained since the 1930s. Such an increase also occurred in four-year-old haddock, whiting and plaice during the same time period, but not in cod. The increment in length of the herring of four years of age was about 1.5 cm per decade, whereas that of plaice and the two gadoids amounted to 2.5 cm per decade. Such increments in length represent one in weight of between one-fifth and one-third, depending upon the initial weight. The phenomenon is widespread and lasted for nearly half a century, yet the cod did not increase in length although the age at first maturity was reduced (Daan, 1978). In an earlier section the increment in length of herring was attributed in part to the effects of more *Calanus* to eat and in part to the decline in recruitment. The remarkable point about the demersal fish is that their recruitment *increased* as recorded, and we have shown that the variation in recruitment does depend to some degree upon the food available.

It is of course tempting to suggest that the increase in length was a direct effect of exploitation as indeed happened to the herring through recruitment overfishing in the late 1950s and the 1960s. However, as far as can be seen there was no detected change in mean length due to relaxation of fishing and increase of stock during the Second World War; it is possible that length was not very well sampled in 1946 and 1947, but that is only another way of saying that the expected change with stock is low. Hence I believe that it is essentially a natural change, partly associated with the availability of food; however, the lack of change in length of four-year-old

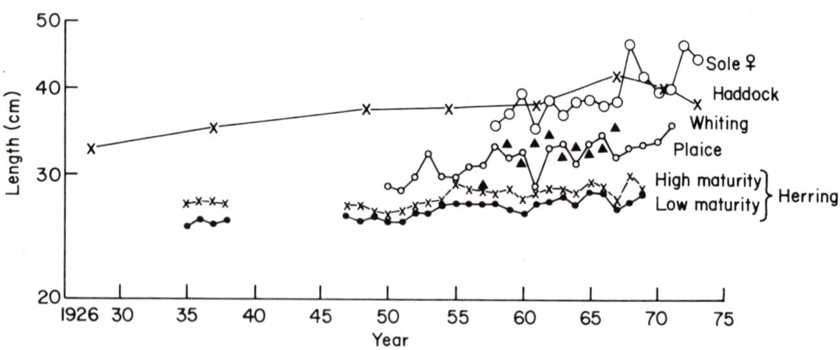

Fig. 79 Changes in length of some commercial fishes in the North Sea since 1930s (Cushing, 1981).

cod suggests that a multispecies component is playing a part. The cod is the most abundant top predator in the North Sea, and because its larvae grow at 10% a day it demands food at a very high rate and pre-empts it from competitors. The upsurge of gadoids in the 1960s indicates a superabundance of food which has gone into increased recruitment and perhaps into the weights of individuals as consequence of exploitation.

CONCLUSION

Considerable changes have taken place in the North Sea in the last few decades. We have classified the results in three parts—on plankton, on pelagic fish, and on demersal fish—because the records exist in this form; there are of course links between the groups. Probably the changes in the plankton are the fundamental ones that generate those in the fish stocks. Since 1948, the quantities of smaller copepods, *Pseudocalanus, Paracalanus* and others, have declined steadily by a factor of four or five, and the season of zooplankton production was shortened by one or two generations. The reduction in the quantities of smaller copepods is inversely related to the number of days of westerly wind over the British Isles, that is, to wind stress from some other direction. In 1950 the input of *Calanus* into the North Sea perhaps from the Norwegian Sea and East Icelandic Current increased by a factor of nearly three. It was probably related to an increase in "arctic influence" (as arctic intermediate water) in the Faroe Shetland Channel. The quantity of *Calanus* remained high in the north eastern North Sea, and so the ratio of *Calanus* to *Pseudocalanus* increased there between 1948 and 1978 continuously. There are two main flows into the Northern North Sea—the Dooley Current between Orkney and Shetland and the Tampen Bank Current on the western edge of the Norwegian Rinne—and we believe that the main entry of *Calanus* into the North Sea since 1950 has been in the Tampen Bank Current.

The changes in the North Sea herring populations were complex but linked partly to the increase in *Calanus*. The first summer's growth of the herring depended upon the *Calanus* anomalies and upon the reduced recruitment. Hence the lengths of herring of each of the three spawning groups increased during the period, partly due to the sustained increase in *Calanus* and partly due to the diminished recruitment. Then the age at first recruitment was reduced, mainly owing to the greater quantities of food. It is possible that the industrial fishery for herring in the eastern North Sea originated in the faster growth of herring with more food available. The predominant event in the 1960s was the sharp increase in the magnitude of recruitment to the Buchan stock of herring in the northern North Sea, and it is tempting to suggest that it was the consequence of more food for less larvae or juveniles which might have improved survival.

The major changes in the demersal stocks are primarily those in recruitment, decline in haddock and increase in sole during the first fifty years of the century, followed by the extraordinary upsurge of cod-like fishes in the sixties. Some changes in distribution occurred which were probably associated with shifts from one spawning centre to another or the expansion of one at the expense of another. The rise of the gadoid stocks in the 1960s occurred during a period of cooling and we know that the haddock, at least, was abundant towards the end of the nineteenth century when the climate was distinctly cooler. Thus the changes in the North Sea appear to have been linked with the recent period of warming and subsequent cooling. In detail, the gadoid outburst was probably associated with the delay in the seasonal production of *Calanus* on which the larval gadoids depend. The remarkable correlation of cod recruitment upon the index of delay of *Calanus* production between 1954 and 1977 both before and after the start of the gadoid outburst in 1962 suggests that it was really part of a more continuous natural change which started in 1950. In its turn the augmentation of *Calanus* in the north-eastern North Sea appears to have been related to the increased arctic influence in the Faroe–Shetland Channel. This event is also part of the long-term climatic changes because Tait (1956) regarded the appearance of the arctic intermediate water in 1949–52 as a new event in the long time series (since 1927).

In many ways the events in the North Sea reflect the climatic changes as clearly as the more dramatic ones such as El Niño or the rise and collapse of the West Greenland cod fishery. They are a little more difficult to disentangle, partly because the ecosystem is somewhat more complex than the simple upwelling one or that in the arctic and partly because the signal is weaker. This is because the North Sea is heavily exploited and the herring and mackerel stocks have been much reduced.

Chapter Eight

The Effects of Cold Winters in North-West Europe

During this century very cold winters have occurred in North-West Europe about once in every two decades, for example, 1928/9, 1946/7, 1962/3 and 1978/9. Such winters are very cold, very long and quite distinctive, when ice floes appear in estuaries or off the coasts of Holland and Germany; in earlier centuries, the Thames froze, but since the eighteenth century the river has not frozen because of improved navigation facilities. In Germany such winters are called *Eiswintern,* partly because the frost endures and partly because the edges of the North Sea and the whole of the Baltic may freeze; indeed in certain ice winters a man could ride from Denmark to Gdansk (Danzig) (Pettersson, 1915). Such winters are associated with a persistent anticyclone over Scandinavia with cold easterly winds flowing across the North Sea and the British Isles. The winds are so cold that the sea freezes off the Danish, German and Dutch coasts of the North Sea and the cold water is blown offshore.

The cold winters played a crucial part in the development of the British trawl fisheries. Houldsworth (1874) wrote that the "productivity of the locality (the Silver Pit) is dependent upon the coldness of the season, which drives fish of all kinds, and especially the soles, into the deeper water". Further

> "The Great Silver Pit was first worked over in a very severe winter [about 1843] [and] in the deeper parts of the western end and near the middle soles were found during that very cold season in almost incredible numbers; the nets were hauled up bristling with fish trying to escape through the meshes, and such catches were made as the most experienced fishermen had never before dreamed of. The discovery soon got wind and a migration of trawlers

from Ramsgate and Brixham took place . . . in subsequent years the Silver Pit has again been found very productive, whenever the winter has been very severe, or as the trawlers call it, in "Pit seasons".

Frank Buckland (1879) refers to the hard winter of 1878–9, when soles were found in the Silver Pit. From evidence of cold winters in North West Europe in Easton (1928) and in the Baltic (Speerschneider, 1915; Betin, 1957), the winters of 1843/4 and 1844/5 were both exceptionally cold.

The migration of fishermen referred to by Houldsworth took place in two parts, first to Barking Creek and Ramsgate from Brixham after the Napoleonic wars and secondly to Grimsby and Hull following the discovery of the Silver Pit. In 1851 the railway reached the north-eastern ports and a market was developed in the rising industrial towns of Yorkshire and the Midlands. On the basis of this rising market, the English trawlermen in their smacks, sometimes in fleets and sometimes single-boating, developed the North Sea trawl fishery and their exploration extended northwards during the second half of the nineteenth century (see Cushing, 1966). Hence the cold winter of 1843/4 (or 1844/5) played a very important part in the development of the English trawl fishery in the North Sea.

Lamb (1959) described the winters in recent climatic changes in the following way. Between 400 and 1200 A.D. there was a secondary climatic optimum when the weather was dry and storm free. Between 1200 and 1400 the weather deteriorated and became very variable, the winters being either very severe or very mild. Between 1400 and 1500 there was a partial recovery when there were fewer harsh winters. Between 1550 and 1600 east wind winters were prominent. But between 1550 and 1850 we entered the Little Ice Age when glaciers advanced to their most extended positions since the last glaciation and the polar pack ice became more extensive and the cold winters more frequent. The evidence for this climatic history is given in Fig. 80 which shows Lamb's winter mildness and severity indices for the three months December to February by decades from 1100 to 1960. The index of winter mildness or severity is the excess number per decade of unmistakeably mild or cold months between December and February; mild winters are scored as positive and cold winters as negative. An analogous procedure is used for the indices of summer dryness or wetness. Instrumental records at a number of stations in Western Europe extend back to 1757 (Easton, 1928), or in central England to 1659 (Manley, 1974). The series has been extended back in time by Lamb and Easton with historical reference. For example, if the sea froze during the instrumental period, good documentary evidence of ice floes at sea in earlier

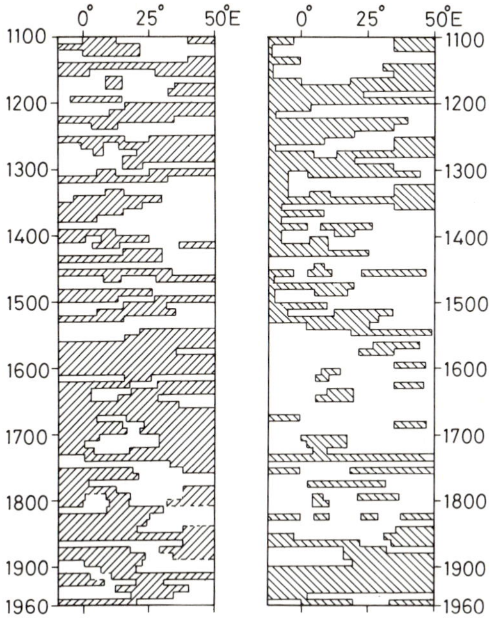

Fig. 80 (*left*) Lamb's indices of summer dryness and wetness (Lamb, 1965). (*right*) Lamb's indices of winter mildness and severity (Lamb, 1965); the excess number in each decade of unmistakably mild or cold winter months (December, January, February), mild being recorded as hatched and cold as blank.

times can give a clue to the air temperature then. But Speerschneider (1915) pointed out some of the early documentary evidence is difficult to interpret. For example the winter in year 1237 is described in Easton as follows:

(*a*) heims humida et calida et in hoc distemperata.

(*b*) Hiver chaud, avec peu de jours de gelée.

(*c*) Ein gelinder Winter, den Winde, Regen und Schnee immer unterbrauchen.

(*d*) Vinteren var mild.

He gives historical references for each sentence. Bell and Ogilvie (1978) have examined the documents as sources of evidence and refer to the unreliability of dates and to the habits of mediaeval copyists in transmitting fiction. In particular they point out that the Icelandic sagas are more literature than history, that the great majority of Speerschneider's sources are derivative, and that there are inconsistencies in Easton's records. The sequence of climatic events may be summarized in the roughest way by indicating the number of occasions in each century on which the Thames froze, up to and including the eighteenth (Lamb, 1959):

900s	1000–1099	1100s	1200s	1300s	1400s	1500s	1600s	1700s
1	1	2	2	0	1	4	8	6

PHYSICAL CONDITIONS

The physical conditions have been described in the North Sea during the winters of 1928/9, 1946/7 and 1962/3. Lumby and Atkinson (1929) noticed that reports of dead fish were first received in early March, 1929, but that the cold period was most intense during the month of February; they also suggested that the flow through the Channel into the Southern Bight of the North Sea might have been reversed. Schnoor (1929) wrote that the wind was easterly in the last week of January and all of February and that it was not very strong; off the Rhine, off Texel Island off north west Holland, and as far north as Skagen the sea temperature was about 0°C or as low as − 1°C. This condition persisted during February but it became warmer by the end of March. Such conditions would have been characteristic of a Scandinavian anticyclone which for example in 1947 persisted from 23 January to 14 March (Vaux, 1953). Ice extended from the Scheldt to the Skagerrak (Simpson, 1953). In the cold winter of 1939–40, there were thick ice floes in shallow water in the Thames estuary (Orton, 1940). At the other end of the North Sea off the Norwegian coast, Dannevig (1930a) suggested that the cold temperatures on the Skagerrak coast lasted only a few days when the Baltic outflow is pressed against the coast by the wind. However, during the cold winter of 1923–4, the water off the Flødivigen hatchery on the Norwegian coast of the Skagerrak remained very cold for three months, the temperature being as low as − 1.0°C down to 17 m.

The cold winter of 1962/3 was the coldest for 223 years (Scherhag, 1963), but it was also cold during the previous summer, i.e. from April 1962 to March 1963 throughout the whole of the northern hemisphere. Ellett (1963) showed that the temperature of the sea had fallen more slowly than in 1946/7 but the abnormally cold water persisted for longer. Crisp (1964) published a typical weather map during the period which showed the characteristic anticyclone with easterly winds across the North Sea; in addition he demonstrated that although the most extreme low air temperatures were experienced on south-eastern coasts, the south-western region of the British Isles suffered the greatest negative anomalies in air temperature and, as will be shown below, the greatest mortalities amongst shore living, intertidal and shallow water animals. Goedecke (1964) established that very cold water (−1°C) at the surface extended as far out as Helgoland at the end of January 1963, but on the sea bed such cold water only extended half way from the coast towards Helgoland. By 12th February, however, the bottom cold water reached the island, but by the 19th March it had retreated. The material is somewhat scattered and is usually collected somewhat late in any cold winter; the difficulty is to decide at the time whether a winter is abnormally cold. The persistent anticyclone over Scandinavia generates easterly winds which are cold enough to freeze the sea on the continental coasts out to some distance from the shore, to cool the water to the sea bed in the German Bight and to freeze the estuaries.

There are three sources of information on the cold winters in time series; Easton (1928) on Northwest Europe since the thirteenth century, Speerschneider (1915) on the ice winters in Danish waters from the time of earliest records until the first decade of the present century, and Betin (1957) on the ice cover in the Baltic between 1720 and 1952. Information before the instrumental period is based on historical records which are sometimes difficult to interpret as indicated above. Pettersson (1915) suggested that cold winters occurred periodically in Danish waters; there were said to be two groups, 1780–1809 and 1875–97, which coincided with the Swedish herring periods. When the Baltic froze between Scania (southern Sweden) and the island of Bornholm, the fishery was said to flourish. In all three sets of independent material, cold winters since the beginning of the nineteenth century occurred outside Pettersson's two periods. In particular Speerschneider shows that the number of cold winters in each decade remained high from 1760 onwards until the end of the nineteenth century, and the same conclusion emerges from Betin and Easton. Hence Pettersson's grouping of cold winters was faulty; however, the frequency of cold winters during the Swedish herring periods might be somewhat higher.

There is considerable evidence on the nature of cold winters, in North west Europe for long time periods. Figure 81 gives Easton's evidence based on historical material from 1205 to 1757, on quasi-instrumental records from 1758 to 1851 and on instrumented measurements from 1852 to 1928. The instrumental records were taken from nine stations: Bremen, de Bilt, Uccle, Paris, Greenwich, Angers, Toulouse, Lyons, and Strasburg. Easton

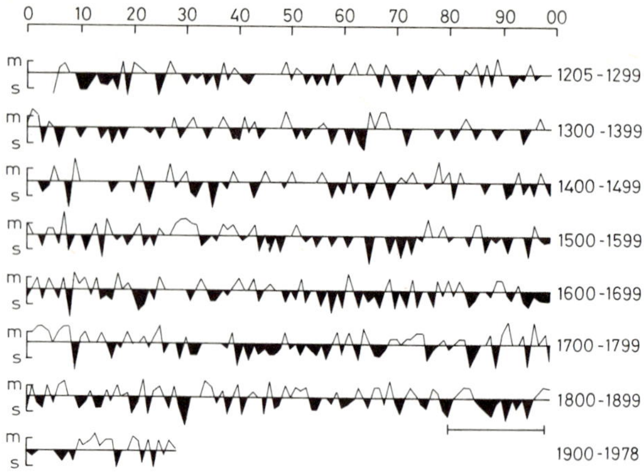

Fig. 81 Frequency of cold winters in north-west Europe from 1205 to 1928. Although the classification since 1851 is based on instrumental records, the main mass of material is based on qualitative observations (Easton, 1928) and is subject to a number of documentary biases. M is mild and S is severe.

developed ten categories across the three groups of material:

Très doux
Doux
Tiède
Normale +
Normale
Normale −
Froid
Rigoureux
Très rigoureux
Grand hiver

In very broad terms Lamb's description of secular changes in climatic terms since the secondary climatic optimum is confirmed, particularly the relative amelioration in the fifteenth century. Figure 82 gives the estimates of Baltic ice cover in thousands of km^2 from 1720 to 1950 (Betin, 1957). The figure shows intense winters from 1780–1790 (part of a Swedish herring period) and the decline in ice cover during the recent period of warming; but the great winters persisted during that period, which suggests that the blocking winter anticyclone over Scandinavia can occur whatever the long term trends if at different frequencies as number/decade. Figure 83 shows four series

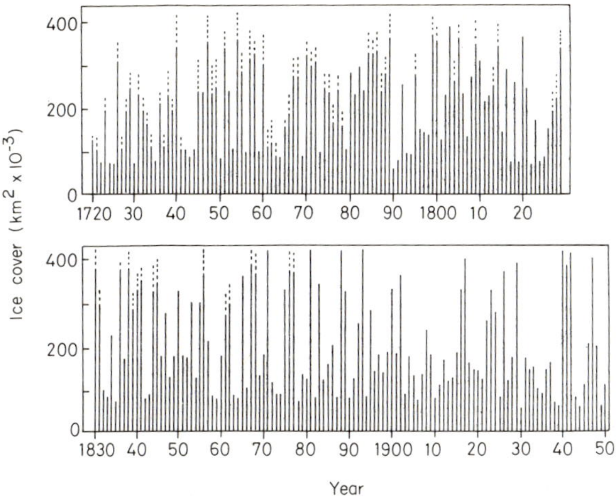

Fig. 82 Ice cover in the Baltic in thousands of km^2 from 1720 to 1790 (Betin, 1957).

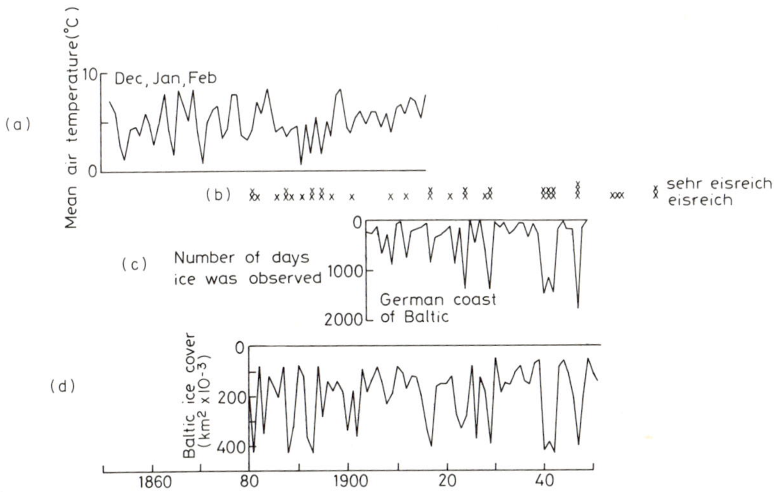

Fig. 83 Four series of observations during the last century or so.

(*a*) Mean of three monthly instrumental observations at Bremen from 1850–1916 (Easton, 1928);

(*b*) Ice winters on the German coast. "eisreich" (icy) x, and "sehr eisreich" (very icy) $\overset{x}{x}$ (Gehrke, 1922, 1923; Goedecke, 1958, 1963, 1964);

(*c*) Number of days on which ice was observed on the German coast of the Baltic (Nusser, 1949, 1958);

(*d*) Betin's (1957) observations of ice cover in thousands of km^2 from 1880 to 1950.

of observations in recent years. Figure 83a is the monthly mean temperature for December to February at Bremen (Easton, 1928) from 1850 to 1916. In Fig. 83b crosses, x and $\overset{x}{x}$, indicate ice winters "*eirsreich*" and "*sehr eisreich*" (Gehrke, 1922, 1923; Goedecke, 1958, 1963, 1964). Figure 83c is the number of days on which ice was observed on the German coasts of the Baltic (Nusser, 1949), and Fig. 83d is Betin's estimates of ice cover from 1880 to 1950. From the four sets of observations in Figure 83, there were more cold winters in the last two decades of the nineteenth century than in the twentieth century (as might be expected from a study of the recent period of warming), which is when the last Swedish herring period occurred. Although the long-term series of historical information, of Baltic ice cover and of instrumental records indicate the general trends of subperiods in climate history since the thirteenth century, the main impression is of the random appearance of cold winters in North West within each of the four main subperiods.

EFFECTS OF THE COLD WINTERS ON PLANTS AND INVERTEBRATES

The earliest detailed record of animals killed in a cold winter in north-west Europe was published by Miller (1855), who found millions of animals dead on the beach below Portobello, near Edinburgh, towards the end of February 1855. Included in the list of dead animals were the molluscs, *Mactra corallina* L., *Tellina fabula* Gmelin, *Donax fabula* (and a few *Spisula subtruncatum* da Costa), large numbers of *Syndosmya alba* Wood and of *Natica monilifera* Lamarck and the crabs *Portunus depurator* L. and *Carcinus maenas* L. Scott (1910) reported that during the winter of 1894–5, the whole Lancashire coastline was covered with ice floes and many hundreds of tons were washed up in the first gale after the frost. It has long been known that the oyster fisheries suffer in the cold winters; Orton (1940) reported that of Essex oysters, all imported from Brittany were lost in the winter of 1939/40 and of the natives, 42%–69% were killed, mostly the larger individuals. Cole (1940) recorded that Whitstable oysters survived well in that winter, but there was 50% mortality on the R. Crouch, 80% on the R. Roach, and 75% on the R. Colne. During the winter of 1962/3, oysters suffered considerable mortality, as did the cockles in the Burry Inlet in South Wales; but the 1963 year class of cockles in the Inlet was outstandingly large, with the consequence that the stock increased considerably in the subsequent decade, as did the fishery and, later, the population of predatory oyster catchers (*Haematopus ostralegus occidentalis* Neumann) (Hancock and Urquhart, 1964; Horwood, 1977). Orton (1931) showed that, in the earlier cold winter of 1928/9 of the drills (oyster predators), *Ocenebra* was almost eliminated, *Purpura* was reduced in abundance and *Urosalpinx* increased. Thus oysters were killed directly, but their predators were changed; as more of the larger individuals were killed, recruitment in subsequent generations may have suffered, because the larger oysters are the fecund ones. The distribution and species composition of the drills changed which may have affected the magnitude of the oyster population and its age distribution.

The best evidence on the death of invertebrates in the coastal zone was collected on European coasts in the early months of 1963. Offshore in the German Bight, Gillbricht (1964) showed that the spring outburst in 1963 occurred much earlier than in 1962 because of the layer of cooler water at the surface; as a consequence, the production of zooplankton started much earlier. Ziegelmayer (1964), in a study of the benthos in the German Bight showed that mortality was greatest among the Lamellibranchiata (except *Macoma baltica* L. and *Nucula nitida* Sowerby) and Echinodermata (especially *Amphiura filiformis* (O. F. Müller)); mortality was less among the Crustacea and least among the polychaetes. Boddeke (1963) wrote that in Dutch coastal waters, cockles and the polychaete *Lanice conchilega* (Pallas) were killed; after the cold winter, shrimp catches increased in the following May, June and July, perhaps because they fed on increased organic detritus. Offshore in the North Sea, Simpson (1953) recorded the deaths of *Phascolosoma vulgare* (Blainville), *Aricia cuvieri* Audouin and Edwards, *Metridium senile* (L.), the hermit crab, *Eupagurus bernhardus* (L.), the worms *Nephthys incisa* (Malmgren), *Nereis longissimma* Johnston, *Glycera gigantea* Quatrefages, the cockle *Cardium edule* L., the polychaete *Aphrodite aculeata* L., *Sipunculus* sp., and the Dublin Bay prawn *Nephrops norvegica.* L.

The material on the deaths of animals on the shores around the British Isles was collected by a number of biologists (Crisp *et al.* 1964; Crisp, 1965; Woodhead, 1964; Waugh, 1964; Hancock and Urquhart, 1964; Newell, 1964; Crisp and Southward, 1964; Moyse and Nelson Smith, 1964; Boaden *et al.,* 1964; de Valera and O'Ceidigh, 1964; Perkins and Williams, 1964; Colman, 1964). Mortalities may be arranged by species in phyla and by position on the coasts from east to west, from Horn's reef off Denmark to the west coast of Ireland; details are given in Table 17. Dead algae were recorded only in the Menai Straits and at Aberfrew on

Table 17

Distribution of mortalities in 1963 by species and by position

	Horns Reef	Dutch coast	Botney Gut	Outer Silver Pit	SW Dogger	Filey	R. Crouch	Essex	Shoeburyness	Herne Bay	Whitstable	S. Side Thames	Kent	Hastings	Isle of Wight	Beaulieu R.	Poole Harbour	Studland Bay	Weymouth	Lyme Bay	Dorset	Exmouth	Torbay	Torquay	Brixham	Devon	Prawle Point	Channel Is.	Helford R.	Scilly Isles
ALGAE																														
Lichina pygmaea																														
COELENTERATA																														
Actinia sulcata																														
Actinia equina																														
Anemonia sulcata																														
Cereus pedunculatus																														
ANNELIDA																														
Andouinia tentaculata																														
Branchiomma vesiculosum																														
Chaetopterus variopedatus																														
Eulalia viridis																														
Lanice conchilega											●																			
Marphysa sanguineum																														
Perinereis cultrifera																														
Pomatoceros triqueter																														
Sabella pavonina								●					●		●		●												●	
Sabellaria spinulosa																														
Sabellaria alveolata																														●
Spirorbis borealis																														
CRUSTACEA																														
Balanus balanoides											●																			
Balanus perforatus																														
Balanus crenatus																														
Cancer pagurus		●		●	●																									
Carcinus maenas								●					●																	
Chthalamus stellatus																														●
Corystes cassivelaunus								●					●																	
Crepidula fornicata																														
Cyprina islandica																														
Diogenes pugilator																														●
Elminius modestus											●																			
Ligia oceanica											●																			
Maia squinado																														
Nephrops norvegica			●	●																										
Pilumnus hirtellus																														
Porcellana platychelus																														
Porcellana longicornis																														
Portunus puber		●																												
Verruca stroemia																														

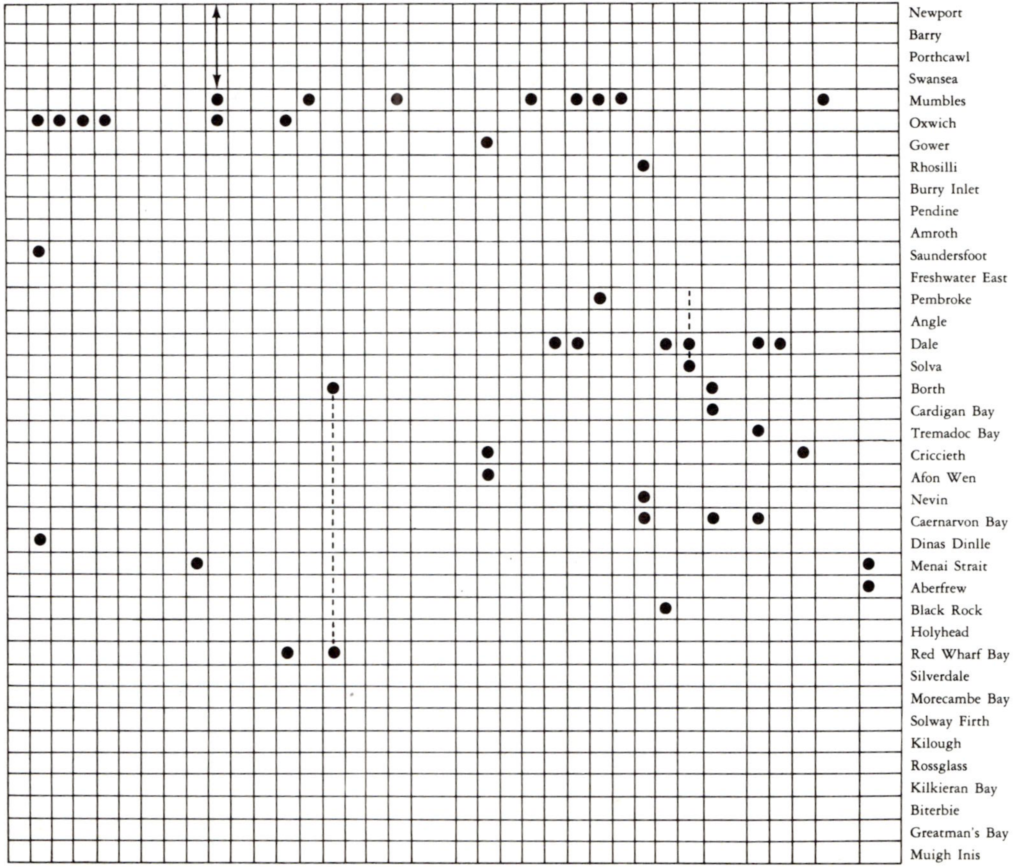

Newport
Barry
Porthcawl
Swansea
Mumbles
Oxwich
Gower
Rhosilli
Burry Inlet
Pendine
Amroth
Saundersfoot
Freshwater East
Pembroke
Angle
Dale
Solva
Borth
Cardigan Bay
Tremadoc Bay
Criccieth
Afon Wen
Nevin
Caernarvon Bay
Dinas Dinlle
Menai Strait
Aberfrew
Black Rock
Holyhead
Red Wharf Bay
Silverdale
Morecambe Bay
Solway Firth
Kilough
Rossglass
Kilkieran Bay
Biterbie
Greatman's Bay
Muigh Inis

Table 17 (*cont'd*)

	Horns Reef	Dutch coast	Botney Gut	Outer Silver Pit	SW Dogger	Filey	R. Crouch	Essex	Shoeburyness	Herne Bay	Whitstable	S. Side Thames	Kent	Hastings	Isle of Wight	Beaulieu R.	Poole Harbour	Studland Bay	Weymouth	Lyme Bay	Dorset	Exmouth	Torbay	Torquay	Brixham	Devon	Prawle Point	Channel Is.	Helford R.	Scilly Isles
MOLLUSCA																														
Anomia epihippium															●															
Barnea candida																														
Buccinum unidatum																														
Calliostoma zizyphinum																														
Cardium echinatum		●		●																										
Cardium edule								●			●	●	●		●															
Chlamys varia																														
Crassostrea angulata											●																			
Crepidula fornicata											●																			
Ensis ensis	●	●															●	●											●	
Ensis arcuatis																														
Gibbula cineraria											●																			
Gibbula umbilicalis																														
Lucinomia borealis																														
Lutraria lutraria																														
Macoma balthica																														
Mactra corallina																														
Monodonta lineatum																														
Mya arenaria																														
Mytilus edulis																														
Nassarius reticulatus																														
Natica catena																														
Ostrea edulis											●																			
Patella vulgata																													●	
Pecten maximus																														
Pharus legumen																														
Pholas dactylus																				●										
Scaphander lignarius																														
Scrobicularia plana											●																			
Solen marginatus																														
Spisula solida																														
Teredo novegica																						●								
Urosalpinx cinerea								●					●																●	
Venerupis decussata																														
Venerupis pullastra															●														●	
Venus mercenaria																														
Venus striatula																														
ECHINODERMATA																														
Astropecten irregularis																														
Echinocardium cordatum																														
Psammechinus miliaris																														
Ophiura texturata																														
PISCES																														
Anguilla anguilla							●																							
Blennius pholis																														
Conger conger														●					●				●					●		
Gadus luscus																														
Gadus merlangius																														
Gadus morhua																														
Labrus bergylta														●					●					●						
Labrus ossifagus														●					●					●						
Limanda limanda																														
Morone labrax														●											●					
Mugil labrosus														●																
Pleuronectes platessa																														
Salmo salar																														
Scophthalmus maximus																														
Scophthalmus rhombus																														
Solea solea																														
Trigla																														
SUM OF OCCURRENCES	1	4	1	3	1		1	4			10	1	6	5	3		2		3	1		1	1	2	1			1	5	3

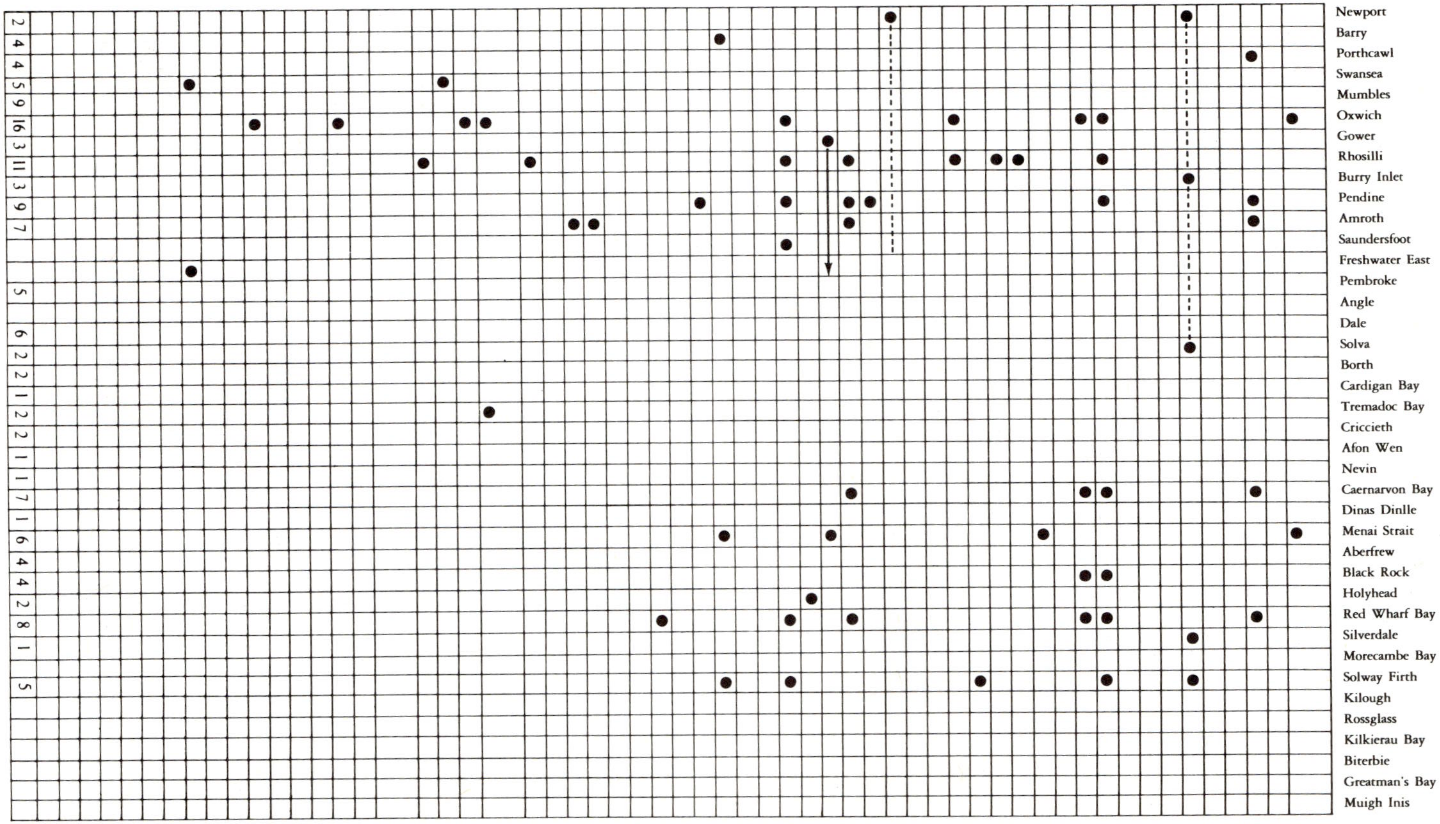

Newport
Barry
Porthcawl
Swansea
Mumbles
Oxwich
Gower
Rhosilli
Burry Inlet
Pendine
Amroth
Saundersfoot
Freshwater East
Pembroke
Angle
Dale
Solva
Borth
Cardigan Bay
Tremadoc Bay
Criccieth
Afon Wen
Nevin
Caernarvon Bay
Dinas Dinlle
Menai Strait
Aberfrew
Black Rock
Holyhead
Red Wharf Bay
Silverdale
Morecambe Bay
Solway Firth
Kilough
Rossglass
Kilkierau Bay
Biterbie
Greatman's Bay
Muigh Inis
2 4 4 5 9 16 3 11 3 9 7 5 6 2 2 1 2 2 1 1 7 1 6 4 4 2 8 1 5

the coasts of Anglesey. Dead coelenterates were found west of the Mumbles on the south coast of Wales and as far as the west coast of Anglesey. Dead annelids were recorded between Essex and Poole Harbour, in the Helford River in Cornwall, at three locations on the coast of South Wales and also between Pembroke and Holyhead; in general, mortalities were more intense on the coasts of Wales than on the east and south coast of England. Crustacea were killed in the North Sea, in the Thames estuary and as far west as Poole Harbour, in the Helford River, on the Mumbles and there were some deaths on the coast of North Wales; dead *Cancer pagurus* L. were found all along the coast from Borth (in Cardigan Bay) to Red Wharfe Bay in Anglesey. Molluscs were more vulnerable in the North Sea, in the Thames Estuary, between Kent and Poole Harbour on the South coast, from the Helford River to Pembroke, and there was considerable mortality on the west coast of Ireland. Echinoderms suffered little except on the coast of South Wales and in Tremadoc Bay. Dead fish were found at Hastings, between Weymouth and Prawle Point (in Devon), and on the coast of South Wales. Although the number of species killed was quite high in different regions, the greatest number suffering mortality was on the coast of South Wales. Indeed the death rate on the coasts was greater in the south west of England where the negative anomaly of the air temperature was greatest (see Table 17).

Crisp (1965) noticed that the mortality was greatest on the "land-influenced" coasts, that is, not on the western coasts of the peninsulas. He tabulated the mortalities by faunal elements in Britain as follows:

Distribution	Severe mortality (%)	Less severe mortality (%)	Unaffected (%)
Southern (21)[a]	57	24	19
Celtic (36)	39	47	14
Northern (26)	0	15	85

[a] Number of species

As might be expected, the southern species suffered most and northern ones least, but most of the Celtic species suffered less severe mortality. Crisp separated the southern species into those that experienced heavy mortality in the intertidal zone on "land-influenced" coasts and those that did not. The first group included *Anemonia sulcata* (Pennant), *Branchiomma vesiculosum* (Montagu), *Audouinia tentaculata* (Montagu), *Sabellaria alveolata* (L.), *Patella intermedia* Jeffreys, *Monodonta lineata* (da Costa), *Gibbula umbilicalis* (da Costa), *Pharus legumen* (L.) and *Balanus perforatus* Bruguières. The less affected group included *Bifurcaria rotunda* (Huds.) Pakeuf, *Cystoseira ericoides* (L.) Ag., *Cystoseira granulata* Ag., *Cystoseira fibrosa* (Huds.) Ag., *Chthalamus stellatus* (Poli), *Xantho incisus* Leach, *Clibanarius misanthropus* Risso and *Paracentrotus lividus* (Lamarck). In general the younger and active individuals were less affected than the older ones; fish and sublittoral invertebrates were killed at air temperatures above zero; the bivalves cannot clear the debris, the topshells become buried in gravel, and the razor shells cannot burrow, nor can the limpets and anemones (Crisp, 1964).

EFFECTS ON FISHES OFFSHORE

The deaths of fishes is reported in three papers—Lumby and Atkinson (1929), Simpson (1953) and Woodhead (1964); brill, conger, cod, whiting, dab, plaice, sole, miller's thumbs, gurnards, pout whiting, eels, mullets, wrasses, skates and rays were all reported dead.

Between December 1962 and February 1963, the Elbe was covered with ice and 300–500 tonnes of freshwater fish were killed (Kothke and Koops, 1963). The effects on soles in the North Sea between February and April (1963) were reported by Woodhead (1964a). Figure 84a shows the distribution of dead soles during the whole period and the areas of heavy mortality lie in the deeper water in contrast to the catches made in 1843/4 in the Silver Pits. Figure 84b shows the distributions of dead fish in the southern North Sea in the three cold winters of 1929, 1947 and 1963; that in 1963 was more extensive than in the other two years. In 1929, reports of dead fish were received first in early March and persisted till the end of April; as much as 40 kg per haul were discarded dead. Crabs were found dead on eastern grounds and off the Shipwash Light in the Southern North Sea; cod were found dead in Tea Kettle Hole. Schnoor (1929) wrote that high catches per effort of sole were recorded in the Clay Deeps, the Silver Pit and in the deep water of the Southern Bight. Ellett (1963) wrote that the highest mortalities were in areas circumscribed by the 2°C isotherm at the surface in early March, a plume of cold water spreading from the continental coast; there was a ten-day interval between the appearance of cold water and the reports of mortality; 10–30 kg per haul of soles were discarded. Inshore there were dead and dying congers, and offshore dead dabs, plaice, whiting and gurnards were caught with occasional brill and turbot and small numbers of cod. Not only were the soles killed in water of 3°C, but they were caught more readily in the cooler water, but inshore catches on

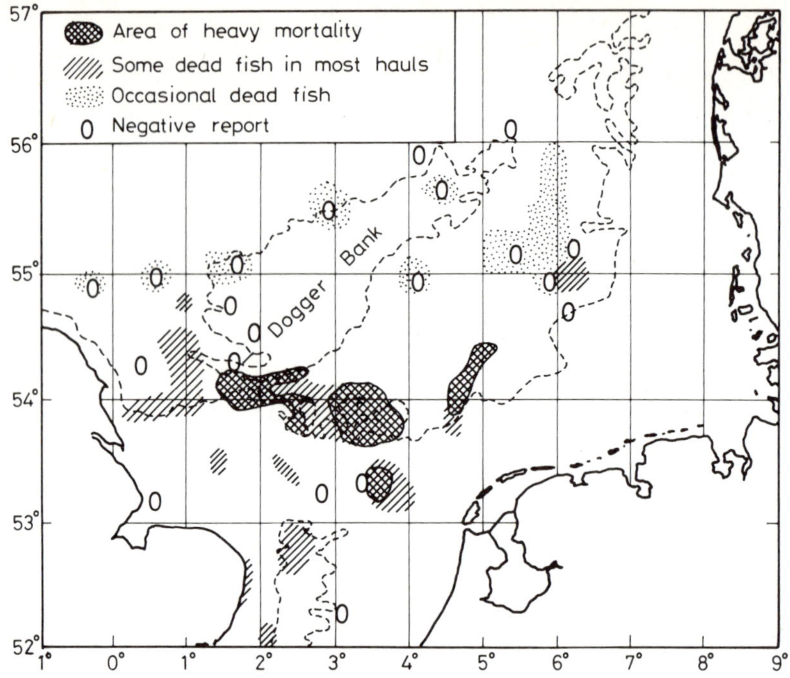
Area of heavy mortality
Some dead fish in most hauls
Occasional dead fish
0 Negative report
Dogger Bank
57°
56°
55°
54°
53°
52°
1°
0°
1°
2°
3°
4°
5°
6°
7°
8°
9°

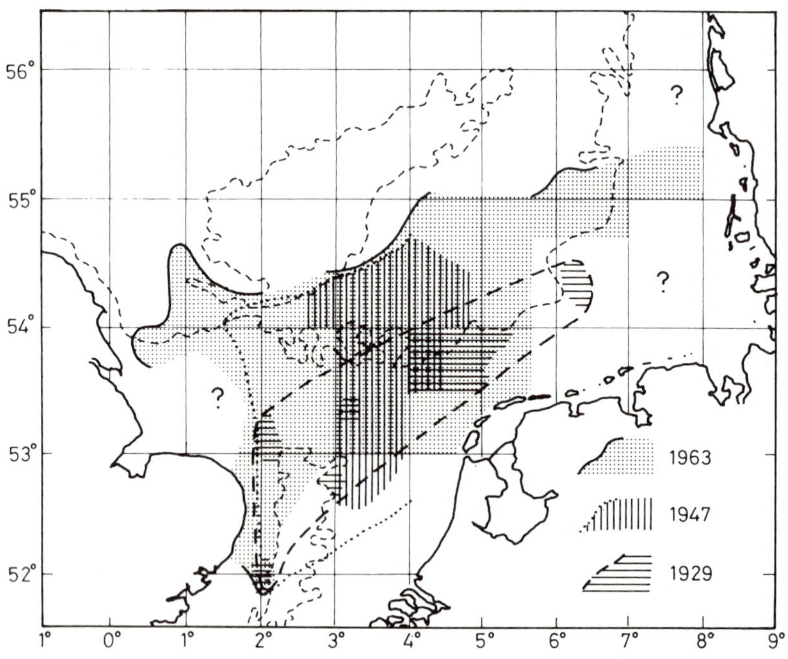
1963
1947
1929
?
?
?
56°
55°
54°
53°
52°
1°
0°
1°
2°
3°
4°
5°
6°
7°
8°
9°

the Dutch and Danish coasts were much reduced. Some flatfish suffered from skin infections. Woodhead (1964b) noted that the stock density of soles (in kg per 100 hours fishing) in the period January to April in 1963 increased by a factor of 3.5; but in some statistical squares (in deeper water) that factor increased to 12. In the Belgian sole fishery, the factor was much higher, up to 20 or 25 in February and March (Gilis, 1963). Further there was no difference between day and night catches perhaps because the fish no longer burrowed into the sand at night (de Groot, 1964). Inshore, dead mullet (*Mugil labrosus,* Risso) were found at Hastings, dead ballan and cuckoo wrasses (*Labrus bergylta* Ascanius and *L. ossifagus* L.) were recorded from Hastings, Weymouth, Torquay and west of Ireland. A large number of eels (*Anguilla anguilla* L.) were found dead in the River Crouch. Offshore, in addition to the soles, turbot and brill were killed; gurnards were frequently caught dead in the Channel; a few dead cod and whiting were found in the North Sea; many pout whiting were caught dead near Eastbourne. Skates and rays, flounders and sandeels were also killed in the cold winter of 1947 (Simpson, 1953). Soles, cod and dabs were caught dead in the Silver Pits and off the Brown Ridges. The numbers dead per 100 landed were:

Soles	Cod	Whiting	Dabs	Plaice	Skates and Rays
13.4	9.1	7.1	2.2	1.7	0.6

Such effects were the consequence of very cold easterly winds blowing persistently over the continental coast, so that very cold water was drifted offshore at the surface. As it was mixed more deeply, fish died and soles in particular migrated into deep water and became more available to capture. In March 1973, my former colleague P. Scholes worked on MV *Cuttlefish* on the Rising Ground and on the Cleaver Bank; of 184 cod caught, 4 were dead (all 70+ cm in length) with necrotic areas over head and body and bloody mucus on the fins; all four were emaciated. This estimate of death rate of about 2% is much less than that observed in the cold winter.

The 1963 year class of plaice was the largest recorded and that of sole was also high (Bannister *et al.* 1974), but in 1947 no cod, plaice or dab eggs were found in water of 0°C (Simpson, 1953). The same phenomenon occurred in 1947 but not in 1929; the cold winter of 1979 in the German

---

Fig. 84 Distributions of dead fish in the Southern North Sea:
(*a*) dead soles in 1963;
(*b*) dead fish in 1929, 1947 and 1963 (Woodhead, 1964c).

Bight has probably generated yet another strong year class of soles (Anon, 1980). The price of soles is very high, and the large 1963 year class encouraged many Dutchman to build new boats with their new technique of double beaming, that is, using beam trawls alternately from each side of the vessel. Such gear is very efficient indeed in the capture of soles. The causes of such large year classes are unknown. However, with persistent easterly winds, the strong and heavy northerlies were precluded; such winds with long fetch makes the water much more turbid and the production cycle is delayed as shown in Chapter 6 (Cushing, 1972). Indeed, an inverse correlation was established between northerly winds in the Southern Bight and plaice year classes for a limited period when the plaice fishery did not extend into the northern North Sea. During the cold winter, eggs and larvae develop much more slowly and perhaps the production of larvae is matched more closely in time to that of their food, an hypothesis put forward in Chapter 11. An interesting point is that the strong year class of soles often characteristic of cold winters compensates for the loss of adults due to death in cold water and by increased availability to capture.

CONCLUSION

The effects of cold winters in European waters have been examined in some detail, partly because fishermen have long detected the deaths of fishes in offshore waters and partly because oyster fisheries have obviously suffered and such effects have been extended in study by marine biologists to other invertebrates. The cold winter effect is interesting because it is the southern sole that suffers most. In northern waters there are records of fish having been killed by the cold, but usually the fish native to such waters survive well in cold water; cod can in fact live in very cold water ($-1.5°C$) which is in fact a super-cooled liquid (Jones and Scholes 1974). The death of soles in the southern North Sea is more like the death of tile fish, perhaps in cold water towards the end of last century—the death of a southern species in the extreme conditions at the northern edge of their range. It is interesting also that cod is sometimes killed in the southern North Sea when cod of another stock is not killed in more northern waters.

The cold winters occur once or twice in two decades. During the Little Ice Age they were more frequent, but it is difficult to extract any periodicity in the records. They occur frequently enough to sustain the plaice and sole stocks in the southern North Sea. If outstanding year classes of these two species do depend on cold winters, one might have expected that in colder periods than the present, perhaps anomalous, seasons the plaice and sole stocks to have been more abundant than they are today. As the recent period of warming has turned a little cooler, the recruitment to the plaice stock has increased somewhat independently of the cold winters.

The fact that southern invertebrates suffered more than northern or Celtic ones should not surprise us, for the sole is also a southern species. However, they suffered more in the south west, where the negative anomalies of air temperature were greater, although it was not as cold as in the east. Similarly, imported oysters in the Essex creeks suffered more than the natives because they came from Brittany. The recruitment to the Burry inlet cockle increased in the cold winter of 1962/3, but it is not known whether the same was true for other invertebrates.

The periodic cold winters give rise to phenomena of general interest which must play a part in any long-term climatic change. The effect of one is immediate, inflicting death on the adults of many species and stimulating high recruitment subsequently in some. If cold winters become more frequent (as during the Little Ice Age), such factors might well have cumulative effects upon the populations. The loss of adults may be less important than increased recruitment, particularly if there is more than one strong year class produced within the average age of the stock. In principle we would then expect the stock to increase, but this would really depend upon the long-term carrying capacity of the food of the adults.

Chapter Nine

Changes in the Baltic Sea during the Twentieth Century

In Chapter 3 the Scanian herring fishery in the Baltic was described in historical terms. It persisted for centuries and provided fish for the Hanseatic ports and others in the Baltic. From the position of the fishery at the southernmost region of Sweden, the fish may have been of North Sea origin, like those which visit the Bohuslån coast during the Swedish herring periods, but this is merely speculation. The Baltic freezes extensively for some months in the winter, and records of ice extend back for centuries (Chapter 3). In this chapter are recorded some physical and biological events which took place in the Baltic during the twentieth century. What is interesting about these events is their links with meteorological processes.

THE STRUCTURE OF THE BALTIC SEA

Although the Baltic has been compared with the Mediterranean, because it has a narrow entrance, and with the Black Sea, because it tends to stagnate in deep water, it has a character of its own. It is a northern sea which freezes extensively in winter, its entrance is very shallow (<20 m), and many large rivers empty into it. The Sound is only 7–8 m deep, and south of the Danish islands there is a sill 16–17 m in depth between Gjedser in Sweden and Darss in the Federal Republic of Germany west of

Rügen. The average depth of the Sea is 100 m, but there is a prominent series of deeps—Arkona (53 m), Bornholm (195 m), Gotland (249 m) and Landsort (459 m)—through which the saline inflows slowly dribble; off this main stream lies the Gdansk deep (113 m) and the Fårö deep (205 m). In general the Sea is relatively shallow in the south and south-west but deeper in the north and north-east.

There is an excess of precipitation and run off over evaporation, so the salinity at the surface is low (<8°+‰), although it increases below 60 m under the influence of the saline inflows. The annual input of fresh water amounts to 630 km^3, 1/34 of the total volume of the sea. The Great Belt is the main outlet of the low salinity surface water, but where it is wide and deep, salt water enters underneath from the Kattegat to the Belt Sea where most mixing takes place (occasionally there is an inflow through the Sound). In this area of transition, the upper Baltic layer flows out into the North Sea and the lower saltier layer flows into the Baltic; between them there is a layer of mixed water. Salinity increases from the surface downwards and from south to north in the transition zone, and transfer occurs in the straits. More water flows out from the Baltic than enters it from the North Sea. When the outflow is least (in autumn and winter) the salinity is greatest, but the salinity of the bottom layer is less in winter and greatest in summer. There are irregular pulses of high salinity water in the bottom layer which maintain the deep salinity beyond the Darss sill. From the Belts the heavy high salinity water flows slowly on the bottom into the Arkona deep, between Bornholm and the Swedish mainland into the Bornholm Deep and thence through the Stolpe Channel into the east and north Bornholm basin. Most outflow from this region is usually prevented by a high sill, so the highest salinity is found in the Bornholm deep. Beyond the sill lies the Gotland deep and beyond that, in the Gulf of Finland, the Landsort deep. It takes up to one and a half years for the heavy high salinity water to travel from the Bornholm deep to the Landsort deep. There is no such flow into the Gulf of Bothnia. In general the Baltic is a sea poor in nutrients (but see below), and those from the deep salty layers are transported inwards and are upwelled near coasts and the inner end of the Gulf of Finland (but not in the Aland Sea or the Gulf of Bothnia).

The two northern Gulfs are covered with ice for two to five months in winter; the temperature of the surface waters is less than 0°C in February in the north and about 2°C in the warmer southern regions. Because the summer does not last long enough to warm the whole layer cooled in winter, there is an intermediate cold layer between 60 m and 100 m,

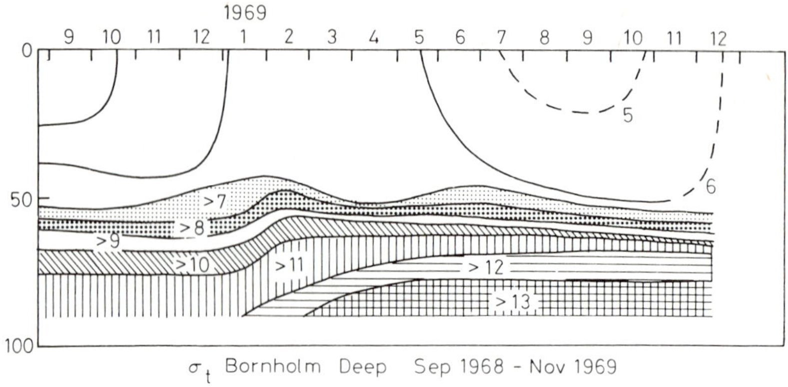

Fig. 85 Seasonal distribution of σ_t by depth, showing the halocline which separates the fresher surface layer from the saltier deep water.

although the water of Atlantic origin below it is warmer, if much saltier. Figure 85 shows the seasonal distribution of σ_t in depth; in the surface waters a thermocline develops in summer in the usual way, quite distinct from the halocline below. Above the thermocline, or more strictly as it develops, the biological processes of production are contained; below it, evolve those of eutrophication. The first are of much interest to the fisheries biologists, whereas the second concern those who fear that the Baltic will become a deep desert like the Black Sea, infested with hydrogen sulphide. This account is based on Segerstråle (1957, 1972), Fonselius (1962), Pickard (1964), and Zenkevitch (1963).

THE BIOLOGY OF THE BALTIC SEA

The biology of the Baltic Sea is much simpler than that of its nearest neighbour, the North Sea, but it is much more variable, spatially between the Sound and the northern Gulfs and temporally from year to year and over many decades. Certain groups are absent or very rare, for example, Porifera, Actiniaria, Madreporaria, Octocorallia, Solenogastrea, Scaphopoda, Pteropoda, Cephalopoda and Echinodermata; of fishes, the selachians are absent.

East of the Darss line, that is in the Baltic proper, the plankton comprises the following commoner organisms:

Blue greens *Anabaena flos-aquae*, L., *Nodularia spumigenia* Mertens, *Anabaena baltica;*

Diatoms *Chaetoceros wighami* Brightwell, *C. danicum* Cl., *Thalassiosira baltica* Ostenfeldt;

Silicoflagellate *Tintinnopsis campanula* Cl. et L., *Helicostomella subulata;*

Rotifers *Collotheca pelagica* (Rousselet), *Brachionus angularis* Gosse, *B. pala* Ehrenberg, *B. bakeri* (Muller), *Anurea aculeata* Ehrenberg;

Copepods *Eurytemora hirundoides* Nordqvist, *Acartia bifilosa* (Giesbrecht), *Pseudocalanus elongatus, Temora longicornis* (O. F. Müller), *Limnocalanus grimaldii* de Guerne;

Cladocera *Bosmina maritima, Evadne nordmanni* Lover, *Daphnia cucullata* Sars, *Chydorus sphaericus* Baird, *Leptodora kindti* (Focke)

In the deeper layers, in the saline water, live *Calanus finmarchicus, Oithona similis* and *Sagitta elegans. f. baltica*. There are two phytoplankton communities: a boreal–arctic one in the south and an arctic one in the north.

The production cycle in the Baltic has been described in chlorophyll in depth and in season by Renk (1972, 1973), Renk *et al.* (1974, 1975) and Yurkovskis and Bramane (1977). In the Gdansk deep, the spring outburst may occur in mid-April or from April to June. Off Gotland and in the Gulf of Finland, the production cycle peaked at the end of April. In Gdansk Bay it occurred as late as September. From a somewhat small sample there is considerable variability from place to place in the Baltic. In the Baltic proper, Fonselius (1972) has shown that the average daily production, estimated by the radiocarbon method, amounted to 0.4–0.8 gC per m^2 in 1964 and up to 0.4gC per m^2 in 1967 (in Danish waters at Albarg Bugt, Anholt Nord and Halsskov Rev, daily production ranged from 0.05–0.13 gC per m^2). Ackefors and Hernroth (1972) have tabulated seasonal production of zooplankton for each period of two months in the Baltic in gC per m^2.

Jan/Feb	Mar/Apr	May/Jun	Jul/Aug	Sep/Oct	Nov/Dec
0.36	0.26	0.39	0.51	0.97	0.38

They have also compiled an annual production budget (in millions of tonnes of carbon) for the trophic levels in the Baltic:

Phytoplankton (including nanoplankton)	20
Benthic algae	1
Zooplankton	1
Benthic invertebrates	0.8
Pelagic fish	0.06–0.09
Demersal fish	0.02–0.03

On the seabed, the whole sea is dominated by the *Macoma* community; indeed, bivalves comprise 90% of the total biomass. In general the productivity of the benthos decreases from the Belts where the biomass in wet weight amounts to some hundreds of g per m^2 towards the Gulf of Bothnia, where it is reduced to some dozens of g per m^2. The following table (Zenkevitch, 1963) summarizes the quantities found in particular places in wet weight:

Species	Region	Biomass (g per m^2)
Cyprina astarte	Kiel Bay	185
Macoma calcarea	Bornholm depression	102.5
Macoma/Cardium	North of Pomerania	67.2
Macoma baltica	Southern Gulf of Bothnia	31.1
Deep water community		3.03

There are not many fish species. There is a Baltic herring, the strømling (*Clupea harengus membras*), and a cod genetically distinct from that in the North Sea (Jamieson and Otterlind, 1971) although individuals of the North Sea stock must be swept into the western Baltic in the deep salinity inflows. There are large stocks of sprats; the flatfish are represented by plaice, dab and flounder. The populations are much less diverse than those of the North Sea. Yet there are relicts such as *Myoxocephalus scorpius* L. and some coregonids and freshwater species such as pike (*Esox lucius* L.), perch (*Perca fluviatilis* L.), eel (*Anguilla anguilla* L.) and the golden shiner (*Abramis brama* L.) are found on the edges of the sea.

The Baltic has been invaded from time to time; the diatom *Biddulphia sinensis* Greville came from Chinese waters in the first decade of the century, probably on a ship. The gastropod *Potamopyrgus jenkinsii* Smith, the bryozoan *Alcyonidium polyonum* Hassall, the copepod *Acartia tonsa*, and the crabs *Rhithropanopeus harrisii* (Gould) and *Eriocheir sinensis* Milne Edwards are other invaders (Zenkevitch, 1963). Lindquist (1960) has published a chart showing how certain animals have extended their range as consequence of the rise in salinity in recent decades (Fig. 86). The open circles show the limit of range of the copepod *Limnocalanus grimaldii* before the increase of salinity and the closed circles indicate the limit after as this brackish water form has been driven northwards. Other copepods, *Acartia longiremis, Temora longicornis* O. F. Müller, *Centropages hamatus* and *Pseudocalanus elongatus* have invaded the Gulf of Bothnia and the Gulf of Finland. Other animals such as the ctenophore *Pleurobrachia pileus* O. F. Müller, the jellyfish *Aurelia aurita* L., the barnacle *Balanus improvisus* Darwin, and the mackerel *Scomber scombrus,* have penetrated the Gulf of Finland. The garfish *Belone belone* L. and the jellyfish *Cyanea capillata* L. have moved further north within the Baltic Sea and the tunicate *Oikopleura dioica* Fol. and the copepod *Oithona similis* have become commoner in the Southern Baltic. The horse mackerel *Trachurus trachurus* has appeared as an invader. Whiting and the long rough dab *Drepanopsetta platessoides* and the cod have increased in abundance (see below).

The biology of the Baltic has been described in some detail because the very remarkable changes that have occurred in recent decades require fairly detailed background in order to be understood. The sea has become more saline, but over several decades there has been an increase in the areas of hydrogen sulphide on the sea bed in the deeps, the desertification of the Baltic. Such changes are linked to the trends in salinity inflows and trends in run off, but there is a fear that a eutrophication has taken place as a consequence of the use of detergents. The second change is the increase in catches of cod, herring and sprat since the mid 1930s; that of the cod is the most remarkable, an increment from a few thousand tonnes in the late 1920s to about 150 000 tonnes in the 1950s.

THE NATURE OF THE SALINITY INFLOWS

The relatively warmer and saltier Atlantic water enters the Baltic through the Kattegat as a deeper layer below the halocline. The inflow is continuous and is compensated partly by the outflow of cooler fresher water in the surface layers. The salinity of the Atlantic water varies in a quasi-periodic manner, reaching a peak every five years or so. The variability in salinity has been recorded in the English Channel, in the German Bight, in the

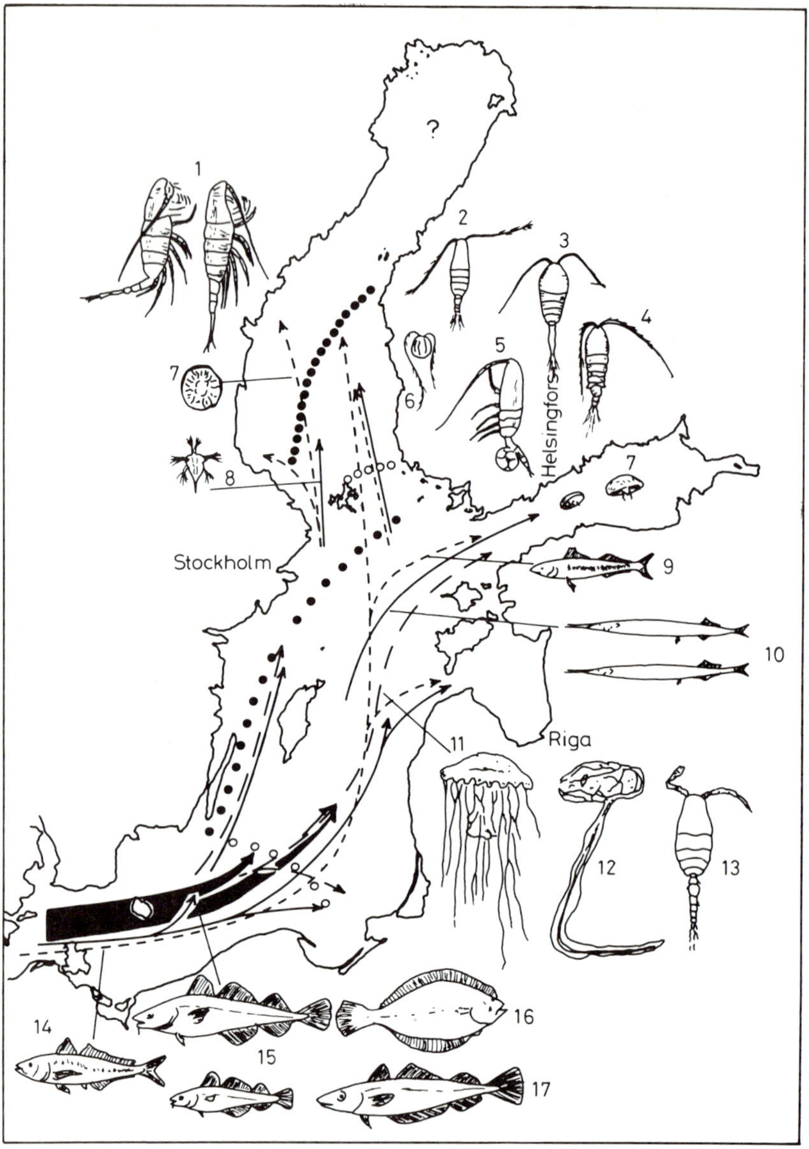

Fig. 86 Changes in range within the Baltic of certain animals as consequence of the rise in salinity (after Lindquist, 1960). Arrows indicate the direction of spread; continuous lines show extension of range with reproduction; broken lines with long strokes show extension of range; broken lines with short strokes show occasional invaders.

Norwegian Sea as well as in the approaches to the Baltic (in the Skagerrak, the Kattegat and in the Belt Seas) (Dickson, 1971). Hence the source of variation must lie out in the Atlantic.

Dickson (1971) has shown that the cause is a warm sea surface temperature (SST) anomaly lying to the east of a cold water pool in the North Atlantic. Depressions entering the area linger in the region of the sharp thermal gradient between the warm and cold water pools as described in Chapter 4. Because of feedback with the ocean surface temperature field, a particular pressure anomaly pattern is generated over the North Atlantic.

Figure 87 illustrates the process: a deep meridional (north to south) trough lies in the atmosphere over the western Atlantic and a strong meridional ridge over north-west Europe. Consequently a cold pool appears in the west Atlantic and a warm pool in the east, and this structure encourages the atmospheric structure to persist. Such processes are considered to generate differences in salinity, due to differences in advection and *in situ* processes throughout the north-east Atlantic. Figure 88a shows the salinity anomalies in the German Bight of the North Sea as running three

---

(1) *Limnocalanus grimaldii*: open circles show distribution of this copepod before the increase of salinity and closed circles show that in 1960.

(2, 3, 4, 5) *Acartia longiremis, Temora longicornis, Centropages hamatus, Pseudocalanus elongatus*: copepods which have become more common in the Gulf of Bothnia and the Gulf of Finland.

(6) *Pleurobrachia pileus*: a ctenophore which has become more common in the Gulf of Bothnia and the Gulf of Finland.

(7) *Aurelia aurita*: a jellyfish which has penetrated the Gulf of Finland and has been observed in the Gulf of Bothnia.

(8) *Balanus improvisus* (nauplii): a barnacle which has become more common in the Gulf of Bothnia.

(9) *Scomber scombrus*: a mackerel found occasionally in the Gulf of Bothnia and Gulf of Finland.

(10) *Belone belone*: a garfish which has penetrated further north into the Baltic.

(11) *Cyanea capillata*: a jellyfish which has been encountered further north.

(12, 13) *Oikopleura dioica*, tunicate, and *Oithona similis*: copepods which have become commoner in the southern Baltic.

(14) *Trachurus trachurus*: a horse mackerel which has appeared in the southern Baltic.

(15, 16, 17) *Gadus morhua*, the cod, *Drepanopsetta platessoides*, the long rough dab and *Odontogadus merlangius*, the whiting, which have all increased in abundance in the Southern Baltic.

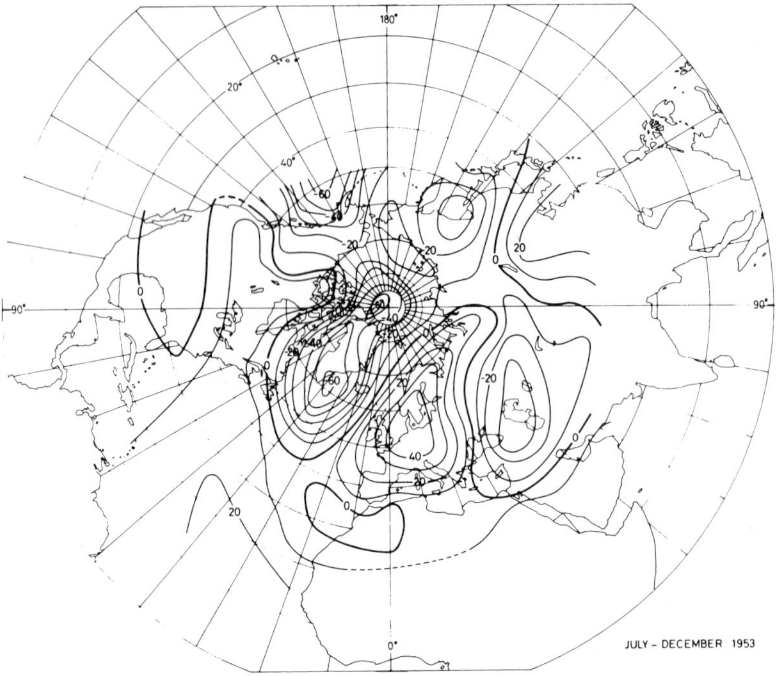

Fig. 87 Structure of depression and anticyclone in the North Atlantic which sustains the cold and warm water pools in the ocean (Dickson, 1971).

quarterly means (from three different sets of fields). The peaks are about five years apart. Similar periodicities can be seen in the temperature records from the Kola meridian in the Barents Sea (Fig. 88b) and in the survival indices of the Arctonorwegian cod which spawns in the Vestfjord and which lives in the Barents Sea (Fig. 88c). The evidence produced in Fig. 88c is of considerable importance because the serial correlation of recruitment is well illustrated and, moreover, it is linked to physical factors. The nature of the link is unknown and needs investigation.

The link between the salinifications in the German Bight, in the Skagerrak, the Kattegat and the Belt Seas and atmospheric indices are shown in Fig. 89. The time series extends from 1947 to 1972, and the two atmospheric indices are the Cyclonicity Index anomaly over the British Isles (which shows the annual tendency for cyclones, +, and for anticyclones, −) and the Meridionality Index anomaly over the British Isles (which shows the quarterly tendency for northerly or southerly winds). During the period there were six main inflows, indicated A, B, C, D, E and F. When there

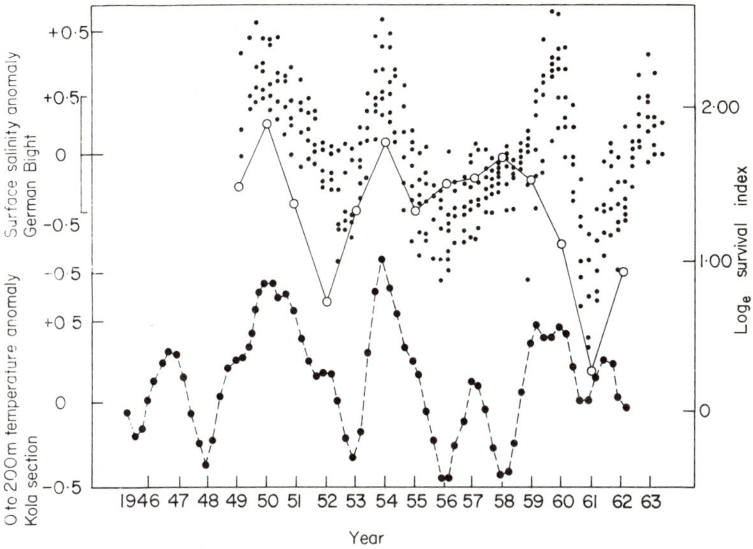

Fig. 88 (*a*) The salinifications in the German Bight (with three different sets of fields) shown in running three quarterly means;
(*b*) the periodicity in the temperature records on the Kola meridian in the Barents Sea;
(*c*) the index of survival of the Arctonorwegian cod stock (Cushing, 1972).

are most anticyclones over the British Isles and southerly winds predominate there, there is high salinity in the German Bight, Skagerrak, Kattegat and the Belt Seas. Figure 89 demonstrates the link between the salinifications, or the saline inflows, and the atmospheric events over the North Atlantic.

Between the light fresh surface layer derived from rain and run off and deeper water of North Sea origin below there is an intense density gradient. This gradient prevents the transfer of oxygen from the surface to the lower layers and so the aeration of the deep basins depends upon the renewal of inflow from the North Sea, rich in oxygen. If the sediments become anoxic, as described below, charged with hydrogen sulphide, the new oxygen in an inflow is used to react with that hydrogen sulphide and in the long run this process tends to lead to anoxic conditions.

Although more water leaves the Baltic than flows in through Danish waters, the warm saline water in the deeps is sustained by irregular inflows from the North Sea. In 1923, 34‰ North Sea water in the Kattegat understream carried an abundance of haddock larvae which were drifted into the Arkona deep where 25 000 kg of two-year-old fish were caught

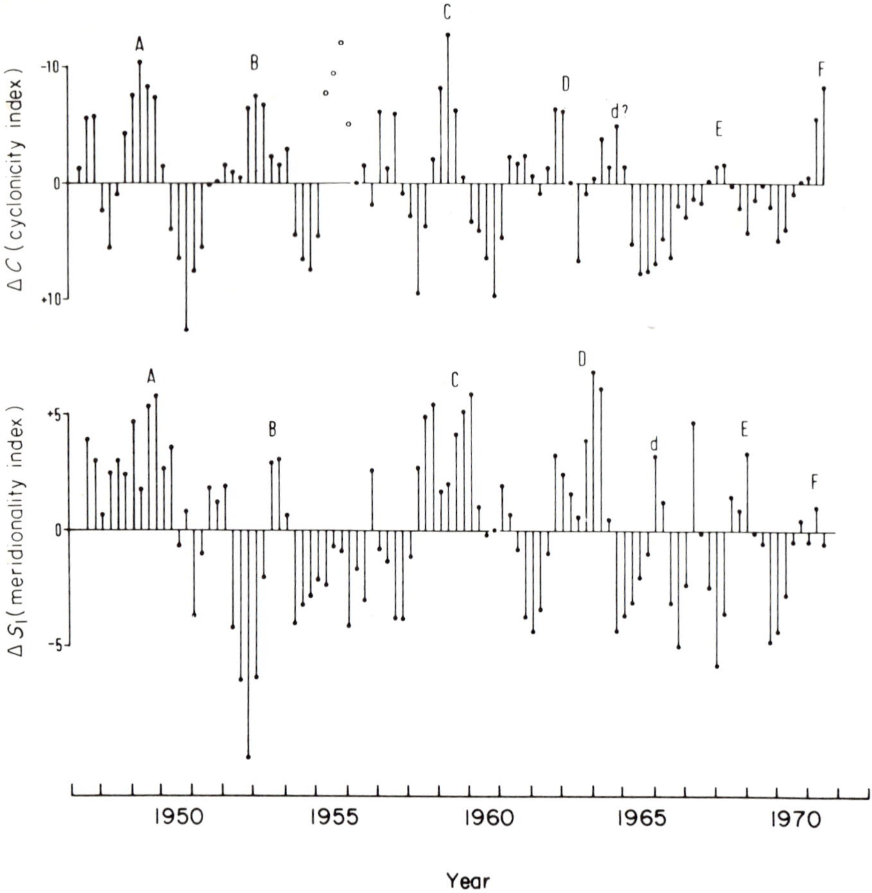

Fig. 89 The link between (*right*) the salinifications, 1947–72, in the Skagerrak, Kattegat and in the Belt Seas and (*left*) indices of cyclonicity and meridionality (Dickson, 1971).

Key to right-hand graph.

Upper	*Middle*	*Lower*
× Vyl	● Skagens Rev	● Kattegat S.W.
○ Elbe 1	○ Laeso Nord	○ Halsskov Rev
	× Anholt	× Gedser Rev

in 1925; in the winter of 1925/6, 500 tonnes were caught, and in March 1926 the fishery stopped because the haddock had vanished (Dietrich and Kalle, 1957). Kalle (1943) described an inflow in 1933/4 and showed that "old water" in the deeps is poor in oxygen but rich in phosphate; the new water in the inflow is poor in phosphate but rich in oxygen, and in that

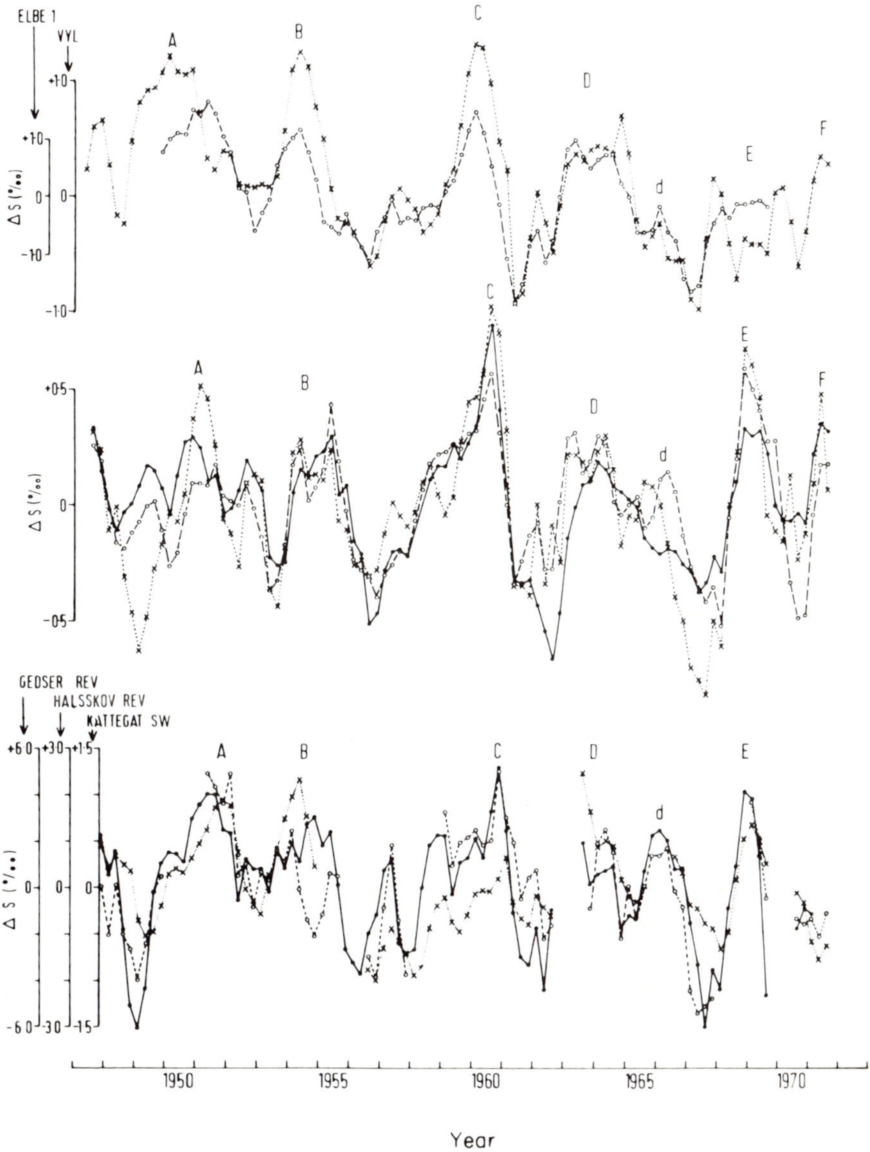

ELBE 1
VYL
+1·0
+1·0
0
-1·0
-1·0
Δ S (°/oo)
A
B
C
D
d
E
F
+0·5
0
-0·5
Δ S (°/oo)
GEDSER REV
HALSSKOV REV
KATTEGAT SW
+6·0
+3·0
+1·5
0
0
-6·0
-3·0
-1·5
Δ S (°/oo)
1950
1955
1960
1965
1970
Year

year 30 000 tonnes of phosporus were released from the sediments by oxidation with low pH on the bottom. In time the water of the inflow loses oxygen and gains phosphorus and "ages". The largest inflow was recorded by Wyrtki (1954) in 1951 under the influence of very strong southerly and southeasterly winds; between 23 November and 17 December there was no Baltic outflow at the surface, only an inflow. The total inflow amounted to 201 km^3 and it was recorded at Drogden, in the Fehmarnbelt and at the *Gedser Rev* LV; a strong increase in salinity was recorded in the Borkum Deep where it had sunk. The weak inflows do not cause big changes in temperature and salinity in the deeps but do renew the oxygen (but not beyond the Arkona and Borkum deeps). The powerful inflows reach the Gotland and Landsort deeps. Notable inflows were also recorded in 1906/07, 1913/14, 1926/7, 1937/8, 1947/8 and 1951/2 (Piechura, 1972). In 1959, 40 000 tonnes of phosphate–phosphorus were released in the Gotland basin and were mixed in winter through the primary halocline to the surface layers, comparable to the release in 1933/4 (Fonselius, 1962). Such a mechanism provides a potent source of fertilization, but it is intermittent and not continuous.

Figure 90a shows the increase in temperature in the Bornholm basin at 80 m between 1902 and 1961; Figure 90b shows the changes in salinity in the Bornholm basin, the inflows including the prominent one of 1951. Figure 91 shows the changes in temperature, salinity and oxygen in the Gotland deep; temperature and salinity increased but oxygen dropped very sharply in 1956. Similar phenomena were observed in the Landsort deep (Fonselius, 1962). Inflows were recorded in 1902, 1907, 1909, 1921, 1925/6, and between 1926 and 1937 there were several minor ones, followed by notable ones in 1948, 1951 and 1959. Hydrogen sulphide was recorded in the Gotland Deep in 1931 and 1958, and the Gotland Deep is the only one that really stagnates; there was a long period of stagnation after the 1951 inflow. Wyrtki (1954) pointed out that the 1951 inflow alone could have raised the salinity of the Baltic Sea by 0.1‰. Segerstrale (1957) recorded an increase of 0.5‰ during the last fifty years at Finnish light vessels and Lindquist (1959) noted a similar change in the Gulf of Bothnia. From Fig. 91, the increase in salinity between 1902 and 1961 amounted to 0.5–1.0‰. Yet there was no long-term increase in salinity in the Bornholm Basin, nearer the source of the inflows.

Fonselius (1969) showed that oxygen in 100 m declined from about 30 ml/l to less than 10 ml/l between 1900 and 1970 in the Gotland deep. Figure 92 shows the regions of deep water contaminated with hydrogen sulphide in September 1968. Figure 93 shows the oxygen and hydrogen

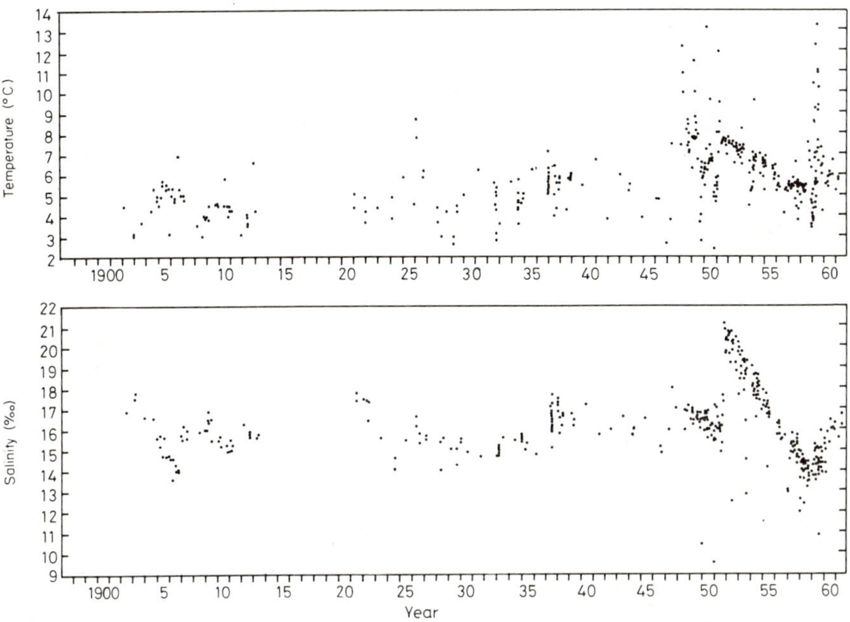

Fig. 90 (*a*) Increase in temperature in the Bornholm basin at 80 m from 1902 to 1961;
(*b*) Changes in salinity in the Bornholm basin at 80 m.

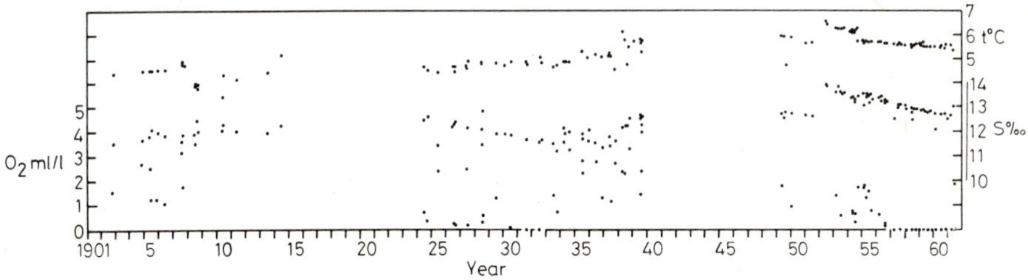

Fig. 91 Changes in temperature, salinity and oxygen in the Gotland deep at 200 m.

Fig. 92 Regions of deep water contaminated by hydrogen sulphide in the Baltic during the September 1968 expedition.

sulphide distributions at 240 m in the Gotland deep between 1953 and 1972: in general there was an increase in hydrogen sulphide from the late 1950 onward (Fonselius and Rattanesen, 1970). During the same period salinity, temperature and σ_t have tended to decline in the Gotland deep as the effects of the 1951 inflow have diminished. Yet over a much longer period, 1893–1970, salinity in the Gotland deep increased from about 11.75‰ to 12.5–13.5‰. Fonselius (1969) showed that in Landsort Deep, the temperature between 200 m and 459 m increased between 1877 and 1966 from about 3.8°C to between 4.2°C and 5.0°C.

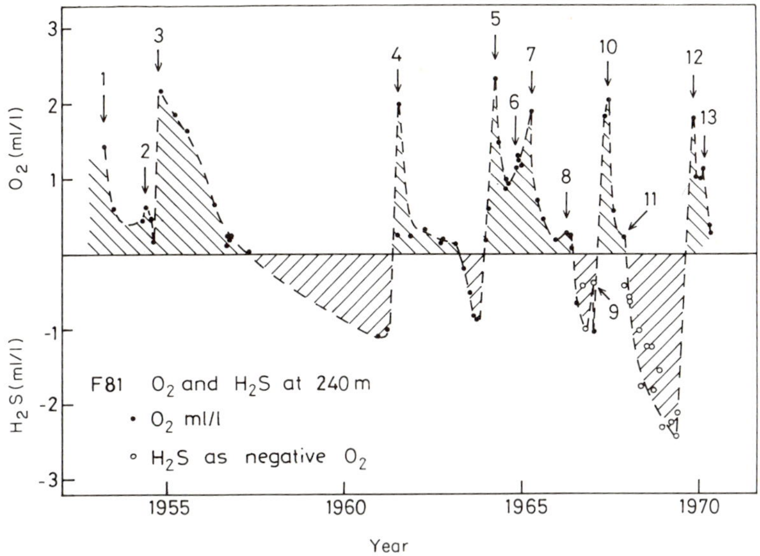

Fig. 93 Changes in oxygen and hydrogen sulphide in the Gotland deep between 1953 and 1972.

Figure 94 (Fonselius, 1969) shows the ten-year running mean values of run off from the rivers Neva, Vuoksi and Kymijoki from the end of the nineteenth century to the mid 1950s. There is a common if variable decline in run off of about one-quarter in about seventy years. If the input of warm high salinity water remained constant if variable, then the general increase in salinity throughout the Baltic might be attributed in part to decreased run off. Three factors are involved: decreased run off, decreased exchange through the Sounds, and the increased intensity of bottom inflows in the Baltic, all of which depend upon a net of meteorological events. An increase in phosphate–phosphorus at the surface occurred in the Central Basin of the Baltic proper from 1951 to 1968, which might originate partly in the inflows of warm saline water in the deeps by release from the sediments, but might also originate from detergent effluent in centres of population around the Sea.

Thus there is a strong link between physical and biological events in the Baltic Sea and meteorological processes across a large part of the North Atlantic. The periodical salinifications maintain the oxygen content of the deep water basins, probably sustain the populations of invading animals and provide fertilization up to the surface by nutrients released from the

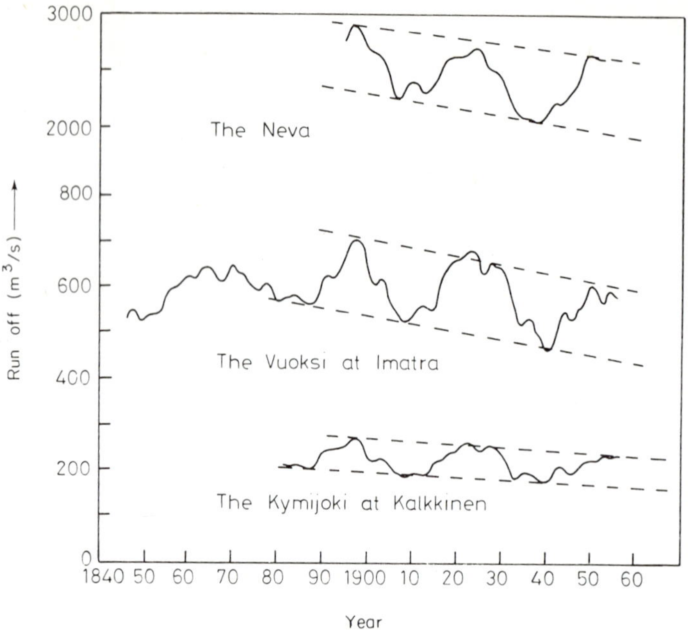

Fig. 94 Ten-year running mean values of run off from the rivers Neva, Vuoksi and Kumijoki (Fonselius, 1969).

sediments. Over a long time period of eighty years the salinity and temperature in the deeps have increased slowly. Although the warm salt water originated from the North Sea and the Atlantic in the salinifications which occurred approximately every five years, the long-term increase in salinity was probably due to a decline in run off from the major rivers, itself a consequence of meteorological events.

THE RISE OF THE COD STOCKS

During the 1970s, the annual catches of cod in the Baltic proper amounted to 200 000 tonnes or more, whereas in the 1920s annual catches were as low as 5000 tonnes (*Bull. Statistiques* of ICES). The Baltic cod, *Gadus morhua callarias*, differs from its North Atlantic counterpart, *Gadus morhua morhua*, and the two subspecies are genetically distinct, both in transferrins and in haemoglobins (Jamieson and Otterlind, 1971). The cod spawns in

April May and June from the Bornholm deep to north of Gotland (Grauman, 1966), and tagging experiments have shown a dispersion as far as the Aland Sea (Thurow, 1974).

The rise in stock was first recorded by Alander (1948), who noticed that the recruitment was unusually high in 1941–4, perhaps associated with the salinity inflows of 1936, 1937 and 1942; he noticed that the 1942 year class was a strong one. A further large year class in 1947 was associated with the inflow of 1946–7. Alander wrote that an increase in salinity made it easier for the cod eggs to float in the Baltic instead of sinking as they often did. He put forward a further hypothesis: "A rise of 10 m or so in the water where the fry hatch gives better access to suitable food".

Meyer and Kalle (1950) wrote that the catches of cod increased in the late 1930s all along the German coast from Lübeck to Danzig (Gdansk) and attributed it to the saline inflows which released phosphorus from the bottom sediments. Figure 95 shows the increase in catches from Thurow (1974) and *Bulletins Statistiques*. There is a steady rise from 1933 onwards until 1940 and another from 1945 onward; increments also occurred in the early fifties and early seventies. Thurow gives information of L_∞, length at infinite age, W_∞, weight at infinite age, (M/K), the ratio of natural mortality to a growth coefficient, and c, a coefficient of trawl selectivity.

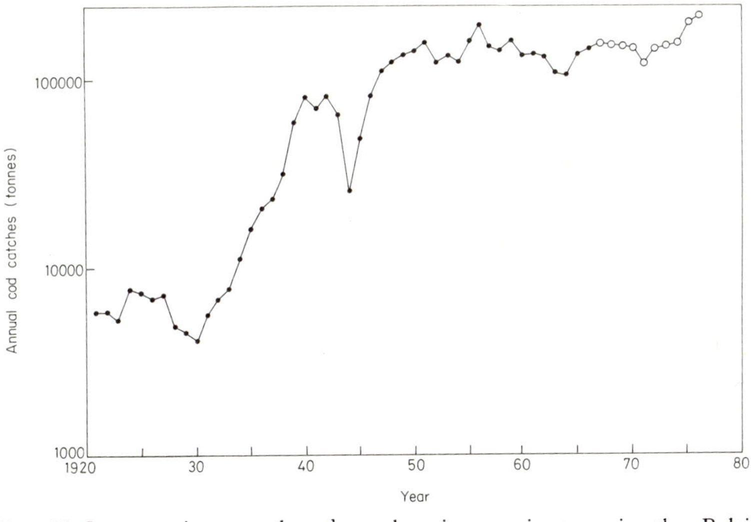

Fig. 95 Increase in annual cod catches in metric tons in the Baltic, 1921–1976. (●———●) Area IIId. (○———○) *Bull. stat.*

With catches, the yield tables of Beverton and Holt (1966) may be used to estimate the change in recruitment from the yield per recruit, Y'.

$$Y' = Y/R_0W_\infty \quad \text{where} \quad R_0 = R \exp M(t_r - t_0)$$

where Y is yield in weight, R is recruitment in numbers at entry into fishery, R_0 is recruitment in the first year of age, t_r is the age at recruitment and t_0 is age in the first summer. The increment is calculated:

	Y'	Yield (kg $\times 10^{-6}$)	$W_\infty Y'$	R_0 ($n \times 10^{-9}$)
1925–8	0.056	7.3	0.035	0.21
1935–9	0.056	20.6	0.035	0.59
1951–7	0.056	130.7	0.035	3.74

From the end of the 1920s to the end of the 1930s, recruitment rose by nearly a factor of 3 and then again in the early 1950s by a factor of more than 6; the total increment in recruitment was by a factor of 17.34.

Grauman (1966) showed that the cod eggs were laid in deeper water; indeed in the Bornholm deep they spawn 4–12 m above the bottom, except where the deep waters stagnate. The following table gives the distribution in depth in three of the deeps:

Depth (m)	Gdansk	S. Gotland	Bornholm
10–20	—	—	6
20–50	2	—	8
50–70	6	2	9
70–100	12	16	8
100–130	—	10	—

Grauman also showed (1974) that buoyancy increases with egg size, so larger eggs reach the 50–70 m layer. The following table gives the percentage distribution of eggs at various depths:

Depth (m)	April	May	June
20–50	5.4	16.0	—
50–70	77.0	25.0	28.0
70–90	17.6	59.0	72.0
% large eggs	58	22	15

In April cod larvae live in the bottom layers, 60–100 m, but by May and June have risen to the top 50 m. Grauman (1974) also investigated the feeding of cod larvae and found that below 70 m their guts were predominantly empty, whereas above 70 m, 80–88% were full. The feeding regime followed a pattern with length:

Food	Length of cod larvae (mm) 4.0–4.5	4.6–7.0	7.1–11.0
Phytoplankton	70%	11%	—
Synchaeta	10%	2%	39%
Nauplii	20%	83%	55%
Copepodite stages	—	4%	6%

As they grow, the larvae start to feed on phytoplankton, change to rotifers, copepod nauplii and thence to copepodite stages, much as they do elsewhere. Renk *et al.* (1974) have demonstrated that the production cycle occurs in April and May, so the cod larvae probably depend upon the food produced then and subsequently. Fonselius (1969) noticed that the depth of the halocline has risen from 80 m to 60 m during recent decades and hence the chance of matching larval production to that of their food has increased. This is Alander's original thesis and is a special case of the match/mismatch hypothesis put forward in Chapter 11.

The increment in cod catch in Baltic since the early 1940s amounts to about 200 000 tonnes (worth about £50 000 000 at present prices) is probably a direct consequence of the slow increase of salinity and rise of the halocline, both of which are linked to a slow decline in run off from major rivers. The direct effect of the rising halocline, which brings the cod larvae closer to the productive layers where their food is produced, is a more likely cause than the intermittent fertilizations from the sediments. The rising halocline is effective on each year class and not on sporadic ones and therefore is a simpler explanation. It is likely that the increments to herring and sprat stocks originate in the same causes (Fig. 96).

The addition of more than one million tonnes of stock of cod herring and sprat to the Baltic system has revealed a gap in the ecosystem which was quickly exploited. The quantities are so large that the whole structure of the ecosystem must have been profoundly changed, and the problem needs multi-species analysis on the grand scale.

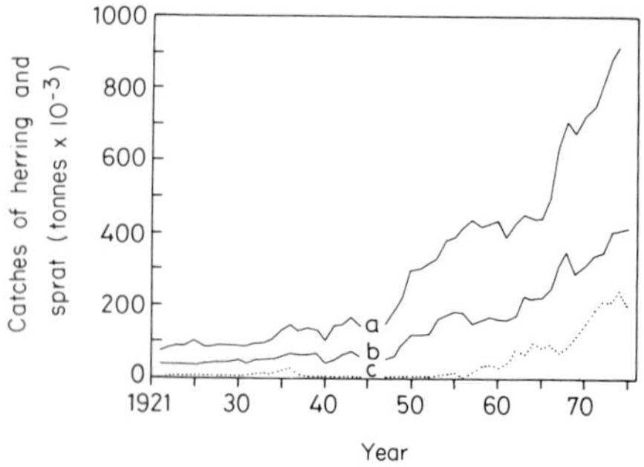

Fig. 96 Increase in annual catches of (*b*) herring, (*c*) sprat, and (*a*) total.

CONCLUSION

The Baltic is a sea of special character, of salinity low enough to be brackish in places. The entrance is so small that the input of saline water is less than the run off from the rivers. Hence there is a sharp halocline and density discontinuity which divides the productive surface layers from the waters below. Because the rivers have released less water into the Baltic it has become more saline and the halocline has risen in the water column, allowing the cod stocks to increase greatly. Brackish water animals such as *Limnocalanus grimaldii* have retreated and have been replaced by other invaders carried in by quasi-periodic salinifications. Such are driven by meteorological events in the North Atlantic and they maintain the oxygen in the deeps against the pressures of stagnation. Phosphorus is periodically released from the sediments and mixed to the surface layers in the winter and this fertilization may assist, but not initiate, the production of strong cod year classes. The remarkable point is that the biological structure is affected so profoundly by meteorological events. It is also remarkable that this structure can accommodate such an increase in the top predator when the herring and sprat stocks also increased.

The salinifications occur every five years or so, but some are more prominent than others. As recorded in the deeps, notable inflows occurred in the years 1902, 1907, 1909, 1921, 1925, 1948, 1951 and in 1959; there were several minor ones in the period 1926–37. They are recorded here to be compared with years notable for other events.

Chapter Ten

The Effect of El Niño upon the Peruvian Anchoveta Stock

The fishery for anchoveta within the Peruvian upwelling system has been the largest fishery in the world, and in 1970 12.28 million tonnes were landed mostly through the major ports of Chimbote and Callão. In the early stages in the mid 1950s catches amounted to less than 100 000 tonnes, but with the increasing demand for fishmeal catches developed, as shown in Fig. 97.

Catches built up rapidly during the 1960s by an order of magnitude to the peak catch in 1970. During the late 1960s, the catches of anchoveta were dried and pressed into fishmeal in such large quantities that the world price was dominated by the landings off Peru. In 1972, catches dropped to 4.45 million tonnes, and since then, large though the fishery remains, it has not recovered. The initial collapse took place during the major El Niño of 1972.

El Niño is a warm current of transequatorial origin which appears irregularly every five or ten years off the coast of Peru and is so called because it appears at Christmas time or just after (*niño* is Spanish for child). Typically, the warm current strengthens from the region between the Galapagos Islands and Ecuador and flows south along the Peruvian coast to between 5°S and 14°S between February and May. It stabilizes for a period, retreats in August, September and October, but strengthens again in the following January, February and March, regressing again in the southern spring after a total life of about eighteen months. The longshore winds persist during

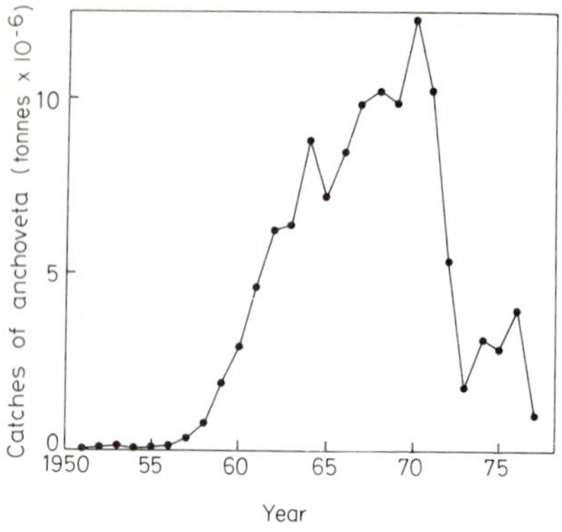

Fig. 97 Annual landings of the Peruvian anchoveta, 1951–77, in millions of tonnes.

the period of El Niño, and the warm water upwells in the same way as the cooler water does in the normal season but it is much less productive by about a factor of three (Guillen, 1976). El Niños appeared in 1891, 1911, 1917, 1925, 1930, 1941, 1957, 1965, 1972 and perhaps 1976; those in 1891, 1941 and 1972 were very strong indeed.

THE PHYSICAL BASIS OF EL NIÑO

The history of physical studies on El Niño started with the work of Schott (1931), who examined events off Peru in the light of changes elsewhere in the Pacific. Figure 98 shows a chart of the major current systems in the ocean. The El Niño of 1891 was a major event, and in March of that year the Cromwell undercurrent surfaced west of the Galapagos Islands between 1°N and 3°S. Within the area 7–10°S and 81–83°W some ships observed zero current and the trades themselves were weak. Schott thought that the warm water in the Eastern Tropical Pacific originated from the west, south of the Galapagos Islands, and that it appeared off Peru as El Niño. The coast of Ecuador is usually dry in the season of upwelling, but during El Niño, rainstorms appear because atmospheric convection and evaporation are no longer suppressed by low sea temperatures (Murphy (1926) gives a full account of the disastrous floods which occurred in 1925 and 1891); in 1891, there was also heavy rainfall in the Galapagos Islands, presumably

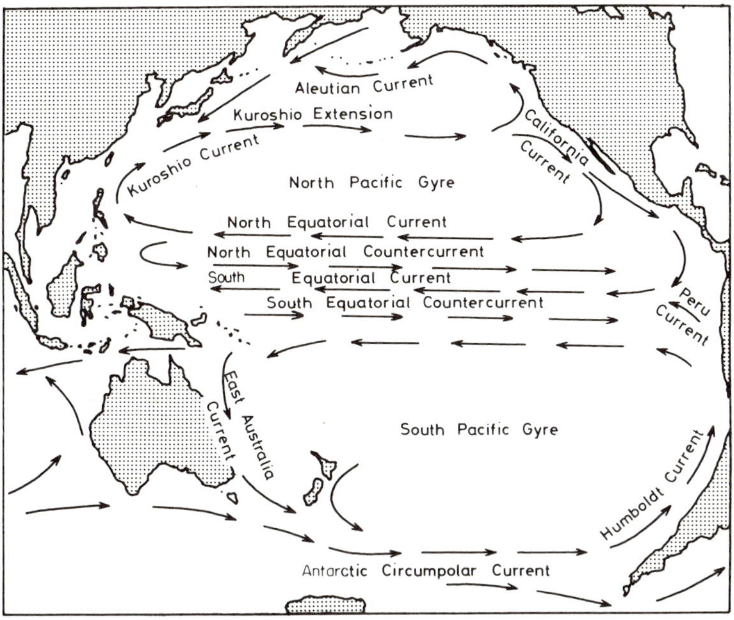

Fig. 98 The major current system in the Pacific; after the Rand McNally *Atlas of the Oceans* (O'Brien *et al.*, 1981).

with no equatorial upwelling. This interpretation by Schott was based upon considerable insight into possible processes.

Bjerknes (1961) studied the Niños of 1939–41 and 1957–8. He noticed that the Eastern tropical Pacific, including the region North of the Equator, was flooded with anomalously warm water and that towards the American coast the thermocline deepened. Then the transequatorial circulation east of the Galapagos Islands could transport more water into the southern hemisphere and by Coriolis force it would be turned to the left towards Peru. Such a movement would be opposed by the South East Trades, so if they were weaker (as happens before El Niño), the appearance of El Niño off Peru would be encouraged. Bjerknes also noticed that El Niño was preceded by weaker trades in the North Pacific. He examined the 1957–8 El Niño (Bjerknes, 1966) and established that the trades corresponding to the atmospheric pressure difference between Apia and Canton Islands weakened enough to eliminate equatorial upwelling in the southern summer of 1956–7, one year before El Niño. Hence he expected upwelling to cease off Peru during El Niño, but as will be shown below, the water continues to upwell but it is warm. These two papers directed attention towards atmospheric events.

The Niño of 1972–3 was described by Wooster and Guillen (1974). The region of warm surface temperature anomalies extended westward along the equator at least to 180°W and the equatorial zone remained unusually warm in January 1973. The normal equatorial tongue of cold water was absent during most of 1972 and southerly equatorial transgressions may have occurred east of the Galapagos, in June, August and November 1972 and in February 1973. In March 1971, surface waters in the equatorial Pacific East of 110°W became significantly warmer than in the previous year. Figure 99 shows the presence of El Niño at stations in Ecuador and Peru as monthly temperature anomalies. Figure 99a (Wooster and Guillen, 1974) shows observations from 1949–72 and Fig. 99b (O'Brien *et al.* 1981) shows another set of observations from 1958–77. The Niños are shown clearly as positive anomalies in 1951, 1953, 1957–8, 1965–6, 1969, 1972–3 and 1976; the major events were in 1957, 1965 and 1972. The regression to normal temperatures is shown in Fig. 99a, but not in Fig. 99b.

Wyrtki (1973) showed that the appearance of warm water off Central America coincided with strong transport in the North Equatorial Countercurrent and the highest peaks coincided with El Niños. Figure 100 shows the sea surface temperature anomalies in the Eastern Tropical Pacific in 1971, a normal year (*above*) and in 1972, an El Niño year (*below*). The area flooded is very large. Quinn (1974) linked El Niño with the Southern Oscillation, the fluctuation expressed by air exchange between the South East Pacific Subtropical High and the Indonesian Equatorial Low. His index of the Southern Oscillation is the difference in atmospheric pressure at sea level between Easter Island (representing the South East Pacific Subtropical High) and Darwin, in Australia (which represents the Indonesian Equatorial Low). When the index is high, the trades blow strongly towards the equator and when it is low, the trades are weak. Quinn used 12 month running means of the indices and found that a large peak precedes El Niño by about

---

Fig. 99 Appearance of El Niño as indicated by temperature anomalies recorded on the coast of Peru:

(*a*) at Paita, Puerto Chicama, Chimbote, Huarmey and Chancay (Wooster and Guillen, 1974);

(*b*) at Baltra, Libertad, Talara, Chicama, Chimbote, Don Martin, La Punta, San Juan, Matarani (O'Brien *et al.*, 1981).

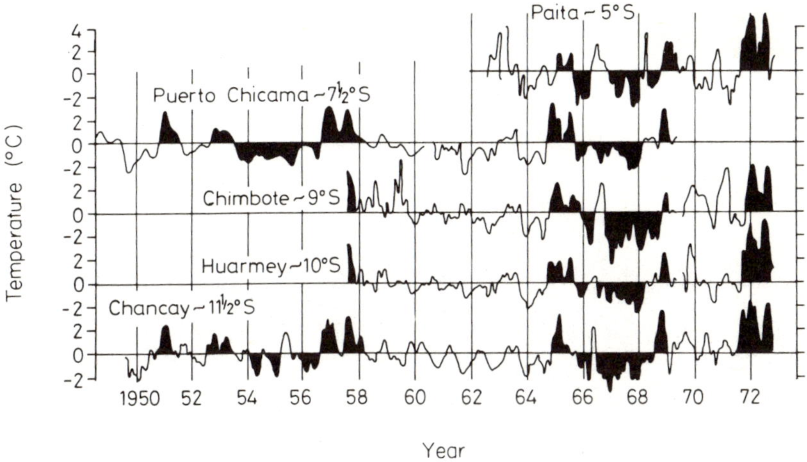
Paita ~ 5°S
Puerto Chicama ~ 7½°S
Chimbote ~ 9°S
Huarmey ~ 10°S
Chancay ~ 11½°S
Temperature (°C)
1950
52
54
56
58
60
62
64
66
68
70
72
Year

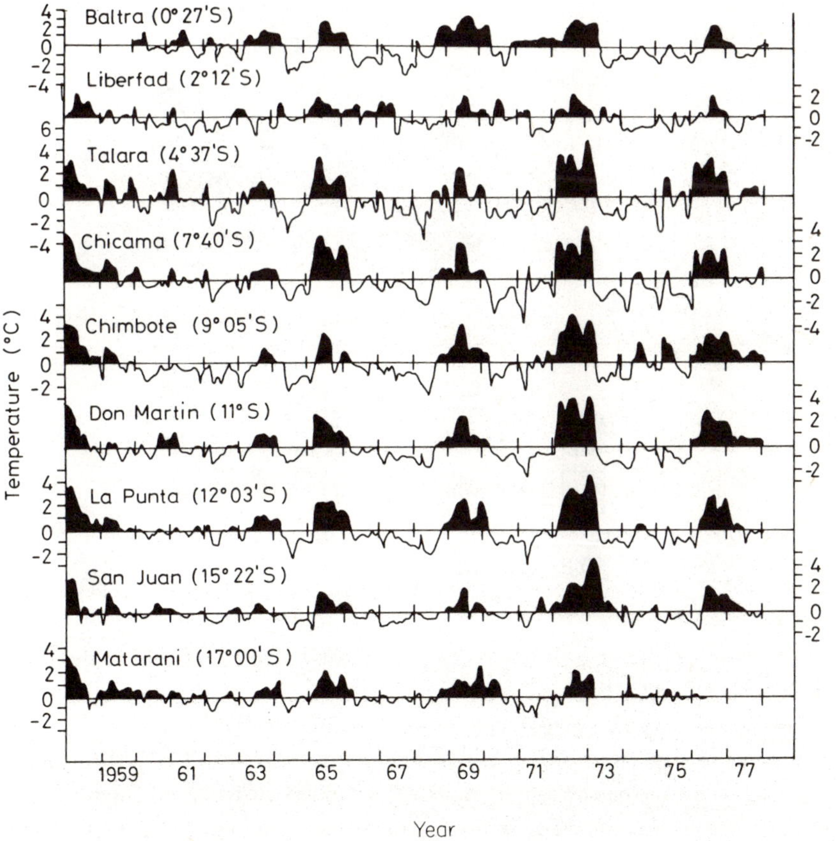
Baltra (0° 27'S)
Liberfad (2° 12'S)
Talara (4° 37'S)
Chicama (7°40'S)
Chimbote (9° 05'S)
Don Martin (11° S)
La Punta (12° 03'S)
San Juan (15° 22'S)
Matarani (17°00'S)
Temperature (°C)
1959
61
63
65
67
69
71
73
75
77
Year

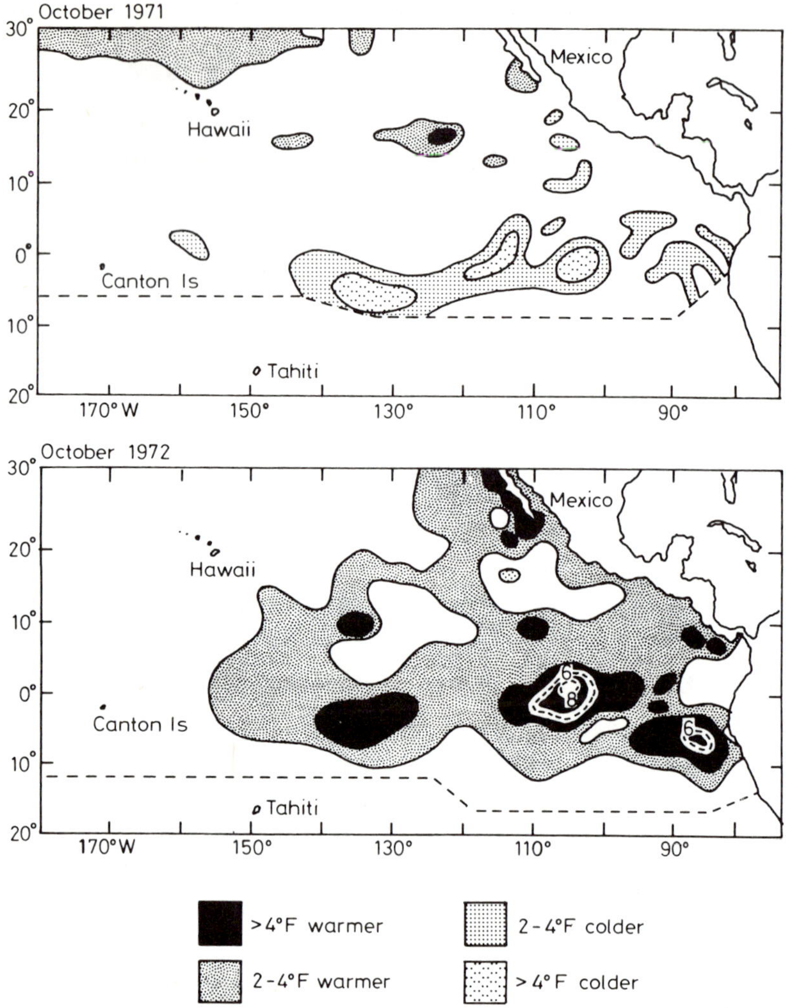

Fig. 100 Sea surface temperature anomalies in the Eastern tropical Pacific in 1971 and 1972 (O'Brien *et al.*, 1981).

a year. This implies that El Niño is preceded by strong easterly trade winds and coincides with their relation (Fig. 101).

These two papers opened the possibility of forecasting the onset of El Niño, but the greater consequence was the ground laid for Wyrtki's (1975) hypothesis which has inspired continuing analysis since. He noticed that in the Central Pacific very strong South East Trades exist for two years

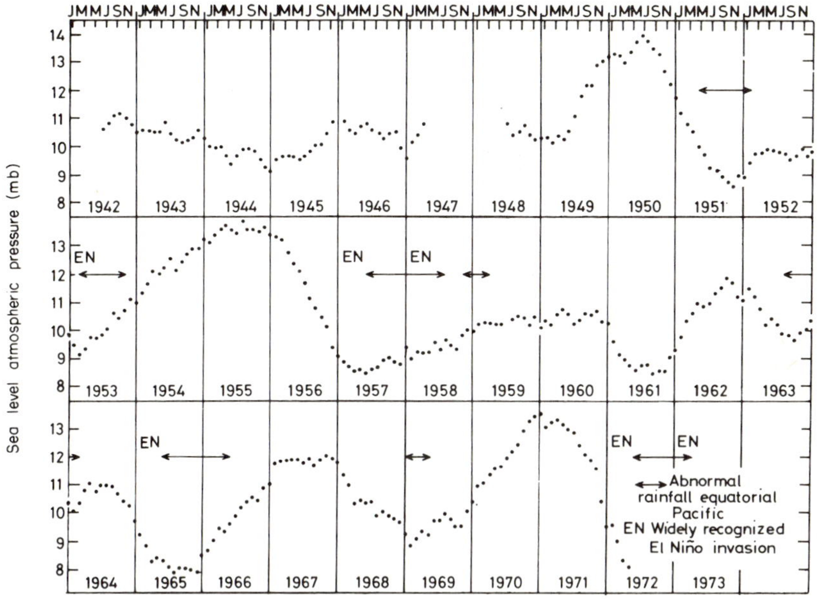

Fig. 101 Twelve-month moving means of the difference in sea level atmospheric pressure (mb) between Easter Island and Darwin in Australia, 1942–72 (Quinn, 1974).

before each El Niño. They intensify the subtropical gyre in the South Pacific and strengthen the South Equatorial current. Because water is stacked up in the western equatorial Pacific (perhaps between Samoa and the Solomons), the slope of sea level from east to west increases, which is observed. Wyrtki's hypothesis states that when the wind stress relaxes in the central equatorial Pacific, the accumulated water flows eastward perhaps by an increase in the transport of the North Equatorial Countercurrent and other equatorial counter currents or in the form of a Kelvin wave. In this case it is a long internal gravity wave travelling eastward along the equatorial thermocline whose velocity, $c = (gh)^{1/2}$, where c is velocity in cm per sec, g the acceleration due to gravity and h the depth of the ocean. The waves are very long indeed and water particles can travel over enormous distances in wave motion. The Kelvin wave travels in the ocean in an eastward direction and the equatorial system (where Coriolis force is zero) provides a wave guide which traps it. At the American coast the Kelvin wave propagates north and south, trapped in the coastal boundary, and it generates a Rossby wave which travels westwards at one-third of the speed of the Kelvin wave.

The most important point of Wyrtki's hypothesis is that it accounts for the rapid warming in the Eastern Tropical Pacific and for the sudden appearance of El Niño off Peru. The hypothesis generated a great effort by the numerical modellers to make computational experiments on an oceanic scale to account for the physical events. Hurlburt *et al.* (1976) showed that when a zonal pressure gradient in a two-layer model of the South Pacific was relaxed, Kelvin waves were generated and there was downwelling in the Eastern Tropical Pacific, deepening the thermocline, as observed. With another two-layer model, McCreary (1976) showed that only relaxation of the zonal wind field is significant in generating El Niño and more important that it is changes in the wind field within 5° of the equator which are effective. It is a somewhat narrow strip on an oceanic scale, and McCreary's results suggest that the effect was more intense in the 5° band north of the equator. Further, the transequatorial flow was stimulated, together with downwelling in the Eastern Tropical Pacific of the right order of magnitude.

Wyrtki *et al.* (1976) reported on a forecast of the 1976 El Niño and on an oceanographic expedition mounted quickly to observe the transequatorial flow. The basis of the forecast was the Southern Oscillation between 1969 and 1974 (Fig. 102) which started to relax at the beginning of 1971; El Niño appeared off Peru in February 1972 and the forecast was based on the relaxation at the start of 1974. El Niño appeared at the beginning of 1976, but was weak; the expedition showed the transequatorial flow and a deeper than normal thermocline in the Eastern Tropical Pacific.

Hickey (1975) correlated sea level, wind stress and sea surface temperature at various lags in the Pacific equatorial region between 10°N and 10°S: the year-to-year fluctuation in meridional shear across many years (as represented by differences in zonal wind stress between 5°–10°N and between 5°S–5°N)

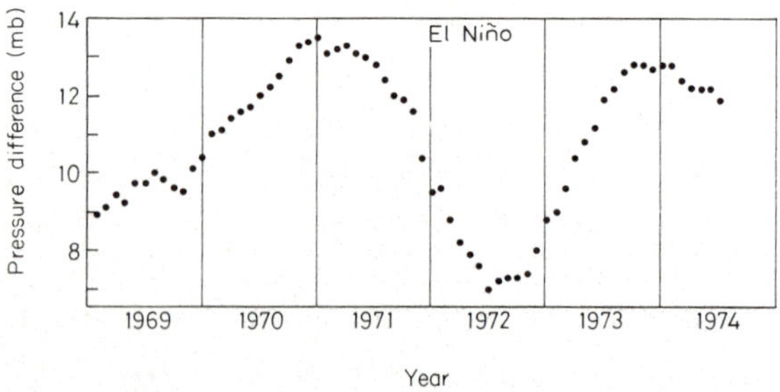

Fig. 102 Index of the Southern Oscillation for 1969–74; the El Niño of 1972 is indicated (Wyrtki, 1977).

in bands of longitude shows the prominent Niños (Fig. 103). A comprehensive analysis of the trade wind field in the Pacific was made by Wyrtiki and Meyers (1976). Figure 104 shows the major result, the annual variations of the area of trade wind and the mean wind stress of the Southeast trades between 2 and 16°S and between 90 and 160°W; in 1955–6, 1964, 1966–7 and in 1970–1, the trades are stronger than normal preceding the major Niños. But no link was established with the north-east trades. Barnett (1977) analysed the same material in terms of the variability of the U (eastward) and V (northward) components. U increased during the early 1950s and early 1970s; the end of this increase preceded a major Niño by

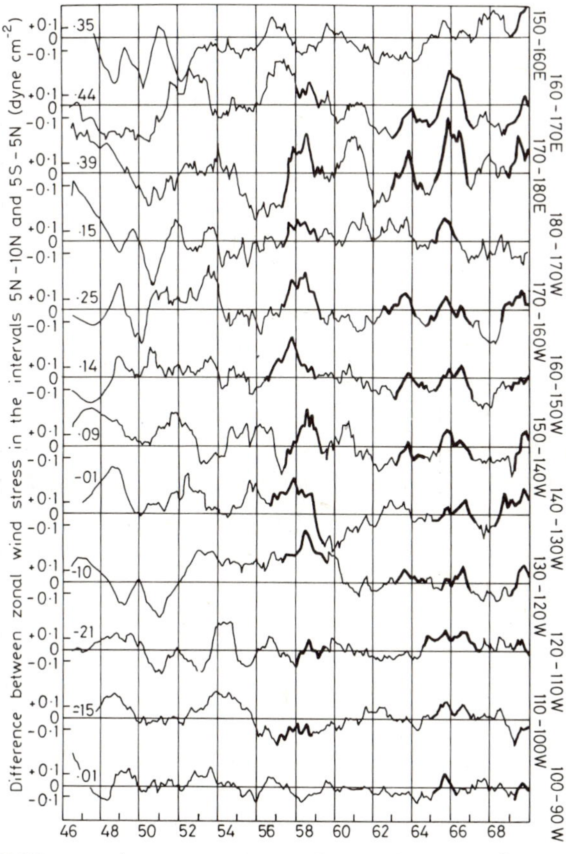

Fig. 103 Differences between zonal wind stress between 5°N and 10°N and 5°S and 5°N at each band of longitude, indices of meridional sheer. El Niño is emphasized (Hickey, 1975).

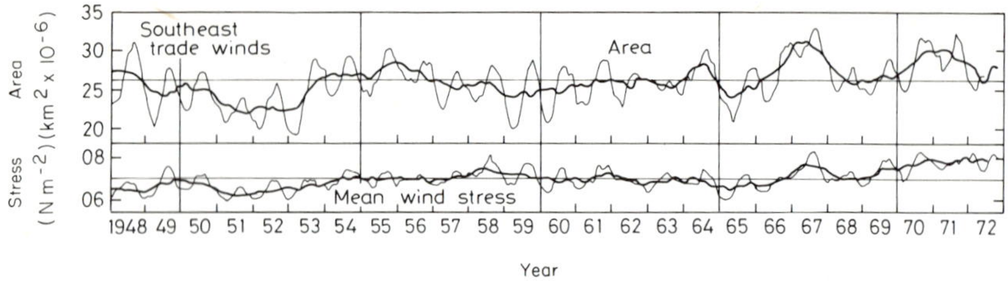

Fig. 104 Area covered by the South East Trades and the mean wind stress of those trades (Wyrtki and Meyers, 1976).

one year. A notable event occurred in the mid 1950s before the Niño of 1957–8, which implies that the northward component of wind was strengthening (or that the southerly component was slackening) 1000–3000 km west and north of the Galapagos Islands. From this intense meteorological study, the occurrence of the major El Niño events is preceded by persistent strengthening of the South East Trades in a narrow region close to the equator in the central ocean.

Wyrtki (1977) elaborated his hypothesis in an examination of the sea level in the Pacific during the great 1972 El Niño. There are five stages: (1) a build-up of sea level in the western equatorial Pacific; (2) a first relaxation which leads to a slow drop of sea level in the west and a first peak in the east; (3) a rapid drop of sea level in the west, a filling of the equatorial trough, while it remains above normal in the east; (4) a minimum of sea level in the west and a second peak in the east; (5) equalization, that is, a very rapid increase in sea level in the west and a return to normal in the east. During the period of build up, the South Equatorial Current is strong, but the Countercurrent and the North Equatorial Current are both weak. This period lasts about a year and the intensity of build up may determine the intensity of El Niño.

When the wind field relaxes slowly, sea level relaxes slowly and there may be no major El Niño. A sudden relaxation, in about two or three weeks, of the trade wind field near the equator will trigger an equatorial Kelvin wave. Within two or three months sea level rises rapidly in the Eastern Tropical Pacific and the thermocline there is depressed. During this second phase, the South Equatorial Current decreases, the Countercurrent increases and the North Equatorial Current remains weak. A weak Niño, like that of 1975, stops at this stage.

The third phase is a long period of weak trades which leads to a further drop in sea level in the Western Pacific. The Equatorial trough is being filled, the South Equatorial Current decreases (and disappears north of the equator), the Countercurrent broadens and indeed may extend south of the equator and intensifies. Sea level off South America remains high for a period of six months. The period of lowest sea-level in the Western Pacific is the culmination of the process and takes place at the second peak of sea level in the east off the coast of Peru. The last phase is the recovery, when sea level drops along the coast of Peru and rises in the Western Pacific and the equatorial currents return to normal; the rapid recovery is caused by the sudden onset of strong south east trade winds.

The sequence of processes across the whole Pacific is summarized in Fig. 105, which shows the events before and after the 1972 Niño within the structure of equatorial currents. There are two important conclusions from this study: (*a*) that El Niño is a panoceanic event, analogous to a seiche in a lake, within which the Kelvin wave is a sensible component, and (*b*) the three stages of El Niño off Peru can be detected in the system described by Wyrtki.

Meyers (1979) analysed the changes in the 14°C isotherm (just below the thermocline) along the equator. He found a two- or three-month lag between the minima of isothermal depth at the eastern boundary, which is about the time needed for a Kelvin wave to propagate eastward. O'Brien *et al.* (1981) were stimulated by this result to write a numerical model of the duration of El Niño, including the intensification of winds for two years before its onset. Then the trades were relaxed, the windstress being halved in two months. A Kelvin wave is propagated in an easterly direction in the model and the thermocline is deepened in the east and a Rossby wave is generated westward from the American coastal region.

O'Brien and his co-authors developed a model of equatorial upwelling which generates equatorial Kelvin waves. This was run for two model years and in the third year the westerly wind west of 140°W was increased by a factor of 1.5 by the end of the year. The winds were relaxed in two months to half their normal value and were kept low for the remainder of that year. At the beginning of the fifth year the westerly winds were restored to their long-term value during a two-month period. Figure 106 shows how the model thermocline in the Eastern Tropical Pacific responds: it is depressed within two or three months of the relaxation of the winds. Two other experiments were made in the computer: one to sharpen the relaxation of the wind and slow its recovery period and one to delay the relaxation by two months. In the first experiment a double peak in the model thermocline is observed; in the second the period of thermocline

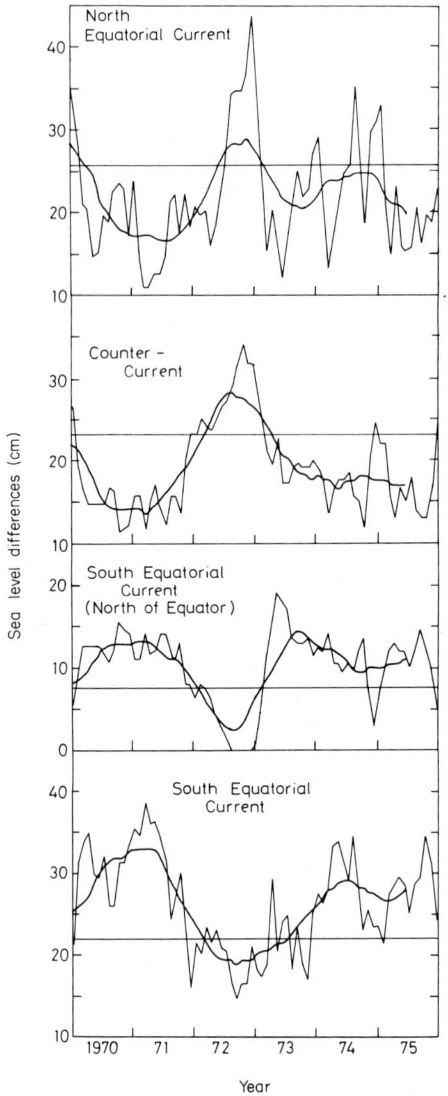

Fig. 105 Sea level differences across zonal currents in the Central Equatorial Pacific, 1970–5, in cm (Wyrtki, 1977).

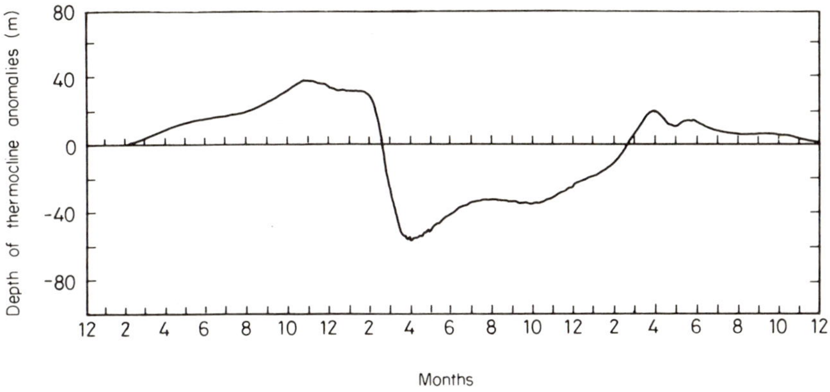

Fig. 106 Depth of thermocline anomalies throughout the period of El Niño as modelled by O'Brien *et al.* (1981).

depression is shortened considerably. Thus not only does Wyrtki's thesis explain the rapid onset of El Niño in terms of panoceanic atmospheric events expressed as a Kelvin wave, but the triple nature of a major event can be adequately simulated by extension of his hypothesis. It should be pointed out that the theory from which the numerical model was constructed was derived from earlier models of coastal upwelling (O'Brien and Hulburt, 1972) and of equatorial upwelling.

The collapse of the anchoveta fishery off Peru in 1971/2 and the failure of the stock to recover stimulated the most remarkable cooperation between meteorologists and physical oceanographers to understand the physical bases of El Niño. It is now possible to forecast El Niño, and perhaps in the future it will become possible to forecast its nature, that is, whether or not it will be a major event. The experimental approach initiated by O'Brien and his students will indicate what observations will be needed at sea in order to advance our understanding of this very complex problem.

THE PERUVIAN UPWELLING AREA

The Peruvian upwelling area lies in the Peru current and extends from Cape Carranza, south of Valparaiso, in Chile, to Punta Aguja. The most important area is that off Peru. The coastline makes a different angle to the prevailing longshore winds in the southern part of the country from that in the north.

Production is high in both regions but the fishery for the anchoveta (*Engraulis ringens* Jenyn) is concentrated in the region north of Callão. The radiocarbon estimates averaged over a broad area are given below, but Walsh (1976) quotes much higher values of 5.6 gC per m^2 per day off the southern coast of Peru at 15°S. Barber and Smith (1981) give average values for 1966, 1969, 1976, 1977 and 1978: in 1966, 1969 and 1978, measurements ranged from 3–10 gC per m^2 per day, but in 1976 and 1977 they ranged from 0.5–3.5 gC per m^2 per day. The region is one of the most productive in the world ocean.

The processes of coastal upwelling have been well described in a number of publications (for example, Barber and Smith, 1981), and the essential physics are given in O'Brien and Hurlburt (1972) and Hurlburt and Thompson (1973). Two-layer models are used: the southerly winds off Peru generate a surface flow towards the equator with a coastal jet and a poleward undercurrent. Due to Coriolis force, an effect of the earth's rotation, the surface layer is driven offshore, and in a thin region nearshore, perhaps 10–20 km broad, water is drawn upwards from below which is the essential process of coastal upwelling. As in other major upwelling areas the longshore wind is part of the South Pacific atmospheric anticyclone at sea level (Thompson, 1977). The coefficient of variation of windstress off Peru is 51%, much less than that off Oregon or north-west Africa (Barber and Smith, 1981), and this variability has necessarily a panoceanic origin. Upwelling continues throughout the year (Wooster and Sievers, 1970), but the long shore winds are strongest between June and August.

The continental shelf off Peru is only 10 km wide. In 1977, the offshore flow was 24 m deep and the onshore flow extended from 25 to 100 m with a peak at 60 m. Consequently the onshore flowing water, about to be upwelled, is in the lower part of the photic layer, and production can start quickly. The depth of surface mixture due to diurnal variation in wind strength is less than 24 m; this may modify production but will not inhibit it. The poleward undercurrent is strong and shallow, possibly because the shelf is so narrow, but the currents near the seabed are weak, so there are considerable quantities of sediment (Barber and Smith, 1981).

An immense amount of work has been done in sampling the biological environment by the Istituto del Mar del Peru (IMARPE) and by the Coastal Upwelling Ecosystem Analysis program (CUEA). Walsh (1976) summarized much of the material:

Mean longshore wind (sec^{-1})	4.6
Mean shelf width (km)	15

Mean irradiance (ly min^{-1})	29
Estimated vertical velocity (cm sec^{-1})	1×10^{-2}
Mean nutrient at 40–50 m	
nitrate (μg-at l^{-1})	23
silicate (μg-at c^{-1})	18
phosphate (μg at l^{-1})	1.8
Mean chlorophyll a (0–20 m) (μg l^{-1})	6
Mean productivity ($gCm^{-2}d^{-1}$)	5.6
Mean zooplankton biomass, dry wt (mg m^{-3})	41

Walsh (1975) distinguished two food chains in the Peruvian ecosystem:

WEAK UPWELLINGS Algae → {anchoveta, zooplankton} → carnivores

STRONG UPWELLING Algae → zooplankton → anchoveta → carnivores

The one-step food chain is characteristic of a restricted upwelling area in the southern Autumn, March and April. During the strong upwelling of the southern spring, when the anchoveta spawn, the two-stage food chain predominates.

Walsh *et al.* (1980) distinguish the area between 6 and 12°S from that further south. There are more than half as many storms in the southern part of the area as in the north, where the anchoveta spawn. With fewer storms in the northern area, the anchoveta larvae would tend to be retained in the coastal area; in both regions, north and south, eggs and larvae were restricted to coastal areas of high chlorophyll and no larval anchovies were collected further offshore in the CUEA observation. It is well known that the food demand of larval fish much exceeds the average concentrations of food organisms (Jones, 1973); dinoflagellates that migrate vertically may concentrate enough to provide food for the first feeding stage of the anchoveta.

The anchoveta live in the upper part of the thermocline and horse mackerel (*Trachurus trachurus*) just below; hake and rosefish (*Sebastes marinus* Linnaeus) live near the seabed. Offshore beyond the dynamic boundary to the coastal upwelling system live the tuna-like fishes and the small myctophids characteristic of the subtropical ocean.

As the algae are passed through the lowest part of the photic layer, they start to reproduce, so that when the upwelling water finally reaches the surface the primary production is well under way. However, because the upwelling system is relatively shallow there may be herbivores in the rising water. Yet the high production must indicate that of herbivores lag the primary production as in temperate waters. As the upwelling plumes are drifted offshore, material is transferred up the food chain to the anchoveta and zooplankton in the surface waters and to the demersal fish below.

THE BIOLOGY OF THE ANCHOVETA

The Peruvian upwelling has been studied for a long time by oceanographers (see Gunther (1936) and Wooster (1961) for a historical account). Zuta and Guillen (1970) and Zuta and Urquizo (1972) give detailed descriptions. Mean temperatures off Punta Falsa, off Puerto Chicama and off Callão were tabulated by Zuta and Guillen; their material shows that the upwelling season lasts from May to November in its most intense form (Fig. 107). Zuta and Urquizo (1972) published the average temperature distributions by months and latitudes for the period 1928–69 and they show the same phenomenon (Zuta and Urquizo, 1972). Three centres of more intense upwelling emerge from the material: (1) Pimentel to Salaverry in 7–8°S, (2) Callão and northwards in 10–12°S and (3) Pisco and southwards in 14–17°S. Of the three subareas, the latter is most intense: this is the region where most of the work of CUEA was concentrated, rather south of the major region of the fishery. Estimates of primary production by latitude are given in Table 18. As might be expected, the higher production tends to be concentrated in the three upwelling foci and from June to November, the biological processes would be expected to be the most developed. The estimates given above are less than those given above from the CUEA work, probably because they were sampled at standard stations rather than in upwelling plumes.

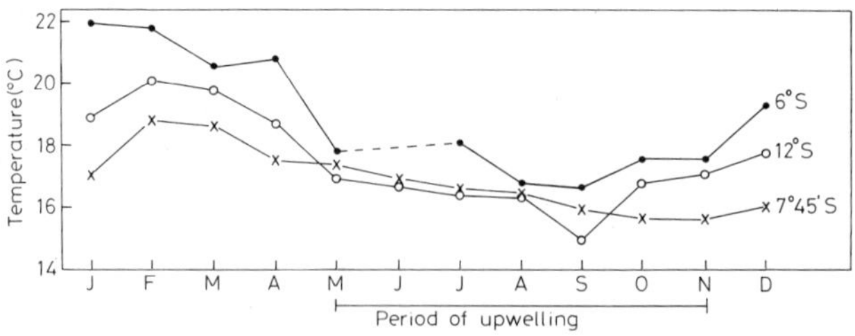

Fig. 107 Duration of the main upwelling system off Peru as shown by the distribution of mean temperatures off Punta Falsa (●———●), off Puerto Chicama (×———×) and off Callão (○———○) (Zuta and Guillen, 1970).

Table 18
Estimates of primary production off Peru (Zuta and Guillen, 1970)

Latitude (°S)	Production ($gC\ m^{-2}\ d^{-1}$)		Mean production ($gC\ m^{-2}\ d^{-1}$)
4	0.20		
5	0.60		0.43
6	0.50		
7	0.55	Pimentel-Salaverry[a]	0.92
8	1.30		
9	0.60		0.60
10	0.55		
11	0.65	Northward from Callão[a]	0.63
12	0.67		
13	0.37		0.37
14	0.60		
15	2.30	Southward from Pisco[a]	1.05
16	0.70		
17	0.62		
18	0.58		0.58

[a] These areas comprise the three upwelling foci

The anchoveta spawn between August and March, with peaks in September and January, as shown in the distributions of maturity stages by months for a number of years (Jordan and Vildoso, 1965; Vildoso and Alegre, 1969; Minano, 1968a). The seasonal distribution of fat content shows an analogous double peak (Lam, 1968). Between latitudes 6°S and 16°S and up to 150 nautical miles offshore in all months of spawning, IMARPE have carried out extensive egg suveys (Guillen and Vazquez, 1966; Guillen and Flores, 1967; Minano, 1968b; Zuta and Mejia, 1968; Valdivia and Guillen, 1966; Flores and Elias, 1967; Flores, 1967; Valdivia and Poma, 1969). The eggs hatch in about two days and are found roughly where they were spawned. Between 9°S and 11°S, spawning starts in July with a somewhat low egg density. By August and September, eggs were found between 5°S and 15°S and as far as 120 or 150 nautical miles away from the coast. Fish which have spawned so far offshore must have been drifted far beyond the dynamic boundary at about 50 km and well outside the area of the fishery, about 30 km from the shore. In November, egg distributions are confirmed to a narrow coastal strip (of about 30–50 km), and later a northern and southern group of spawners may be separated in their egg distributions.

In September, spawning begins three months after the start of the strongest upwelling in June and ends three months after the end of that period. The first zooplankton generation might be expected to appear about six weeks after the start of intense upwelling in May or June. The second zooplankton generation, with much augmented numbers, might be expected to supply nauplii to anchoveta larvae spawned after September. Further a generation of zooplankton hatched after the period of intense upwelling might well support the summer spawners which mature in January and February. If the larvae drifted in the Peru coastal current towards the equator at 10 or 30 km per day (Zuta and Guillen, 1970), they would drift out of the area in one to three months. But the population must remain within the upwelling system; the larvae would only have to sink below 25 m to find themselves in the poleward undercurrent of which an onshore component contributes to the upwelled water.

The growth of the anchoveta has been studied by the progress of length modes in time. There are two length groups of recruits (Saetersdal and Valdivia, 1964), and fish of 5 cm in length appear in two groups the first in January and the second in May (Boerema *et al.* 1967). The spawning of August, September and October generates the first group and that in January gives rise to the second; the latter grows outside the period of major upwelling and it is not surprising that it grows more slowly. Figure 108 shows growth for the first two and a half years (after Saetersdal and Valdivia, 1964); also shown are the periods of upwelling and those of spawning. Length is displayed on the ordinate and the dashed lines represent five important events: (*a*) 5.5 cm, the length at which the smallest fish have been recorded in the fishery (in January and in May); (*b*) 8.5 cm, the length at which maturation starts; (*c*) 11.5 cm, the length at which fish may spawn for the first time; (*d*) 14.5 cm, the length at which the anchoveta spawn for the second time; (*e*) 17.0 cm, the length at which the third spawning might occur (by 1974 the 1971 year-class had reached a length of 17 cm (Valdivia, 1976), and indeed the results of tagging experiments have shown that the fish can live for as long as five years). The growth curves of both spawning groups are given with ages, number of rings on the otoliths and the maturation stages; vertical arrows indicate the part contributed made by each group to the spawnings. As the fish grow older,

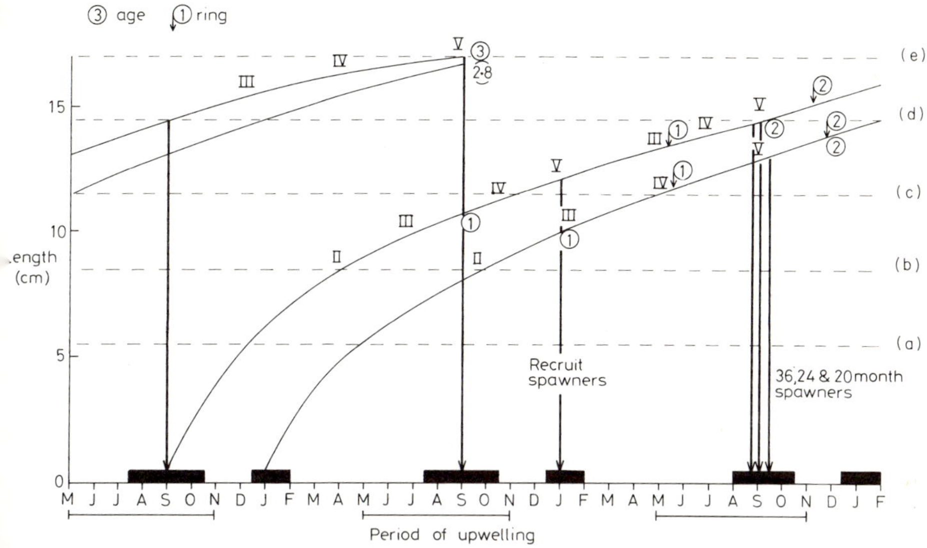

Fig. 108 Growth of anchoveta; periods of spawning (as blocks on the abscissa) and upwelling (as bars below the abscissa) are shown. Five critical lengths are given in the diagram

(*a*) the length at which the smallest fish have been recorded in the fishery (5.5 cm);

(*b*) the length at which maturation starts (8.5 cm);

(*c*) the length at which the fish may first spawn (11.5 cm);

(*d*) the length at which anchoveta spawn for the second time (14.5 cm);

(*e*) the length at which they may spawn for the third time (17.0 cm) (adapted from Saetersdal and Valdivia, 1964)

Ages are indicated as for example: (1), number of rings on the otolith are indicated; (2), maturation stages are given in Roman figures.

their distributions in length will tend to merge, so the separation of the two spawning groups becomes artificial with age. This means that in the real population the products of the two spawning groups become mixed in the unexploited state and so the separation of spawning season in this case should not be evidence of stock separation. But heavy fishing for a very long period indeed might lead towards a real distinction between stocks.

The ages of the fish are estimated from the rings on their otoliths; Vildoso and Chuman (1968) examined otoliths collected between October 1963 and December 1964. The remarkable result emerged that two rings were probably laid down each year on the otolith structure, but it should be pointed out that only half were legible. The first ring is laid down one year after spawning and the second about six months later. The sharpest change in ring structure (from opaque to hyaline, or vice versa) occurs when the period of stronger upwelling starts in May and when it ends in November. The changes in intensity of upwelling appear to be reflected in the appearance of the biannual rings. Any checks due to upwelling are not recorded in the first year of life, perhaps because the little fish may have grown at constant temperature in the cool upwelling water, as suggested from their retention within the upwelling area.

The upwelling production cycle resembles that temperate waters in that production starts from zero and there is a lag between primary and secondary production (but see Barber and Smith, 1981). In temperate waters, fish spawn for a long season, older fish spawning first and recruits last, and this is also true of the anchoveta. Hence the January spawning comprises recruit spawners, and although they are very abundant, their egg production is low (Minano, 1968). The spawning in September is more productive because the majority of spawners are adult. The construction of Fig. 108 assumes that recruits spawn in January and again in September or November, but in later life the anchoveta are assumed to spawn once a year. The figure subsumes three lines of evidence, on growth, maturation and age in a consistent manner.

As indicated in Fig. 108, the recruiting fish appear first in the fishery at a length of 5.5 cm, 3–5 months after spawning. The little fish appear at Chimbote and Callão (180 miles apart) at about the same time (Boerema *et al.* 1967). Hence it is reasonable that the recruiting fish appear at the coast in the upwelled water. The Peruvian coastal waters are flooded with recruiting fish between January and July, which is why the *peladilla* (small fish) season is closed in February and March and during the winter months.

The biology of the anchoveta depends upon the physical structure of the upwelling region in which the fish population survives. Spawning occurs during the period of upwelling in the region of coastal upwelling and, because of the offshore drift, up to 240 km outside it. The larvae grow mainly during the upwelling period and the fish start to mature at about nine months and spawn for the first time at about one year of age. They spawn again in their second year and perhaps in their third. The stock maintains itself in the upwelling system; the adults must drift offshore in the upwelling plumes but they may return in the upwelled water in the coastal region. But far offshore outside the dynamic boundary, they may be lost to the system as food for tuna-like fishes. Within the coastal system, up to 50 km from the shore, the population must live in the physical structure which endows it with drift at different depths in all four directions. It can move offshore and equatorward in the top 24 m and inshore and poleward below that depth. Hence by merely varying their depth at different times or seasons the fish have full freedom in their potential migrations.

THE BIRD POPULATIONS

In each of the major upwelling systems there are colonies of guano-producing birds—cormorants, boobies or pelicans—and there is a Cabo Blanco or Cap Blanc covered with guano in each. One of the first signals of a major El Niño is a premature migration of cormorants from Peru before the juveniles are fledged. In normal years they migrate north or south after breeding and return in the southern spring. In January or February if El Niño appears the cormorants fly away to Chile or Ecuador, having abandoned their nests. And the reliable and popular signal of a major El Niño is the appearance in February of starving pelicans on the streets of Guayaquil and Lima.

Three species of guano bird feed on the anchoveta: the *guanay* (*Phalacrocorax bougainvilli* L., a cormorant), the *piquero* (*Sula variegata* Tschudi, a booby) and the *alcatraz* (*Pelecanus occidentalis thagus* Molina, a pelican). The cormorants fish in daylight in broad bands of large flocks and dive to depths of nearly 12 m; the booby is less gregarious and dives in daylight to a depth of not greater than 15 m; the pelican fishes both day and night in small flocks and feeds at or just below the surface (Jordan, 1967). All three species breed on the guano islands and headlands in November, and the chicks are hatched and reared in January and February. After breeding, the adults migrate to Chile or Ecuador at the end of February. Between February and May, the growing juveniles feed on their own and fly away south or north in April, May or June (Jordan and Cabrere, 1960).

All three species breed on capes and islands off the coast of Peru where there are considerable deposits of guano. It is excavated and exported by the guano companies for fertilizers. Jordan (1967) suggests that there has been an "improvement of nesting sites and protection of various headlands where important centers of nesting were successfully established"; certain headlands were walled off to exclude predators. The number of birds increased from about 3 000 000 in 1910 to 8 000 000 in the 1930s. Numbers fell sharply to 4 000 000 after the major El Niño of 1940–41 and did not recover until the late 1940s. Subsequently, numbers continued to rise and reached a peak of 20 000 000 in 1956; then the numbers of birds declined as a consequence of competition with the fishermen and effects of El Niño, as will be shown below. By 1972–73, the numbers were reduced to about 1 000 000. Numbers were estimated from the production of guano, and Fig. 109 shows a slow build up in numbers from 1910 to 1956, before the fishermen had started to compete with the birds for anchoveta. The care of the colonies by the guano companies may have augmented the increase but did not stimulate it. The 1940–1 El Niño delayed the increase considerably. The question arises whether Fig. 109 illustrates a recovery after the great El Niño of 1891.

Jordan (1967) estimated that each bird takes 430 g per day of 3–16-cm anchoveta (but they tended to take the larger fish). Although the cormorants can fish as far as 48–64 km from shore, they prefer to work inshore near the headlands and islands perhaps because the fish are more abundant and shallower there. Between February and June, the recruiting anchoveta appear inshore all along the coast in abundance and this may be why the cormorants gather inshore.

The population (estimated from guano production) increased from 4 000 000 in 1909 to 20 000 000 in 1956 but in 1957, numbers fell to 7 000 000 after an El Niño and subsequently recovered to 18 000 000 by 1962. The collapse was due to the 1957–58 El Niño. In 1965 after the 1965–66 El Niño the numbers fell to 4 000 000 and did not recover. After the El Niño of 1972–3 the population declined to 1 000 000 (Jordan

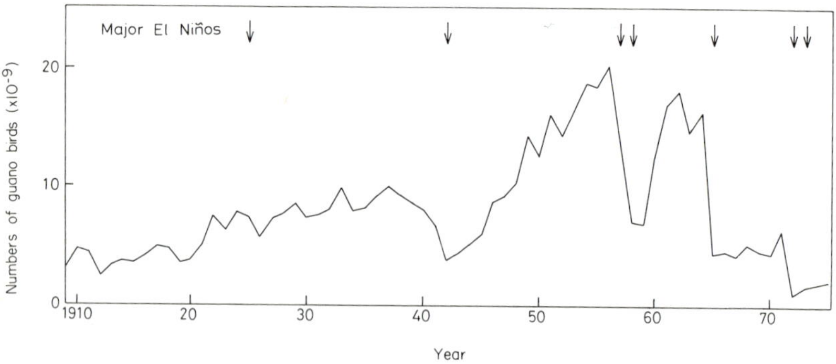

Fig. 109 Numbers of guano birds between 1909 and 1964 (Jordan and Fuentes, 1966).

and Fuentes, 1966; Jordan, 1976) (see Fig. 109). During major El Niños the population collapses because the adult bird abandon their chicks at the onset of El Niño and fly to Ecuador or Chile long before their normal migration. Consequently, the bird population lacks good recruitment for one or two years, and if major events occur close to each other, the population may well face collapse. As El Niño presses quickly to the south, the anchoveta population may be carried beyond 15°S and the adult birds do not have enough to eat because they do not follow the fish. When the population of birds recovered from the 1957 El Niño the catches of anchoveta were less than 2 000 000 tonnes. In 1965, catches had increased to more than 7 000 000 tonnes and the guanay birds have probably suffered in competition with the fishermen.

Let us suppose that the young bird population was twice that of the adults and that they eat 300 g per day for 150 days. From 1960–64, the adult bird stock amounted to 15.8×10^6, or 31.6×10^6 juveniles (2 juveniles per adult) which may have eaten $(31.6 \times 300 \times 150 \times 10^6)$g or at 5 g per fish, $(31.6 \times 300 \times 150 \times 10^6/5)$ fish or 0.28×10^{12}. During this period stock amounted to 13.45×10^6 tonnes or 1.12×10^{12} fish (at 12 g per fish) (Anon., 1972); the quantity of recruits eaten may have comprised 25% of the stock. A similar calculation for the period 1965–71 when the number of birds was reduced to 4.89×10^6, suggests that the quantity eaten was 0.9×10^{12} or 6% of the stock. In Anon. (1973) (and Csirke, 1980) is given a stock/recruitment curve, numbers of recruits per unit of effort and adult stock in weight per unit of effort. Csirke suggested that this stock/recruitment curve was generated by differences in density, themselves stock related. I put forward an alternative explanation, but of course the two possibilities are not mutually exclusive. From the observations we construct Table 19.

Between the two periods, recruitment increased by 63% as the bird stock was reduced after the El Niño of 1965–66. The sums of recruits and

Table 19

Effect of bird predation upon anchoveta recruitment

Period	Stock in weight per unit of effort	Numbers of recruits per unit of effort	Numbers of recruits eaten	Sum of recruits and numbers eaten
1960–64	583	270 ± 85	146	416
1965–71	368	438 ± 73	33	468

numbers eaten were not very different, which suggests that the stock/recruitment curve *without bird predation* is not the dome-shaped one shown in Anon. (1973) but is an asymptotic one; so far as is known dome-shaped stock and recruitment curves are characteristic of some gadoids and some Pacific salmon stocks (Cushing and Harris, 1973). The numbers of recruits eaten were estimated from the percentage of stock as estimated above.

THE 1972–73 EL NIÑO

The 1972–73 El Niño was almost certainly the most intense observed since 1891; it started in February 1972, had stabilized by March of that year and in August, September and October it retreated and strengthened again in the following January, February and March and then died away in May 1973. During the regression of August–October 1972, the sea surface temperature was more than 1°C warmer than the average between 1925 and 1973 (Miller and Laurs, 1975). Usually during the regression, the water is cooler and normal productive upwelling takes place. Wooster and Guillen (1974) noticed some early signals. In March 1971, the equatorial waters between Ecuador and the region of the Galapagos Islands were already warmer. They also observed a patch of warm water offshore from Peru in 16–18°S in August 1971, as if there had been an offshore southerly and transequatorial flow in the early months of that year; the pelagic crab (*Euphylax dovii* Stimpson) was caught in this patch, an invader from Central American waters (Vildoso, 1976).

When El Niño started in February 1972, the adult birds left Peru as is usual during a major event. The population declined to 1 000 000 and by 1974 the stock had not recovered. The presence of exotic marine invaders in Peruvian waters was recorded in some detail during this El Niño. They included animals native to the waters of Central America, Mexico and Southern California, but one remarkable exotic, the milkfish (*Chanos chanos* Kluntz) may have come from Indonesia. It appeared for the first time off Peru in 1972–73 (Vildoso, 1976) but is sometimes recorded in Central American waters; it may have travelled on the Equatorial Countercurrent. During this El Niño primary production was reduced by a factor of three (Guillen, 1976) as compared with a period of normal upwelling. A third event was the dispersal of the anchoveta to the south and the consequent disruption of the fishery. By the autumn of 1971, the shoals had migrated beyond 15°S and were found close inshore by the fishermen as if they had been pushed south by the warm current (Valdivia, 1976). In earlier El Ninõs during the life of the highly developed fishery (in 1957–58 and 1965–66), the fish reassembled after dispersal and neither the recruitment nor the parent stock appeared to be affected. The recovery of the bird population after the 1957–8 El Niño was considered a justification for the rapid expansion of the fishery.

In the spring of 1969, the stock of anchoveta was very low, but the spawning in January 1970 generated high recruitment. The spawning in the spring of 1971 comprised second-time spawners hatched in September 1969 and first-time spawners hatched in January 1970. The year class hatched was the lowest recorded during the history of the fishery, five months before El Niño appeared. The two subsequent year classes hatched during the El Niño years of 1972 and 1973 were also low probably because the primary productivity in the upwelled warm water of 1971–2 and 1973 was sharply reduced. Either all three year classes were determined within five months or the 1971 year class suffered a different history from its two successors.

In the spring of 1971, the proportion of fish in maturity stage V (nearly ripe) was reduced from a normal 70% to 28.6% at Chimbote and from a normal 60% to 12.8% at Callão; at the same time, the fat content remained high, instead of declining as it usually did in the advancing season (Valdivia, 1976). The year classes hatched in 1969 and 1970 were the highest recorded (Valdivia, 1976). In August, September and October 1971, fish of 13 cm in length may have comprised second time spawners (spawned in September 1969) and first time spawners (spawned in January 1970). If growth were density dependent, the first time spawners (spawned in January 1970) did not mature. Then the egg production between August and October 1971 might have been reduced considerably. Such an interpretation would disconnect the observation on maturation in the spring of 1971 from the subsequent El Niño in the following February.

In 1974, nearly half the catches comprised large fish 15–18 cm in length and they were survivors of the 1971 year-class at three years of age; until 1974 no fish larger than 15–16 cm had been observed in the fishery and fish of three years of age had not been seen since 1961 and 1962 (Boerema *et al.*, 1967). Three year classes, contributed to the length distribution of the catches in 1974, 1971, 1972 and 1973 (Valdivia, 1976); because the year classes of 1971 and 1972 were of the same magnitude in 1974, Valdivia concluded that that of 1971 suffered little fishing mortality. Even more remarkably, Jordan (1976) reported the recovery of fish of 14–15 cm in length tagged in July 1970 as follows:

1970	1971	1972	1973	1974
4448	422	6	44	27

Not only did the fish live for five years or so but they were effectively absent from the fishery in 1972, when in fact the exploitation rate was apparently high. Perhaps part of the stock was shifted south during the intense 1972 El Niño and evaded the effects of fishing.

During the period of study of the population dynamics of the anchoveta, that is, since the mid 1950s, only one very strong El Niño had appeared, that of 1972–73. In 1965–66 the phenomenon was less intense. During the El Niño regression the temperatures off Peru in the spring of 1965 were normal or just above normal (Wooster and Guillen, 1974); cool productive water probably upwelled during the major spawning season and recruitment of the subsequent year-class was not reduced at all (see Fig. 99a). During the retreat of El Niño in 1972–3 there was no period of cool water upwelling, for the temperature of the water remained high. This is the period when the main anchoveta spawning takes place and perhaps the larvae suffered from lack of food.

In the unexploited state, the stock of anchoveta may not be affected much by El Niño except by enhanced recruitment after the most intense events because of reduced predation by young birds. Presumably in 1940–41, the stock recovered from a decline after some years, and recruitment may well have failed in those years. However, an immediate effect of El Niño is to drive part of the stock south into waters that remain fairly productive. The main conclusion is that in an unexploited state the stock survives El Niño. During the ordinary events recruitment is perhaps normal because of normal upwelling during the retreat of El Niño. In the major events cool water is no longer upwelled during the spawning seasons and predation by young birds may be reduced.

When the stock of anchoveta became exploited, fishermen and birds competed for the fish available. The effect of El Niño reduced the bird population dramatically and indirectly, recruitment was augmented. We have speculated on the effects on the fish stock particularly upon the 1971 year class. Whatever the mechanism, the sharp fall in the recruitment of three successive year classes (1971, 1972, and 1973) reduced the stock so severely that under continued exploitation it has not recovered.

EVENTS SINCE THE 1972–73 EL NIÑO

In January and February 1976 an El Niño appeared, but by July a normal upwelling system replaced it and lasted right through the main spawning season, September, October and November. However, the 1976 El Niño was remarkable for very dense blooms of the dinoflagellate *Gymnodinium splendens* which produced a dense red tide (or *aguaje*). Large quantities of jellyfish, called *malagua,* were observed for the first time since 1917. The red tide extended from 15°S to 15°S, a very extensive area, and the chlorophyll concentrations reached 50–500 μg per l, a very high density indeed (in the 1972–73 El Niño the quantities of chlorophyll were as low as 0.2–0.6 μg per l). Off San Juan in the south of the region hydrogen sulphide was recorded below the poleward undercurrent. In September, October and November, with normal upwelling of cool water, no *malagua, aguaje,* or hydrogen sulphide was recorded. In 1977 the water was a little warmer than usual, as if the El Niño had resurged, but there was no red tide or jellyfish. This account was taken from Kesteven *et al.* (1977).

The El Niño of 1972 concentrated the adult fish in some cooler areas close to the shore and large catches were made and by September 1972, the stock was reduced to 2 000 000 tonnes. In 1973 and 1974, recruitment remained low, and catches reached as much as 1 500 000 to 3 500 000 tonnes. Such catches were only one-eighth or one-quarter of the maximum catch of about 12 000 000 tonnes, but they were a large proportion of the stock. However, by September 1975 the spawning stock had reached 4 000 000 tonnes and with a strong recruitment in the spring of that year total stock (not spawning stock) rose to 10 000 000–12 000 000 tonnes. During the first half of 1976 the growth rate of the anchoveta was reduced, the fat content from March to June was very low and there was some suggestion that the natural mortality increased. Catches of 4 000 000 tonnes reduced the spawning stock size to 6 000 000 tonnes in September 1976 which declined to 2 to 3 000 000 tonnes by the beginning of 1977, with very poor recruitment. A further decrease in stock size occurred when 1 000 000 tonnes of anchoveta were taken in the fishery in 1977. An examination of the stock/recruitment relationship shows that the best recruitment is produced from spawning stocks of 10 000 000 tonnes or more and if the spawning stocks are reduced below 5 000 000 tonnes, recruitment will be reduced. Thus the stock has not been allowed to recover. It would seem reasonable to close the fishery whenever the spawning stock falls below 5 000 000 tonnes. Another useful rule might be to close the fishery whenever El Niño occurs until it has been established that the recruiting year-class has not been reduced. In view of our present knowledge of the dependence of recruitment upon parent stock, it might have been desirable to have closed the fishery completely in the early 1970s.

Other changes have been taking place. Between 1966 and 1972 sardine catches ranged from 1100 to 6000 tonnes; in 1973, 130 000 tonnes were taken and between January and April 1977, 300 000 tonnes were caught. In 1972 there was a great increase in numbers of sardine larvae in the plankton samples and there were larger catches of mackerel and jack mackeral in 1973, hence a true increase in these stocks has probably taken place. In March 1976 and March 1977 acoustic surveys revealed a stock of 6 000 000 tonnes of pelagic fish, predominantly sardine, mackerel, jack mackerel and saury. This evidence taken together suggests that stocks of other pelagic fish have risen as the stock of anchoveta had declined. It need not surprise us, because the anchoveta stock has suffered from recruitment overfishing since the early months of 1972 and we expect stock switches as the result of recruitment overfishing—as anchovy succeeded sardine after the collapse of the sardine stock off California (Ahlstrom, 1966) and off South West Africa (Newman, 1977).

The most remarkable of all the changes has been noticed by Walsh (1981), who compared measurements made in 1966–9 with those in 1976–9. The most important observation was that the carbon content of the sediments on the Peruvian shelf had doubled during the decade; oxygen levels near the seabed had fallen from 0.2–1.0 to <0.1 ml O_2 per litre. Thus there was a "larger than normal organic loading" on the bottom of the shelf and on the upper slope, which implies an increase in the sedimentation rate by an order of magnitude. Because the quantity of copepods remained the same between the two periods, the reduction in the stock of anchoveta switched part of the primary production from food to sunken detritus. Euphausid production doubled, and the sardine yield increased considerably. What prevented the copepods from exploiting the primary production more fully? Dramatic changes have taken place since 1971, yet the response of the ecosystem has been slow. This is the first demonstration of the effect of fishing upon an ecosystem as distinct from a stock.

CONCLUSION

The stock of the Peruvian anchoveta, on which the former largest fishery depended, is governed always by panoceanic forces. The productivity of Peruvian waters depends on persistent coastal upwelling which itself derives from the position and endurance of the South Pacific atmospheric anticyclone. The periodic appearance of El Niño alters the productivity of Peruvian waters dramatically, and the transequatorial flow at the root of El Niño finds its origin in atmospheric and oceanic events on a transpacific scale. El Niño appears on average every five or seven years. The difference in pressure between the Indonesian Low and the South Pacific High, expressed as the Southern Oscillation, augments for two years or so with stronger easterly winds, and as a consequence water stacks up in the Western Pacific. When the winds abate, a Kelvin wave is generated in the equatorial thermocline, so the water stacked in the Western Pacific crosses to the East within the equatorial countercurrent system, floods the Eastern tropical ocean, and flows across the equator between the Galapagos Islands and the Americas. The fact that the wave travels quickly accounts for the rapid appearance of El Niño. The elucidation of the physical processes is a triumphant vindication of the beliefs of Schott and Bjerknes, extended and pinned with very great skill by Professors Wyrtki and O'Brien.

Until 1972 it was thought quite reasonably that the stock of anchoveta survived immune from the potentially adverse effects of El Niño. All that appeared to be affected was the availability of fish to the fishemen; serious though this was, capture continued after El Niño from an undamaged stock. Amongst other differences, an important one between a major and a minor event is that during a major event the intermediate retreat of El Niño leaves warm and unproductive water off Peru and recruitment of anchoveta may fail. During the intermediate retreat of a minor event, the water welling up off the coast of Peru is cool and productive. This is the major spawning period of the anchoveta, and if recruitment is determined at an early stage of the life history, the stock survives. Because stock was reduced in 1891 and in 1940–1, recruitment may have failed in those years.

When the fishery became heavy, the bird population was reduced in each of the 1957–8 and 1965–6 El Niños, presumably because there was no longer enough food for the birds, adults and juveniles. Remarkably, between the 1965–6 and 1972–3 El Niños, recruitment was enhanced, as if the predation by young birds had been reduced. Speculation suggests that the failure of the recruit class of 1971 might have been an indirect effect of this enhanced recruitment. The failure of the stock of anchoveta to recovery is probably due to poor management; the only secure way of obtaining recovery is to close the fishery down until it appears. As a consequence there is not enough food for the birds and the bird stock remains low. One of the most remarkable discoveries was the change in the ecosystem, the failure of the herbivores to respond to a greater availability of primary production.

Since the days of the great Scandinavian oceanographers at the turn of the century and the International Conference in Christiania in 1899, fisheries biologists have always hoped to establish links between physical processes and biological events. As might be expected, the links are complicated, but the physical point on which the biological processes depends is simple—the presence or absence of normal upwelling of cool water during the intermediate retreat of El Niño.

Chapter Eleven

The Biological Mechanisms

In general it is the recruiting year classes of fish populations that are most affected by climatic factors. We should, however, recall the death of the tilefish off New England and the deaths of fish and invertebrates in the waters around the British Isles during the cold winters. Fish have been killed in enormous numbers for unknown reasons in certain years in the north-west Indian Ocean, or Arabian Sea. Part of the anchoveta population off Peru is probably pushed to the south from time to time by El Niño; populations have moved further into the Baltic as the salinity there has risen and indeed others have retreated; during the recent period of warming, individuals and in some cases populations were shifted northward. The proximate causes of such events are disparate, but some are the consequence of changes in the current structure of the ocean.

However, the most dramatic changes in fish population in response to climatic factors are those in the magnitude of the recruiting year classes. As shown in Chapter 6 the variability of recruitment is high and that of the stock is much less. Indeed the function of a multi-age stock is conservative and that of the recruitment is exploratory, to sample the environment. Then the mean recruitment represents the success of the population in exploiting the environment. Trends in time of the magnitude of recruitment upward or downward represent the changes in fortune of the stock. This chapter will investigate the consequences of the effects of climatic factors upon the recruiting year classes of fish stocks.

SIMPLE RELATIONSHIPS

The recruitment to the West Greenland cod stock increased when the water offshore became warmer, perhaps because they might not have survived easily in the cooler waters characteristic of the period before the augmentation. However, the increment was probably also associated with increased flow of the Irminger Current and a greater chance that juvenile cod would cross the Denmark Strait between Greenland and Iceland. The simple event was probably the advection of relatively warm water from Iceland which also warms the West Greenland offshore waters. As shown in Chapter 5, there are many cases in which recruitment is related to temperature, directly or inversely. When an animal lives at the poleward edge of its range it can only reproduce satisfactorily in warmer water; in the *Octopus* plague in the western English Channel in 1950, the water was warm enough in summer off Brittany for the *Octopus* to breed, a direct and simple effect of temperature. As part of the match/mismatch hypothesis described below, the inverse dependence of recruitment upon temperature might be a function of delayed development; that of egg and larval stages is an inverse power function of temperature and very cold water delays development considerably and perhaps increases the chance of finding food in the variable production cycle in temperate and high latitude waters. Many inverse dependences of recruitment upon temperature have been described and they may indicate the match/mismatch relationship.

The apparently simple correlations of recruitment upon some convenient physical factor are probably always indices of more complex relationships. However, one relationship is genuinely simple, that relating recruitment to primary production. If recruitment is determined at an early stage in the life history, it might well depend directly upon food closely related to the primary production. Longhurst (1971) reported that certain clupeids spawned in the upwelling plumes in certain systems, *Cetengraulis mysticetus* Günther in the Gulf of Panama, *Sardinella anchovia* Valenciennes in the Gulf of Cariaco, *Sardinella aurita* Valenciennes off Ghana and *Sardinella longiceps* Valenciennes on the Orissa coast of India. They may well do so to maximize recruitment as their poleward analogues do by spawning at a fixed season close to a production cycle which is variable in time of onset. The array of correlations in the Californian upwelling area penetrate many parts of the ecosystem. High recruitment of tuna-like fishes depends on the open ocean divergences of the Pacific, and El Niño reduces the recruitment of the Peruvian anchoveta when the productivity of that warm current is diminished. There are three current theses on the dependence of recruitment upon parent stock: (1) that larval and juvenile mortality is density dependent (Beverton and Holt, 1957); (2) that density-dependent mortality is caused by the aggregation of predators (Ricker, 1954, 1958);

(3) or that density-dependent mortality is generated by food lack (Beverton and Holt, 1957) or density-dependent growth (Shepherd and Cushing, 1980). If theses (2) and (3) were right we would expect recruitment to depend upon food and perhaps upon primary production at a constant level of stock.

Steele's (1974) model of a North Sea food web (Fig. 110) is based on observations of primary production, records of fish catches, and good estimates of the transfer rates. The essence of the model is that material is transferred from primary production to carnivores by herbivores and that the herbivore faeces are transformed directly and quickly to the benthos. Thus the herbivores play a central part in the system. All the larval and juvenile fish are grouped as pelagic fish, although the numbers used by Steele were based on catches of adult fish. The point is that the food chain, primary production, herbivore and pelagic fish (with larval fish of all species) is a simple one, although an array of species would constitute a more complex chain. To some degree, differences between species are predetermined in the egg production of the conservative stocks, so the food chain tends to retain its simplicity.

Cushing (1982) suggested that there are three sources of variability in Steele's model (1) the initial variation in primary production (2) a change in the ratio of invertebrate to vertebrate carnivores following change in the ratio of *Pseudocalanus* to *Calanus*, and (3) changes in the species composition of the fishes, as the time of onset of the production cycle changes in time

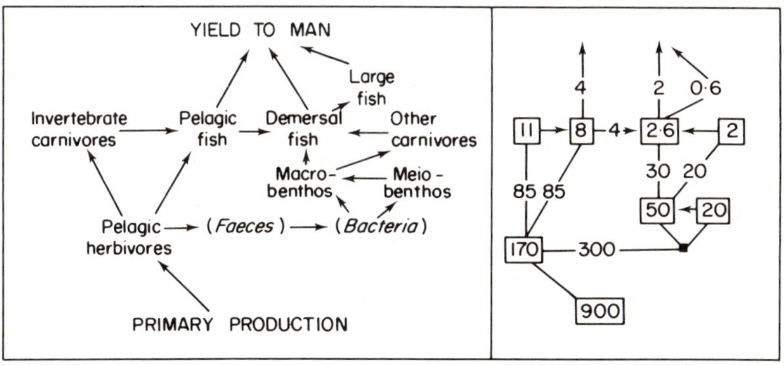

Fig. 110 Steele's (1974) model of the North Sea food web, based on observations of primary production, records of fish catches and good estimates of transfer rates; the numbers in the squares represent production in kcal $cm^{-2}yr^{-1}$ and the numbers across the arrows represent the quantities transferred.

in a systematic way, as will be described below. From recent experience in the North Sea (see Chapter 7), the two latter sources of variation take place slowly on a decadal scale (see Figs. 65 and 75), whereas the differences in primary production of importance are annual, even if trends might occur between decades. Figure 59 showed the annual variation of primary production in the western English Channel for a decade, variation by a factor of three. Smith and Eppley (1982), with a correlation between primary production and temperatures off Scripps' Pier for 16 years, have used the much longer time series of temperature observations to construct a derived series of primary production in which some trends may be seen. The variability of primary production will differ from region to region, by depth of water or by latitude. In the upwelling areas and in the divergences of the open subtropical ocean, there are plumes and patches where recruitment may depend directly, but locally, upon primary production. Unless fish find ways of aggregating on to such plumes and patches, the dependence of recruitment upon primary production may have also a component of random variability. Conversely, high correlation in the upwelling areas might suggest that fish do aggregate on to patches of their food. Thus in the subtropical ocean recruitment may be a direct function of primary production merely because the food chain is simple and the food web of competitors is sustained by their relative population fecundities. Recruitment remains variable, and one source of variation may originate in the serial spawning complex of aggregations as fish search for food in patches.

THE MATCH/MISMATCH HYPOTHESIS

Cushing (1970, 1975) put forward the hypothesis that the production of fish larvae was matched or mismatched to the production of their food in temperate waters. In subtropical seas, the spawning of fishes is linked to upwelling areas or open sea divergences because they feed and spawn at the same time. In temperate seas, fish spawn at a fixed season in order to maximize the chance of larval survival in the face of a highly variable time of onset of the production cycle. Cushing (1967) had associated the distribution of herring spawning grounds in their particular seasons in the North Sea and the waters to the west of the British Isles with the detailed distribution of the production cycles in time and space. In other words, a stock releases its larvae into the local production cycle at the best time to secure good survival on average. The time of onset of the production cycle in spring may vary, for example, by as much as six weeks in British waters (Colebrook, 1965). The evidence for the variability in timing of the spring production cycle is taken from the material collected by the Continuous Plankton Recorder Network. Colebrook and Robinson (1965) and Robinson (1970) related the variability in time of onset of the spring outburst in the North Sea and the north-east Atlantic to estimates of the degree of stabilization of the water column. Colebrook (1979) extended the observations to the whole of the North Atlantic. The average date of production for each region between January and June, weighted by abundance, was taken as the time of onset of the spring production cycle. In the North Atlantic between 50 and 60°N in the open ocean and between 40 and 65°N near the continental shelves, the time of onset varied from the end of March to the end of May. This result, a second form of variability, is the consequence of systematic differences across the ocean between different regions. Then the delay in production between 1948 and 1972 in the North Sea and in the north-east Atlantic is the third form of variability in the timing of the spring outburst. The time of onset of the spring outburst in temperate waters varies from year to year at one position, varies from position to position, and may trend earlier or later in time across the decades.

The evidence that fish spawn at the same season in temperate waters each year is given in Cushing (1969). The spawning seasons lasts for up to three months, but the peak date varies little from year to year. But the long spawning period is itself an essential part of the process. If the time of onset of the production cycle varies by six weeks and the spawning season lasts for three months, then some larvae must survive in the most adverse arrangement.

A total of twelve egg surveys of the plaice spawning ground in the Southern Bight of the North Sea have been carried out since 1911. The peak date of spawning was January 19 with a standard deviation of less than a week. Grab samples of herring spawn were taken for some years on spawning grounds of the Atlantoscandian stock south of Bergen on the coast of Norway during the 1930s. Each ground is small and discrete and can be classified as a "late February", "early February" or an "early March" ground. From the series of observations made, the standard deviation of the peak date of spawning was low, less than a week. The sockeye salmon spawns on its native redds up to 1300 km from the sea in the Fraser river network, in British Columbia. Spawning was seen by the river bailiffs and recorded for a set of 51 sites for a period of seventeen years; the standard deviation of the peak date of spawning was yet again less than a week. The

Arctonorwegian cod stock spawns in the Vestfjord in Northern Norway between late January and early April; catches were recorded by weeks and the peak date of catch was taken as the peak date of spawning. Figure 111 shows the peak date of catch for each year between 1894 and 1967. Within a period of about a week or ten days, the peak date has been retarded, and the period of retardation is very roughly that of the recent period of warming. Thus there is slight evidence that a fish stock can alter its peak of date of spawning, but the nature of the change remains obscure. Deviations of peak date from the moving mean (of tens) are low, less than a week. The trend itself is less than ten days, probably a small proportion of the variation in timing of the production cycle in that area; it is unknown but it is probably greater than that in the North Sea. This is the evidence that commercial fish stocks in temperate waters spawn at a fixed season each year. Although the evidence is detailed, only four stocks were examined because evidence of the time of spawning in time series in temperate waters is lacking. The essential point is that the variation in peak date of spawning is low as compared with the variability in timing of the spring outburst. Then if the production cycle becomes delayed (Fig. 112) we would expect recruitment to a fish stock to increase or decrease as it became matched or mismatched to the production cycle. When the spawning of the sardine, *Sardinops caerulea,* and the spread of the northern anchovy, *Engraulis mordax,* were examined, the peak date of spawning was found to extend for months, not for a period as short as a week. Further, the larvae of tuna species are distributed right across the subtropical Pacific for nine months of the year

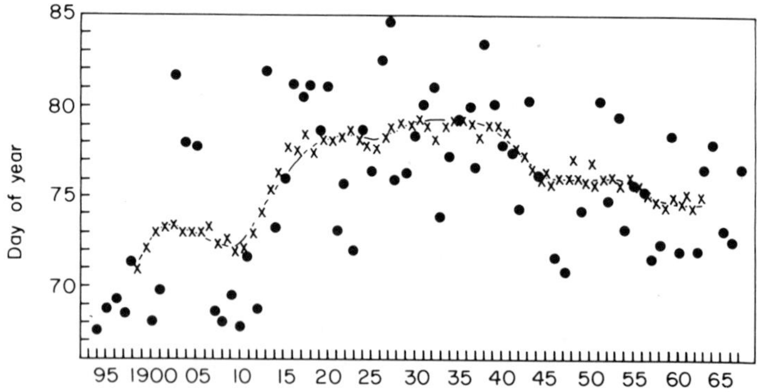

Fig. 111 Peak dates of catch of skrei, or large cod, in the Vestfjord in Northern Norway between 1894 and 1967 (Cushing and Dickson, 1976), fitted by moving means of tens.

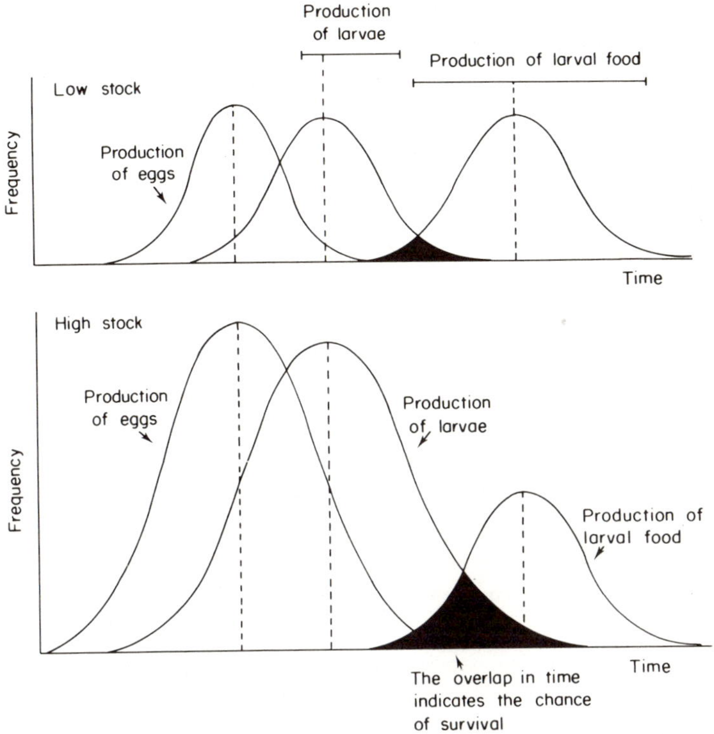

Fig. 112 Principle of the match/mismatch hypothesis. The error of production of larvae is asymmetrical in time because development is an inverse power function of temperature. That of larval food is symmetrical but, in time, broad. The overlap of the two distributions indicates the chance of match (or mismatch) of larval production to that of their food. The distributions are presented at high stock and at low stock; the chance of overlap is greater at high stock, where the variance of recruitment is observed to be less.

(Matsumoto, 1966). Hence, the match or mismatch of larval production to their food may be confined to waters poleward of 40° latitude.

There are two distinct processes: (1) the production cycle occurs at a given time in response to physical factors as described in Chapter 6; (2) the production cycle is advanced or delayed as the same physical factors combine on a time scale of decades. In the first process the production of fish larvae in time should be matched or mismatched to that of their food; if matched, recruitment would be high within the limits of variation of the primary production. If mismatched, recruitment will be low—more so if primary production is low, but less so if it is high. The principle is illustrated in Fig. 112 at high stock and at low stock. The error band on the peak date of the production of larval food estimates the variation in time of onset of primary production on the assumption that the latter is linked directly to herbivore production. That on the production of larvae is that of larval development as an inverse power function of temperature: in warm water development is slightly quickened, but in cold water it is sharply delayed. In the cold winter of 1963, the year class of plaice in the North Sea hatched in that year was the largest on record; in the other cold winters which occurred since records on plaice of the southern North Sea were collected, the year class of 1929 was also high, but not that of 1947. If plaice spawn on January 19, on average their larvae are hatched in mid February and the production cycle occurs between then and April. Then if the production of plaice larvae is delayed considerably, there is a greater chance of their production matching that of their food. Lastly Fig. 112 shows that at high stock the overlap between larval production and that of their food is greater than at low stock, implying that the variability of recruitment is lower at high stock than at low stock, as is observed (see for example, the array of stock/recruitment curves given in Cushing and Harris, 1973).

In the second process, the production cycle is advanced or delayed in a decadal trend such that match or mismatch of production of larvae to that of their food is approached across the years. Figure 78 shows the dependence of North Sea cod recruitment upon the index of delay of *Calanus* production in the North eastern North Sea. In 1963 cod recruitment increased sharply compared with previous years and has remained high since, yet the correlation extends from 1954 to 1978 across the time of the gadoid outburst. Because recruitment is correlated with an index of delay, it is possible that the production cycle was delayed enough in 1963 to provide a better match of larval production to that of their food and in subsequent years the delay has remained. This suggestion cannot be demonstrated until the physical factors throughout the period have been analysed with the methods developed by Woods (1980).

Another of the great events in the twentieth century was the rise of the cod stock in the Baltic from perhaps ten thousand tonnes or so in the 1920s to several hundreds of thousands of tonnes in the 1960s and 70s. Salinity had slowly increased above and beneath the pycnocline which as a consequence became somewhat shallower. In the Baltic the cod lay their eggs near the bottom and they rise towards the surface. Hence the larvae probably appeared near their food somewhat earlier in their lives than they had in earlier decades. That is, the production of the cod larvae was matched more closely to that of their food, merely by a change in the depth of the halocline. But the production cycle in the Baltic must vary in time of onset much as it does elsewhere. It remains possible that the physical control of the time of onset of the production cycle was independent of the increase in salinity and that there was an improved match for other reasons.

The match/mismatch hypothesis implies a phenomenon restricted to temperate and high latitude waters, that is, poleward of 40° latitude. In subtropical seas, recruitment may be linked directly to primary production as indicated for the upwelling areas and for the offshore divergences in Chapter 6. Spawning seasons are much longer (up to nine months for some tuna-like fishes) and fish spawn serially; feeding and spawning proceed at the same time. Spawning is possibly linked to food. Animals may aggregate in the upwelling plumes and in the divergences far offshore, and if they spawn there, they ensure the survival of their offspring for a limited but critical period in the future. In temperate and high latitudes, production stops in winter and is concentrated in the spring outburst; although it continues in summer particularly within the thermocline and in the autumn, the greatest transfer of material takes place in spring, when most of the important commercial fishes spawn (exceptions are the autumn spawning herring and the summer spawning sole). The variability of the production cycle in time of onset may be considerable, and the spring outburst may advance or delay across the decades and the centuries. Therefore to maximize the chances of exploiting such variation in time of onset of the production cycle a fish stock should perhaps spawn for a long season, comprise long-lived individuals, and spawn at the same time each year.

THE DEPENDENCE OF RECRUITMENT UPON PARENT STOCK

The dependence of recruitment upon parent stock or the stock/recruitment relationship is often little more than a statistical arrangement to establish whether fishing has affected the magnitude of recruitment. It is much more than that, if only because the variability about any fitted curve is almost certainly governed by climatic factors. Further, it expresses the way in which the stock survives by converting the food available to growth, in the face of death. The logarithmic form of the Ricker curve is $\ln(R/\alpha P) = \beta P$, where P is the stock of eggs produced, R the number of recruits, α the coefficient of density independent survival and β the coefficient of density dependent mortality. Rounsefell (1958) used it in this linear form to analyse the deviations from the regression of $\ln(R/P)$ on P to investigate the effect of climatic factors upon the Karluk River sockeye salmon stock. He found that the autumn water temperature affected egg survival, that larval survival was affected by the length of growing season, that spring temperatures governed the emergence of the fry, that rainfall affected their survival, and that the numbers of fry were governed by numbers of two- and three-year-old fish. These are possibly not the crucial factors, but the method may reveal those which deserve further analysis.

The coefficient of density independent survival, α, estimates the average effect of climatic factors upon recruitment during the period of data collection. Figure 54 shows upward and downward trends in time, probably distinct from any effect of exploitation (with the exception of the Downs stock of herring after 1955 and the Atlantoscandian stock of herring after 1958). Such natural trends in the magnitude of recruitment must indicate changes in density-independent survival. In the Ricker equation any decrement in α must lead to a decrement in β; if density-independent survival is reduced, so is the density-dependent mortality. It is a truism that the fecundity of the population represents a mortality between the stock of eggs and subsequent recruitment. Hence if one component of mortality is increased, the other must be reduced. A change in α alone alters the slope at the origin but does not alter the shape, which is governed by β, the coefficient of density-dependent mortality. Cushing (1979) described the changes by generating recruitment from stock and then composing stock from recruitments over a model period of fifty years. The effect of the fixed fecundity was simulated by passing all curves through a central point in stock so that any change in α generates a change in β. Figure 113 shows the result; as the initial slope, α, is reduced, the coefficient of density-dependent mortality is reduced. A family of curves pass through the central point and the vertical bars represent the standard errors of the means of fifty years of model observations. Thus with fixed fecundity, any change in density-independent survival, due for example to climatic factors, alters the shape of the curve. The dome-shaped curve represents the behavior of a stock which responds quickly to a change in recruitment, whereas the near linear one represents that of one which responds slowly. The first is resilient because it is to some degree insulated from environmental changes, or indeed from changes due to fishing. In contrast, the second responds

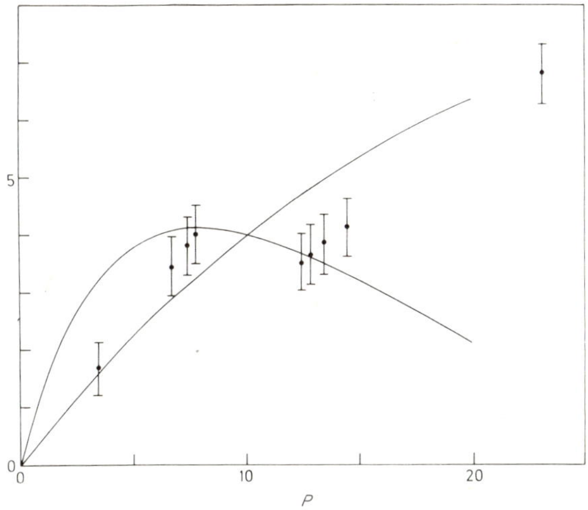

Fig. 113 The change in shape of the Ricker stock and recruitment curve as the initial slope is reduced; with reduced initial slope, the density-independent survival, density-dependent mortality is reduced, because fecundity, the capacity for total mortality, is fixed. Fixed fecundity is represented in the model by passing the family of curves through a central point in recruitment as a central point in stock (Cushing, 1979).

to environmental change and is vulnerable to recruitment overfishing. Cod-like fishes, which are very fecund are examples of the first, the resilient group; herring-like fishes, which are considerably less fecund, are examples of the second, the responsive group. The remarkable point is that a decrease in larval survival makes the stock less resilient, less able to resist environmental change, and only able to respond to it. Further such a stock becomes more vulnerable to recruitment overfishing. Such are potential changes. It is not yet known what changes in density-independent survival are in magnitude and what their links are with climatic factors. All we can say is that such changes are proportions of the fecundity of the adult.

Cushing (1979) expressed similar results with the use of the logistic curve ($N_{t+1} = N_t \exp[r(1 - N_t/K)]$, where N_t is the number in the first generation and N_{t+1} that in the second, r is the rate of natural increase, and K the carrying capacity of the environment, the food). The relation of numbers to carrying capacity is given by $(1 - n_t/K)$ and the exponent $(-rN_t/K)$ expresses density dependence. The same result appears—as the rate of natural increase is decreased, so are the density dependence and the shape of the curve; in the same way the stock becomes less resilient. As r is reduced from 1.5 to 0.22, the characteristic return time T_R ($= 1/r$) increases from 0.67 to 4.55. The return time expresses the capacity of the population to return to equilibrium and an increase in return time implies a loss of resilience. It is a shift from a K-strategy, a self-stabilizing system, to an r-strategy, an opportunistic system. If r is reduced by climatic factors, the system becomes less resilient and more opportunistic. To some degree the result is that given with the Ricker equation but expressed in a different framework. However, climatic factors might affect K as well as r. The results are the same, but differences in r can only be modulated by changes in K, even if food is probably that at a critical stage in larval or juvenile life history and not in the adult stages at all.

Figure 114 shows another application of a stock/recruitment curve. It is a Ricker curve of recruitment upon parent stock about a bisector. Where the curve intersects the bisector, the stock replaces itself in the unexploited state. The variability of recruitment is indicated at two stock magnitudes. Normally any recruitment will tend to return the stock to the point of replacement. At low stock, however, a sequence of low recruitments will tend to reduce stock far below the average levels. At high stock, a sequence of high recruitments will tend to augment stock far above average levels. Thus the system of stabilization which we call the stock and recruitment curve contains within it the capacity for large changes of stock magnitude like the rise of the Atlantoscandian stock of herring with the 1904 year class. The Plymouth herring stock collapsed naturally in the 1930s. I

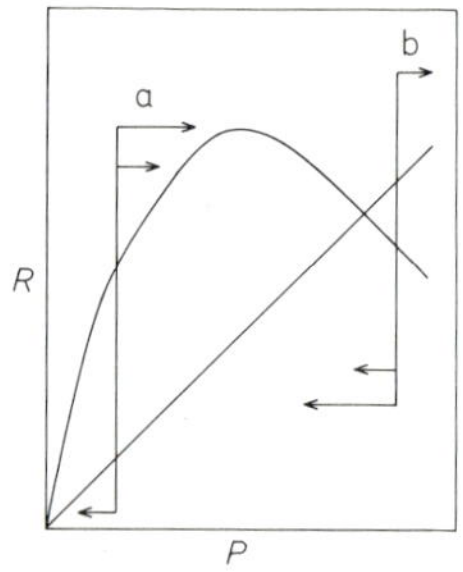

Fig. 114 Changes in recruitment at extremes. If a sequence of recruitments occurs below the bisector at low stock, the stock must decline dramatically; if such a sequence takes place above the bisector at high stock, stock must increase dramatically. In this way great changes in stock can occur as if the system were being rectified (Cushing, 1969).

believe that climatic factors can generate sequences of high or low recruitments which induce the very large changes in stock. In this way the very long term changes characteristic of herring-like fishes can take place. Because changes take place quickly after long periods of apparently stable stocks, Cushing (1966) used the word *rectification* to describe them. The relationship given in Fig. 114 puts the phenomenon in a framework of population mechanisms, but the real question is why low or high recruitments can appear in long enough sequences to cause the rectified changes.

CONCLUSION

The response of a fish stock to climatic factors and their changes is also its form of stabilization. The high variation of recruitment in an unexploited stock is its capacity to sample the food available. Because recruitment is usually determined at an early stage in the life history (see, e.g., Rauck and Zijlstra, 1978 on the southern North Sea plaice), it may depend on food available for larvae and postlarvae just after metamorphosis. These food organisms are those that grow in the production cycle, the spring outburst of temperate and high-latitude waters, and in the upwelling areas and divergences of subtropical seas.

If the magnitude of the recruiting year class is determined by the amount of food available at an early stage in the life history, the stock is stabilized by that process. The effect of climatic factors upon the quantity of food is really part of the same system. There is a single process by which the quantity of food is modified by climatic factors and at the same time determines the year class strength and hence the stability of the stock at one level or another. This is because recruitment is probably determined at an early stage of the life history when the quantity of available food is governed by the physical and biological factors at work in the production cycle.

In the upwelling areas and in the divergences of the open ocean there is some evidence that recruitment is linked to primary production. If such a link is real, the food of the larval or post-larval fish depends upon primary production and to some degree it determines subsequent recruitment. The variation of recruitment in such stocks may be relatively low, by a factor of three or so. In the seas poleward of 40° latitude the variation of recruitment is much greater, by an order of magnitude or more. The increment in variability may be no more than a consequence of match and mismatch.

The rise in the cod stock in the North Sea may be associated with the delay in the production of *Calanus* in the north-east North Sea (see Fig. 78); it might even be a demonstration of the hypothesis that recruitment depends upon the match of larval production to that of their food. Common to both systems on either side of 40° latitude is the possible dependence of year class strength upon primary production. That in subtropical seas may vary in the spatial distribution of upwelling plumes and divergences whereas that in the waters of temperate seas and high latitudes may vary most in time.

Consequently the physical factors which govern recruitment are those which determine the magnitude of primary production, the places where it is found in subtropical seas and the times when it occurs poleward of 40° latitude. The positions of upwelling are governed by the strength and direction of the longshore winds, which are themselves a function of the position and magnitude of the subtropical atmospheric anticyclones. The angle made by the coastline to the axis of the longshore wind is also important, as is the presence or absence of capes along that coastline because upwelling always appears off a cape. The important factors in the seas poleward of 40° latitude are those that determine the critical depth and the subsequent depth of mixing, that is the wind strength and direction together with the degree of heating averaged on a diurnal basis, added to which is the rising angle of the sun as spring advances.

Common to both systems, that of subtropical seas or those poleward of 40° latitude, is the generation of the primary production from very low levels. In this point the production cycles of the upwelling areas and of temperate seas resemble each other. The quantity produced is governed partly by the physical factors indicated above and partly by the degree of control by the herbivores. If there is no control by the grazers, production is stopped by nutrient lack, whereas if control starts early in the cycle, the reduction in nutrients is the consequence of herbivorous activity. In any particular production cycle, prediction of the magnitude produced is somewhat complex. More importantly, it can only be attempted when the precise role of the physical factors becomes established. Many models have simulated production cycles, but none is yet well enough described for predictive purposes.

To forecast the recruitment of fish stocks, components of variance have to be added to the system, spatially and temporally as described above. The fact that correlations can be obtained, such as those in upwelling stocks or as in the North Sea cod stock referred to above, suggests that the differences from year to year are enough to account for part of the variation in recruitment. To do so more fully, however, would require the assembly of a theoretical apparatus more extensive than a simple correlation.

One of the most interesting results from this study was the effect of climatic factors on the resilience of the stock. As the rate of natural increase is reduced by environmental factors, the stock becomes less resilient, less self-stabilizing, and more opportunistic. If the rate of natural increase is enhanced by environmental factors, the stock acquires more capacity for self-stabilization as density-dependent factors become more pronounced. This is the common sense of perturbations lasting for periods up to a decade. Changes in amelioration or deterioration, however, may endure for a number of decades. In periods of deterioration at a given position, stocks will become opportunistic and some may be reduced in magnitude below the noise level. However, they may become replaced by other species emerging from that noise level of numbers, which themselves become more resilient in the modified environment. Such may be the exchanges between anchovy and sardine recorded by centuries in the anoxic sediments off California.

There are three interlinked processes: (1) the generation of recruitment within the complicated structures of the cycle of primary production; (2) the match or mismatch of larval production to that of their food in waters poleward of 40° latitude; and (3) the replacement of species by more resilient ones in decades of climatic changes. Species can be replaced in the match/mismatch process, but this is a distinct one based on the stabilization mechanisms rather than upon the variation in time of onset of the production cycle. To gain a detailed understanding of the generation of recruitment the effect of each process must be disentangled and the cogent physical factors isolated.

Chapter Twelve

The Effects of Climatic Changes upon Management

At the present early stage of development, decisions on management are made from year to year and climatic changes are elucidated by decades, so climatic changes do not modify management decisions. On the other hand, an opinion on the possible nature of an observed change may make managers cautious or optimistic. In this chapter past events will be reinterpreted somewhat in terms of contemporaneous climatic events. Often they were not recognized until later, and sometimes (but not always) a management decision might have been different. The events will be recounted in terms of the vital parameters of mortality growth and recruitment.

MORTALITY

Had the tilefish been under management control in the 1880s, the catastrophic death of perhaps all the population on the *Lopholatilus* bank in 1882 would have raised no problems at all. Fishes can be killed in masses by oxygen lack, by red tides, or by low temperatures. Although the event is usually sudden and the only problem facing the managers is to assess the proportion of stock killed, the death of adult soles during a cold winter in the Southern North Sea, the Pits winter, may have particular consequences for management. Many fish are killed directly, but others become more available to capture as they migrate into deeper water. Hence both sources of mortality—fishing and natural causes—might be affected and so potentially the Total Allowable Catch (TAC). Locally the death rate was high, up to 13% for perhaps a month or so; if the winter was intensely cold, the area of dead fish was extensive (Fig. 84). For a third of the year, however, stock density of soles increased by at least a factor of three and probably more in some localities; the fish were available to capture both during the day (when they are usually caught) and during the night. Whereas the natural mortality might have been increased above its conventional annual value by only a little, the increment of fishing mortality was probably significant. A cold winter is a little difficult to recognize in the initial stages, but the measurable effects on the sole population do not appear until the event is obvious. One might expect that in the future, investigations on the death rates of sole during the cold winters might be needed for management purposes. Other fishes die during the cold winter, but so far as is known their catchability does not increase.

Cold winters kill oysters and other shellfish in European waters. Fish and shellfish are killed wherever the water becomes much too cold for them, but if their stocks are to be managed adequately the death rates should be estimated. Other forms of mortality, such as death by predation, will become important in future investigations: predation might be reduced during the cold winters.

GROWTH

The growth of fish depends primarily upon the amount of food available, although gross differences in temperature modify it positively or negatively. In management today, growth is described in annual increments of weight taken from the sampling systems. Consequently growth is sometimes not studied for its own sake, and it is possible that significant changes are sometimes not observed. Of course the study of growth remains a proper one for interested biologists, but it is no longer an essential part of the processes of management, although it would need only the smallest tweak of a programme to set that right.

In the Central North Sea, the quantity of *Calanus* available to the juvenile herring increased by a factor of three in 1950, as shown in Chapter 7. Burd (1962) showed that the mean length of three-year-old fish, depended positively upon the numbers of *Calanus* per unit volume (averaged during the three years of growth). There were two consequences to the Downs stock of herring: (1) l_∞, the length at infinite age, increased from about 28 cm in 1933–49 to about 31 cm in 1958 (Burd, 1962) and (2) between 1950 and 1955, recruitment changed from one partly at three and partly at four years of age to one wholly at three years of age. The increment in weight caused an increase in yield per recruit of nearly 20% (Figure 115a shows yields per recruit and Fig 115b theoretical stock density, yield per recruit per unit of fishing mortality for five year classes). The effect of the reduction in age of recruitment was to expose the year class to fishing mortality six months early and hence to increase the effect of fishing on the stock to a small degree. During the early 1950s the fishing mortality on the Downs stock increased with the augmentation of growth, so the catch per effort curve would have shifted across the year classes shown in Fig. 115. Hence the decline in catch per unit of effort with time or with increasing fishing mortality masked the true decline in numbers. The point is that the yield per recruit model is a steady state formulation and changes in parameters with time should be accounted for, when management decisions are made.

In fact, however, at least from the mid 1950s onwards, the Downs stock suffered from recruitment overfishing (Cushing and Bridger, 1966; Cushing, 1968), and the subsequent decline in recruitment dominated the reduction in stock density. The rise of the industrial fishery for young herring was probably associated with the growth change and hence the full potential of recruitment was never realized. The collapse of the Downs stock of herring by recruitment overfishing was probably due to the combined effect of adult and juvenile fisheries, initially masked to some degree by the natural changes in growth and recruitment but in fact exacerbated by them.

RECRUITMENT

Changes in recruitment, up and down, are the largest changes in a fish population; we recall that the last Norwegian Herring period started with the 1904 year class of the Atlantoscandian herring. The gadoid outburst, however, was not really recognized by fisheries biologists for some years, perhaps because the recruitment to the haddock stock remained highly variable; the high year classes were those of 1962, 1967, 1974 and 1980, but it was the cod stock that sustained a sequence of strong recruitments. But in general an increment of recruitment which is sustained poses few problems for management: the TAC is joyously adjusted upward. However, there are difficulties. Cushing (1981) analysed the changes in the cod stock at Iceland and was able to separate the Iceland and Greenland components in the fishery at Iceland; see Chapter 5 for a description of the two components. Figure 49 shows the Iceland recruitment (as numbers of fish at

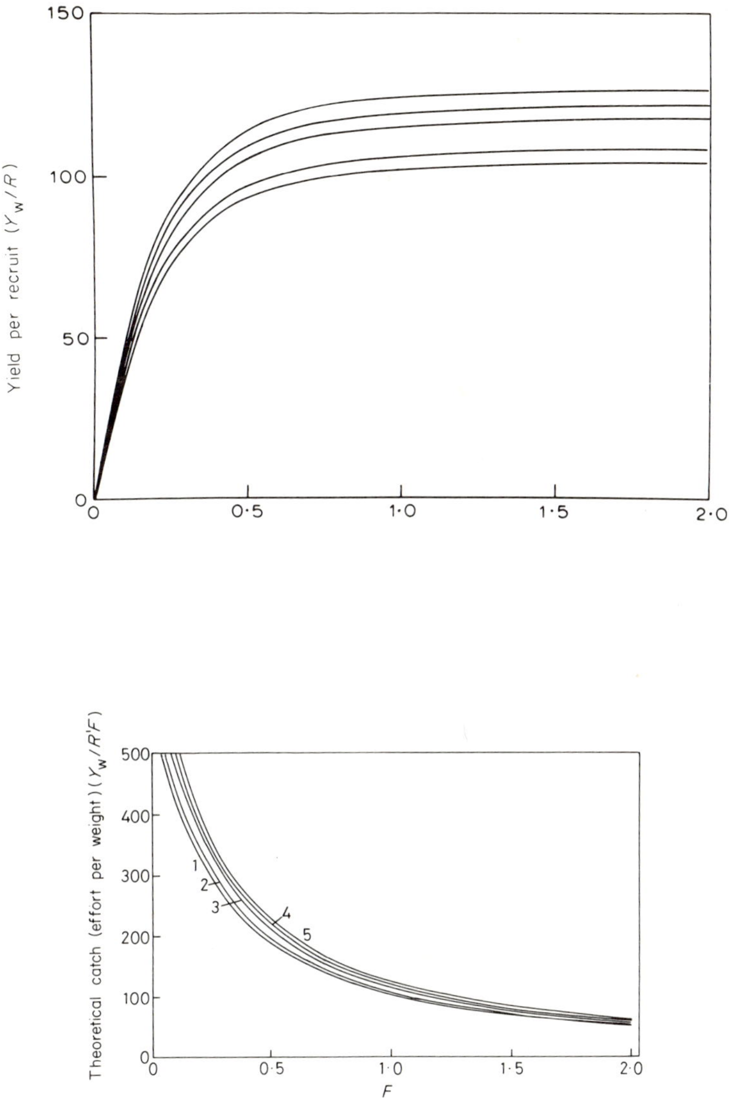

Fig. 115 Yields per recruit (*a*) and theoretical catches per unit of effort (*b*) for five year classes ((1) = 1932; (2) = 1943; (3) = 1953; (4) = 1947; (5) = 1950) of the Downs stock of herring (Cushing and Bridger, 1966).

age four) and Figure 48 shows the immigrant recruitment from Greenland as outstanding year classes in the spawning fishery at eight years of age, distinct from those spawned off Iceland (see pp. 143). Figure 49 shows the remarkable increment of year classes, 1931, 1932, 1933, 1934 and 1935 at Iceland, and this succession must have enhanced the stock for more than a decade afterwards. The important point about Fig. 48 is that the predominant year classes of 1942 and 1945 at Greenland, might have originated in the high Icelandic stock generated by the five successive high year classes about a decade earlier. The two figures show the response of the two cod stocks at Iceland and Greenland to the recent period of warming (see Chapter 5).

After 1945, the highest year classes no longer appeared. Indeed although the lesser recruitments from East Greenland may have increased a little, those at Iceland were perhaps a little lower. Hence during the late 1950s and early 60s, the stock decreased naturally and independently of fishing. Between 1940 and 1970, the exploitation rate increased by a factor of 2.5 (Cushing, 1982a). Icelandic fishermen looked back to a high period in the 1940s and 50s. There were disputes between the Icelanders and the British concerned with the right to fish in Icelandic waters which were defined in different ways successively, 6.4, 19.3 or 322 km from the baselines. The differences might have been less severe had either side known that the stock had declined naturally during the recent period of exploitation after 1945. Thus, during the recent period of warming there was a golden age well recognized by Icelandic fishermen, and their expectations were never sustained. Climatic changes affect management directly as shown above, but also indirectly through the opinions held by the managers.

A decline in recruitment poses a particular problem for managers. Figure 116 shows the decline in recruitment and in subsequent catches per unit of effort of the Plymouth herring. The failure in recruitment from 1926 onwards is part of the Russell cycle described in Chapter 5. Obviously, had the stock been managed, the correct procedure would have been to exploit the stock mercilessly from 1930 onwards. But this interpretation is based on hindsight from the interpretation of the Russell cycle made decades later. In 1930, a manager would not have been able to distinguish the natural decline from one due to fishing. Nor could he today, when recruitment declines and when fishing mortality is high at the same time, but the distinction sometimes emerges subsequently. Of course a prudent manager must suspect recruitment overfishing and frequently he is right.

Figure 117 shows the decline in recruitment of the cod stock at West Greenland during the 1960s which led to a zero TAC being enforced by

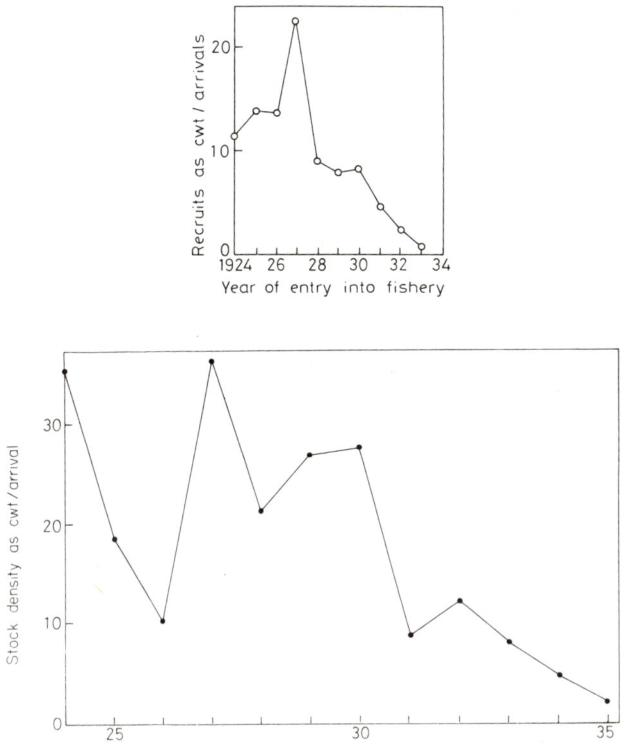

Fig. 116 Decline in recruitment (*a*) and in stock density (*b*) (as cwt per arrival) of the Plymouth herring (Cushing 1961).

ICNAF in the late 1970s. Fishing mortality during the 1960s ranged from 0.3 to 0.6, but recruitment declined after 1963. No dependence of recruitment upon parent stock was published, and the important point was that the managers concluded that the decline in recruitment was not the consequence of a decline in stock due to fishing (but after 1974 they tried to increase the biomass of spawning stock). The zero TAC was an attempt to preserve the stock until recruitment recovered. The opinion revealed by the action taken was that it was a short term fluctuation and that the stock would recover.

Recruitment has declined in many stocks, often under the pressure of fishing. The fishermen who reduced the stock may claim that a natural change has occurred. However, this is only part of that cut and thrust towards the establishment of good management. We are concerned here with cases where the decline in recruitment was of natural origin and was shown at the time or afterwards, even if long afterwards.

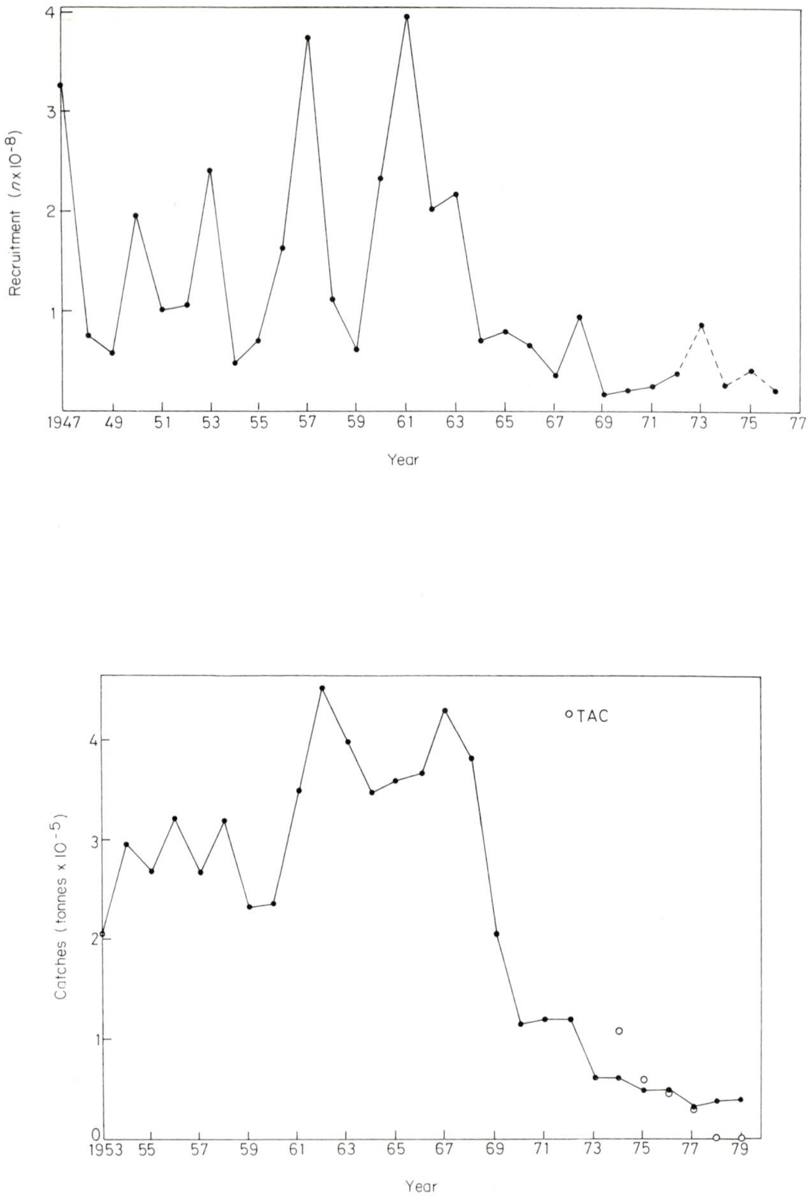

Fig. 117 Decline in recruitment to the West Greenland cod stock (Horsted 197 , 1976) (*a*) and decline in catches. TACs are indicated (○); zero TACs existed in 1978 and 1979, but some fish were landed as by catches and the Greenlanders were permitted small catches.

The most dramatic fall in recruitment was that of the Peruvian anchoveta during the 1972–3 El Niño. As recorded in Chapter 10, there was also a failure in 1971, which may have been independent of El Niño, but the point cannot be decided. Figure 99a shows the course of El Niño as changes in surface temperature off Peru with time in 1957–8, 1965–6 and in 1972–3. The advance, regression and the return in each El Niño is clearly shown. In 1957–8 and in 1965–6, the water during the regression was of normal temperature which means upwelling was productive. In 1972–3, the regression was weaker, the water being warmer than normal because El Niño water was being upwelled, which was unproductive. The anchoveta spawn during the regression and if the water is productive recruitment may remain near the average; however, if the water is unproductive, recruitment probably declines, as indicated in Chapter 10. The time of onset of El Niño can be forecast, and in the future it might become possible to predict whether the event would be as intense as that of 1972–3. The collapse of the anchoveta stock was predominantly due to the natural failure of the 1972 and 1973 year classes.

The action taken by management at the time was unsuccessful in the sense that the stock has not recovered. A manager might be faced with a problem: to be cautious on the basis of a belief about the nature of the recruitment failure and then to decide whether to ban catches during El Niño. From the stock/recruitment curve published by Csirke (1980) one might conclude that catches should have been banned, but the managers knew very well that the failure of recruitment in 1972 and 1973 was of natural origin. This is very like the West Greenland problem, where the managers acted sensibly on the basis of a belief that the recruitment failure was natural. Perhaps in the light of Csirke's relationship, catches should have been banned for a period after the El Niño in the hope that recruitment would have recovered.

THE STOCK AND RECRUITMENT PROBLEM

The problem of dependence of recruitment upon parent stock may have a fairly simple solution which is difficult to reach merely because recruitment is so variable. However, the high variability of recruitment is an essential part of the population structure, as it samples both food and predators; contrast the relative stability of the stock with many age groups. It has long been hoped that an analysis of the control of recruitment by environmental factors would help to elucidate the dependence of recruitment upon parent stock.

Many stocks throughout the world ocean have collapsed through recruitment overfishing. Amongst them are the North Sea herring, the Atlantoscandian herring, the Californian sardine, the Peruvian anchoveta (after collapse) to name only the most prominent for which the cause of decline is known. Not only is recruitment highly variable but that variability tends to be greater at low stock: a very strong year class may appear which lulls the fishermen into a false sense of security. Such an event may also increase stock temporarily, and at low stock levels there may be quasi-cyclical events generated in this way. Indeed, environmental events from year to year may be serially correlated. The high variability makes it very difficult to derive the right relationship from the data with enough certainty to convince the obstinate but if that relationship were obtained independently, then the variation need not trouble us so much and the chances of different courses of action could be estimated.

Cushing's (1982) study of the Icelandic cod stock was used to show how the constants might vary by fitting Ricker curves by decades (Table 20). It will be recalled that there are two possible forms of exploitation. The quantities of α, the density independent survival differ in the two treatments, but the coefficients of variation are the same. The density-dependent

Table 20
Exploitation as on mature stock

Decades	α	SD	β	SD
EXPLOITATION AS ON MATURE STOCK IN EARLIER PERIOD				
1	0.392	0.365	0.00103	0.00139
2	0.191	0.164	0.00095	0.00115
3	0.318	0.197	0.00172	0.00119
4	0.908	0.266	0.00459	0.00105
EXPLOITATION AS ON IMMATURE STOCK IN LATER PERIOD				
1	0.568	0.530	0.00103	0.00139
2	0.277	0.238	0.00095	0.00115
3	0.318	0.197	0.00172	0.00119
4	0.908	0.266	0.00459	0.00105

survival in the fourth decade differs significantly from those in the third and second, but not from that in the first. Further those in the first, second and third do not differ from each other. Thus the density-dependent survival increased during the 1950s, and as a consequence the coefficient of density-dependent mortality decreased. In other words the stock became less resilient and more opportunistic during the recent period of cooling. From the point of view of management this means that at the same level of fishing mortality the stock becomes somewhat mroe vulnerable to recruitment overfishing.

In the history of fisheries research there have been many conflicts between proposed effects due to fishing and those due to natural causes, fishing mortality and natural mortality, recruitment failure due to fishing or to natural causes. When recruitment fails for three years in succession during a period of heavy fishing pressure, it could well be a direct effect of fishing, but it may equally be a consequence of natural or climatic factors. During the mid 1960s there were four successive poor year classes in the Barents Sea cod stock; again the last three year classes (1977, 1978, 1979) in that stock were very poor indeed (Anon., 1980). The first may well have been due to fishing because fishing mortality had increased sharply in the years just before, but the second may well have been natural in origin because the fishing mortality had remained constant for some years. More rigorously, however, the two possible causes cannot be distinguished, and we return to the same point: that the stable relationship of recruitment upon parent stock should be elucidated independently.

As suggested by Beverton and Holt (1957) and in Chapter 2, recruitment might be determined during a critical period. Rauck and Zijlstra (1978) published a good correlation between the numbers of 0-group plaice in the Friesian Island shrimp fishery and the subsequent recruitment: the fish were sampled on their way into the nursery ground and so the animals had metamorphosed only a short time before. Cushing and Shepherd (in press) have suggested that the critical period lies between some larval stage and an age just after metamorphosis; the work was done on the Southern North Sea plaice and the evidence from Rauck and Zijlstra is critical.

The interesting question is how we might expect climatic factors to affect recruitment. In Chapter 6 it was suggested that year class strength depended to some degree upon primary production, based on the high correlations between recruitment and indices of upwelling in the Californian upwelling area. The argument was extended to temperate waters with an extension of the match-mismatch hypothesis. The explanation of the gadoid outburst in the North Sea (Fig. 78) depends on the delay of *Calanus* production during the 1960s, that is, it depends on the food available to the larval fish, which of course must depend upon the primary production. Primary production varies by a factor of three, its onset varies in time by as much as six weeks, and in the North Sea and north-east Atlantic it has been delayed by as much as a month since 1948. Thus there is great capacity for variation and differences in the food available during the critical period can potentially affect recruitment. Given the quantity of food available to the larvae and the numbers of their predators, it is possible that some of the greatest variability in recruitment will be accounted for, that the effects of climatic changes could be incorporated and that a curve describing the dependence of recruitment on parent stock will be developed with somewhat less variability.

This extension of argument is speculative, but if it were established there would be considerable consequences in management practice. Much management today is executed with yield per recruit models where it is assumed that the magnitude of recruitment is not affected by an increment in fishing mortality. The threat of recruitment overfishing may only be met by general production (or Schaefer) models in which yield is estimated as a function of stock or of fishing mortality based on an inverse dependence of catch per unit of effort on fishing effort. It is possible, however, that the index of stock density, or catch per unit of effort, is biased upwards as fishing increases, so the maximum sustainable yield is overestimated. This is because the coefficient relating stock density to abundance, the catchability, is itself inversely related to abundance (in a purse seine fishery for herring, Ulltang (1976) and in a trawl fishery for cod, Garrod (1977)).

Recently, Shepherd (in press) has devised a method by which the variability of recruitment in a stock/recruitment curve is exploited to make provisional estimates of yield before there is any decline of recruitment by heavy fishing. If the variability about the dependence of recruitment on parent stock could be reduced as indicated above, yield curves would be derived from the stock/recruitment curves directly, much as Shepherd has done already. Then stocks would be secure from recruitment overfishing much as they are now secure from growth overfishing with the yield per recruit model. Although techniques are available for most eventualities in stock management, the real advance would be in confidence, in that steps can be taken with a little more certainty.

In pursuing this line of argument we look into the future because the essential information need to establish such a dependence of recruitment on parent stock has not yet been adequately collected. If the magnitude of recruitment depends upon primary production in the upwelling areas and in the offshore divergences of the open subtropical ocean, then the critical period must occur during a brief period of perhaps two or three months, because in any longer period the little fish will have grown too far from the base of the ecosystem. The information required is the number of predators on and the quantity of food available to the larval and post larval fishes, but it must be collected for many years in many places and the costs of research vessels increase each year.

Evidence has been produced that the magnitudes of recruitment to fish stocks increase or decrease in time, and some of the changes are considerable, of an order of magnitude or more. Such increments or decrements are of course uncommon. Forecasts of recruitment are made today from groundfish surveys or beach surveys as part of the management procedure. The numbers of recruits are usually estimated at about one year of age, and estimates differ from stock to stock according to the length of the time series.

If recruitment is linked to primary production and with the time at which it occurs, a forecast of recruitment might be made a year earlier. As most fishes recruit to the fishery or join the adult stock at the age of three, four or five years of age or more, the additional year may not matter very much. Managers, however, need to know about the outstanding year class, high or low, as early as possible, so there is a case for using the effort in sampling one-year-old fishes to sample the primary production at the right time and place each year. Again, such decisions lie in the future.

By the perversity of events doubt is now cast upon present measures of primary production. In the oligotrophic ocean, Gieskes (1979) produced evidence that the radiocarbon method carried out in small bottles may underestimate production by a factor of four. On the same cruise Tijssen (1979) confirmed the higher magnitudes of production by a factor of four. Postma (1979) came to the same conclusion from diurnal estimates of particle numbers. More recently Schlumberger and Reid (1981) have shown that estimates of production with the oxygen method below a shallow density cap across much of the North Pacific are considerably greater than with the radiocarbon method, which extends the argument from the oligotrophic ocean to the waters of higher latitudes. The arguments presented above on the link between recruitment and primary production rest on physical estimates of upwelling and divergence and upon differences in primary production off Peru. In the future, however, it would be desirable to measure primary production with new methods.

CONCLUSION

Strictly the management of fish stocks proceeds from year to year as if their vital parameters do not change from year to year. This is an eminently reasonable procedure because the annual differences are small, except for the occasional outstanding year class, large or small. In the future, if and when fish stocks become managed more effectively, part of the management procedure will provide long-term indications of the tonnages in the stocks. These would not be forecasts but merely indications to allow the industries to think ahead to a greater degree. When a stock is heavily exploited fishermen's representatives often complain that the variation from year to year, which is due to recruitment differences, is too great. As management improves, such variation would be reduced considerably, and some sort of long-term quota say for five years or so, would become possible. Within such a system a series of annual TACs would of course still be needed because of the yearly changes in recruitment.

At the present time weather can be forecast for a few days ahead, so reliable forecasts of climate are probably a little remote. However, changes in the vital parameters do occur which are linked to climatic factors, the predominant example of which is the effect of El Niño. A slow change is now occurring in fisheries science in that data, particularly gut contents, are being collected to study the interactions between species, to monitor changes in natural mortality and recruitment, and the latter is affected by climatic changes. Thus in the future the right material will become available.

In the history of fisheries research there has often been disagreement whether an event, usually a decline in stock, is due to fishing or to natural causes. For example Burkenroad (1948) disputed Thompson's (1936) claim that the recovery of the Pacific halibut stock was due to reduced fishing as a consequence of a close season. Skud's (1975) more recent conclusion is that the immediate increment of stock density in 1932 may have been due to a good year class and that Thompson was correct, if lucky. One of the many steps forward made by Beverton and Holt (1957) was an independent estimate of natural mortality from the transwartime year classes of the Southern North Sea plaice. Other methods, such as tagging, can be used to estimate fishing mortality, and it is now standard practice to separate fishing from natural mortality, an advantage not available to Thompson and to Burkenroad.

When a stock suffers from recruitment overfishing, the decline in recruitment takes place when fishing mortality is high. There is then a presumption that the decline was due to fishing, although it might have been due to natural causes, as probably happened to the Plymouth herring as part of the Russell cycle during the 1930s. There is only one solution, and that is to stop fishing on the stock. If it recovers, the decline in recruitment was due to fishing, as was shown when there was a recruitment failure in the British Columbian herring stock. The same has now happened with the North Sea herring stock in that the recruitment is now recovering somewhat after a ban of four years. If the recruitment had not recovered, its decline would have been attributed to natural causes. The Californian sardine stock collapsed dramatically in the late 1950s when the fishing mortality was high, and the failure was attributed to recruitment overfishing (Clark in Clark and Marr, 1956; Cushing, 1971; Murphy, 1966). After the collapse, the sardine fleet went out of business, so fishing was stopped for economic reasons. The sardine stock did not recover but was replaced about a decade later by the stock of northern anchovies (*Engraulis mordax* Girard) (Ahlstrom, 1966). Subsequently, Soutar and Isaacs (1974) showed from the scales left in anoxic sediments that sardine and anchovy had alternated in abundance for very long periods. The question arises whether the anchovy replaced the sardine stock because it had been reduced by fishing or whether a natural succession was due at that time in any case. The question cannot be answered decisively, but the study of Bakun and Parrish (1981) on the Californian sardine reveals that with Murphy's estimates of recruitment a multiple regression was established with upwelling indices, the curl of the wind stress and the stock. The regression shows something of the structure of processes involved, that recruitment depended upon upwelling and upon stock, but nothing is revealed about the crucial events in 1949 and 1950 when the stock collapsed and when it was subsequently succeeded by the stock of northern anchovy.

It is convenient to assume that the variation of recruitment is randomly distributed at any given level of stock. Figure 88c showed that the recruitment to the Arctonorwegian cod stock is correlated with the salinifications which enter the Baltic every five years or so (see Chapter 9). Indeed the link is close enough to show quasi-periodic changes in recruitment. In other words, a poor recruitment may be often followed by another poor one, or a high one by another high one. There is thus a component of serial correlation in the time series shown in Fig. 88c. It is not a necessary component of the variability of recruitment and obviously the degree of seriality will differ from stock to stock. Recruitments to the Arctonorwegian cod stock, to the West Greenland cod stock and to the Californian sardine all evince three or more poor year classes in succession. If the old conflict between the effect of fishing and that of the environment is to be resolved in the proper management of recruitment overfishing, the degree of seriality in the time series of year classes and its dependence upon climatic factors must be deeply investigated.

The study of climatic factors upon the vital parameters of the fish stocks is the study of annual changes in their environment. On the long time scales, such as the alternation of Swedish and Norwegian herring periods since 1760, the changes are obvious but still demand explanation. The same is true of the recent periods of warming and cooling, for the immigrations into northern waters at that time remain unexplained in the physical sense. The changes in recruitment that dominate the fisheries are probably related to the availability of food for little fish for a relatively short period, and it is these differences from year to year and through the decades that are of importance to fisheries biologists and ultimately to the managers of fish stocks.

Chapter Thirteen

The Effect of Climate upon the Fisheries

Men have long thought that fisheries were affected by climatic events, partly because fisheries often lasted for a few decades and partly because they sometimes appeared and disappeared suddenly. In more recent years, as estimates of recruitment have become available by decades, the high variability of fisheries has been attributed to the type of phenomenon that modulates climatic change. It has also been ascribed to events of random origin. Obviously there are elements of truth in both views, perhaps ordered by the magnitude of recruitment; the outstanding 1904 year class of Atlanto-scandian herring might have originated in specific and novel factors, but there is probably a noise level in the magnitudes of recruitment below which the pursuit of determinant factors is not worth while. The late Dr. W.B. Chapman was one of those who believed in the predominant effect of climatic factors upon the lives of fish stocks and he played a part in encouraging the late Professor Jakob Bjerknes to study El Niño. This final chapter summarizes the material, and in it some conclusions are drawn.

THE RECENT PERIOD OF WARMING

The recent period of warming from about 1880 to about 1945 was a sustained and pronounced physical event shown as an increment of 0.5°C in air temperature throughout the globe (Dickson and Lamb, 1972). Between 1925 and 1935, southern animals invaded waters off France and the British Isles, and between 1935 and 1945 boreal animals reached high latitudes. The cod recolonized the banks off West Greenland during the

1920s and 30s. The onset and return of the Russell Cycle were recognized, as were the eastward spread and subsequent regress of southern barnacles in the western English Channel. The biological events are linked in time but are quite distinct, suggesting that factors other than temperature may have been involved.

Figure 118 (Trenkle, 1956) shows the strength of the upper westerlies at 500 mb in the region 50°–60°N, 60°W–60°E in December, January and February from 1881 to 1965. From 1880 to 1898 they blow at less than 11.3 m s^{-1}, whereas between 1898 and 1925 they were stronger than that average; after 1925 they relaxed to a wind strength less than 11.3 m s^{-1}. Dickson and Lamb (1972) noted that the circumpolar vortex contracted somewhat as the upper westerlies increased and that they shifted northward. A similar intensification of the upper westerlies took place in the southern hemisphere. Figure 42 shows the changes in number of days of westerly winds at sea level over the British Isles each year and they peaked also in 1925 after which they declined. Perhaps the surface winds followed the trend of the upper westerlies.

During the recent period of warming many of the prominent events occurred between 1925 and 1935, the dramatic decade. In addition to the southern invaders off the coasts of France and Britain, the first stage of the Russell Cycle was completed between 1926 and 1935, the West Greenland cod stock received its first great impetus and that at Iceland gained five strong year classes in succession, 1931 to 1935. There was a second period of change between 1935 and 1949 before the cooling started, which followed the recent period of warming. There were no changes in the Russell Cycle in the Western English Channel, and the barnacle distributions may well have been stable. However, major changes in distribution occurred in the waters off Iceland and off East and West Greenland; indeed some of the animals seen had never been recorded before, such as the swordfish at Iceland, a prominent and exotic invader. This was the period when boreal animals reached the Barents Sea and when Atlantic benthic animals migrated north in the West Spitsbergen current. During this second decade the Greenland year classes of 1942 and 1945 generated the period of highest stock. Hence the trend of change in the upper westerlies coincided with the biological events prominent during the recent period of warming.

Iselin (1940) suggested that as the circulation intensifies under an increase of wind stress, the Florida Current flows more quickly. The whole anticyclone contracts, and the thermocline deepens in the Sargasso Sea. Sea surface temperatures rise there and fall in northern waters because of reduced

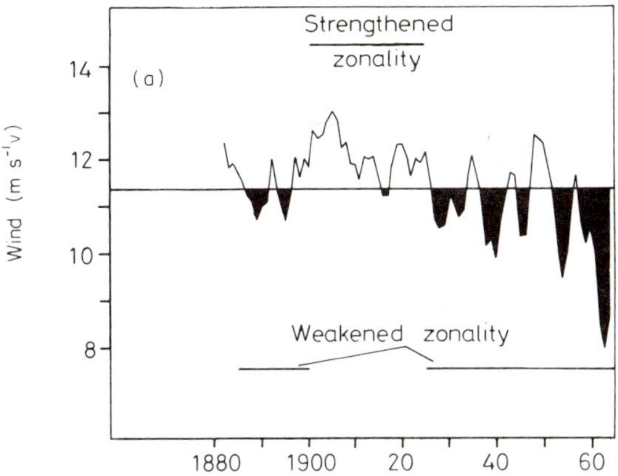

Fig. 118 Changes in the strength of the upper westerlies graphed as three-year running means from 1881 to 1965 (Trenkle, 1956).

advection. Stommel (1965) showed that between 1925 and 1955 there was an increase of about 2°C in the surface temperatures in the Norwegian Sea and that the depth of the 10°C isotherm in the Sargasso Sea may have decreased by perhaps 50 m. However, he thought that the anticyclone should not expand or contract with long-term differences in wind stress. Taylor (1978) compared the difference in sea surface temperature between the periods 1894–8 and 1920–4 and found an increment of heat in the Sargasso Sea. Taylor produced an adaption of Needler's (1967) model of the oceanic thermocline (which strictly applies to the interior of the ocean) in which an increment in the intensity of the subtropical high (and the subpolar low) produces higher temperatures in the anticyclone and lower in the cyclone, an increase in the depth of the Sargasso thermocline and an increase in advection towards the north-east Atlantic.

Today physical oceanographers make great efforts to model the oceanic circulation numerically (see, for example, Holland, 1977), but such models have not yet been used to find the effect of perturbation. There is, however, enough information to suggest that the biological consequences of the recent period of warming owed much to a movement of water into the north-east Atlantic—how else would swordfish find their way to Iceland? (Migrating fish are assumed to move with the ocean currents because they must rarely have external referents (Harden Jones (1968).) However, until the advection in the north-east Atlantic can be modelled fully, the links between wind stress and swordfish must remain mysterious in a physical sense.

FOUR SYSTEMS

Four other systems were examined: (1) the cold winters in North West Europe; (2) the salinifications and long-term changes in salinity in the Baltic; (3) the changes in the North Sea; and (4) the origins of El Niño. They are quite distinct, both in their origins and in their manifestations, but their effects on the fish stocks, particularly on their recruitments, are common to some degree.

Cold winters have appeared every one or two decades in the present century but were more frequent during the Little Ice Age. Fishes and coastal invertebrates are killed in significant quantities, but the more important consequence is that the recruitment to the plaice and sole stocks in the North Sea is augmented and the same may be true for some commercial invertebrates. The increment in year class strength more than compensates the losses of adults in the cold winter. During the Little Ice Age, when the winters were more frequent, the flatfish stocks may have been more abundant than during this century. It was suggested in Chapter 11 that the stronger yearclass arises as consequence of improved match of larval production to that of the larval food because of delayed development in cold water, but this has not been confirmed.

In the Baltic the relatively frequent salinifications probably maintain the oxygenation of the deep water sediments, and the long-term increase in salinity is perhaps associated with slackened flow from the rivers. The very large increases in the cod and herring stocks may well be associated with the quicker rise of eggs and larvae from the seabed or near it towards the euphotic layer where food for the larvae is abundant (Alander, 1940). His suggestion is a special case of improved match of larval production to that of the larval food. If this is true, there are two important consequences: (*a*) that an increment of cod recruitment of an order of magnitude follows simply from more available food and (*b*) that the Baltic ecosystem can accommodate a considerable increment of cod and herring, very roughly 0.12 gC m^{-2}. Either the increment of fish displaces an equal decrement from another part of the system on demand, or the ecosystem is a loose enough structure to expand or contract.

In the North Sea, the switch from the smaller copepods to the larger *Calanus* in 1950 (since sustained) generated changes in growth rate, age at first maturity, and recruitment to the herring stocks. Because *Calanus* is the preferred food of the adult herring (Cushing, 1964), there was an increment of about 20% in stock due to growth alone, another instance of accommodation within an ecosystem. The greatest change, however, was the increment to the gadoid stock, which started in 1962 and which has been since sustained. The increased recruitment to the cod stock was linked to the delayed production of *Calanus* since 1962 and separately to the decline in sea temperature in the North Sea. As the production of cod larvae became more closely matched to that of *Calanus* (their food), recruitment increased. Again, recruitment is perhaps linked to food at an early stage in life history and the gadoid stocks expanded readily into a gap in the system which might be expected in one as heavily exploited as the North Sea.

Off Peru the anchoveta year classes of 1972 and 1973 were reduced by an order of magnitude when the primary production was cut by a factor of three during El Niño. The stock collapsed entirely because there was not enough food for the recruiting year class, presumably at an early stage in its life history. The subsequent history of the stock is less simple because it was fished before it had recovered. If Walsh (1981) is right the copepods did not exploit the gap left by the adult anchoveta perhaps because enough larval and juvenile anchoveta remained to restrain the development of the copepod populations.

The climatic events in the four systems differ profoundly. The cold winters are generated by a persistent anticyclone in winter over Scandinavia. The Baltic salinifications occur when cyclone and anticyclone feed each other over a broad warm-water pool in the North Atlantic. The changes in the North Sea probably owe their origin to a change in the pattern of input in the North, but the hydrodynamic basis of this change has not yet been established. El Niño originates in the relaxation of the winds which had sustained a great head of water in the western Pacific. Yet from all four the fish stocks extract their advantage in food. They have had the endless course of time to explore their changing environments, but they emerge with the information needed to sustain their great numbers.

THE SOURCE OF RECRUITS

The fish stocks have been able to obtain food for their recruits under a number of climatic changes, each of which recurs irregularly over long periods of time. Two questions arise: which stage in the life history is critical in establishing the magnitude of recruitment and how is that quantity determined? The conclusion from Ottestad's (1942, 1960, 1969) correlation between cod catches and tree rings is that some components of variance may be common, and this restricts the choice of factors. The growth of trees is probably governed by air temperature, irradiance, and humidity (Fritts, 1974). Differences in air temperature indicate other climatic changes, particularly in wind strength and direction. With irradiance, such factors are those that determine the time of onset of primary production and perhaps to some degree its magnitude. Once the little fish have metamorphosed and live on the sea bed or near the beaches, the link with elemental climatic factors must become attenuated.

Templeman (1972) demonstrated that most year classes tended to be high in particular years throughout the North Atlantic, a conclusion confirmed by Garrod and Colebrook (1978) with a more critical analysis. If the factors that determine recruitment are pervasive enough to spread across an ocean, they are probably of the same nature as those which determine primary production. The time of onset of primary production is governed by the date at which the critical depth exceeds the depth of mixing, itself governed by wind stress and convection. Subsequently as primary production proceeds, its development is controlled for a time by the ratio of compensation depth to the depth of mixing. Once the algal reproductive rate reaches its maximum, production is controlled by biological processes. With limited information from temperate waters and upwelling areas, primary production may vary from year to year by a factor of three. The physical and biological processes which determine magnitude cannot be distinguished, but both in temperate waters and in upwelling areas the time of onset of primary production is determined by physical processes.

There are three independent lines of evidence that link recruitment to primary production. The first is that the recruitments to pelagic and demersal stocks are linked to indices of upwelling. The second is that the good year classes of tuna species are associated with cool water in the open ocean, possibly the product of offshore divergences. The third is the failure of recruitment to the Peruvian anchoveta stock during the El Niño of 1972–3, when the primary production was reduced by a factor of three. In the first two phenomena it was probably the time of onset that was crucial, but in the third it appears to be the magnitude of production. In subtropical seas, where coastal upwelling or divergence is determined by physical factors, the processes are relatively simple. Yet recruitment cannot be determined solely by physical factors, because where species compete, the essential processes probably occur during the same early life stages when recruitment is determined. In an environment where sardine and anchovy compete off California it is then surprising that links with physical processes do emerge.

In water poleward of 40° latitude production stops in winter and starts again in spring. Because the physical processes, which control the production cycle, are variable, the time of onset changes from year to year by as much as six weeks. As the peak date of spawning of spring spawners in temperate waters is fixed to within a week, the production of fish larvae may be matched or mismatched in time to that of their food. It is possible that there are two small points of evidence in support of the hypothesis: (1) there is a correlation between cod recruitment and the delay in *Calanus* production in the North Sea and (2) the increased cod recruitment in the Baltic may be associated with the decreased depth of halocline. The production of larval food probably depends upon the magnitude of production, but perhaps not on its time of onset. Match or mismatch would depend upon the time of onset of production and perhaps indirectly upon its magnitude.

The description of such processes is complex. The physical processes determining the start and early stages of the production cycle have not yet been understood fully. The production cycle itself has been simulated many times independently of the necessary physical processes, and the biological processes are not yet well enough known to estimate the magnitude of production very well. This somewhat discouraging view is perhaps more pessimistic than it need be, because considerable advances are in train (e.g. Goldman *et al.* (1981) on radiocarbon measurements). Perhaps in the course of the next decade solutions to the problem of the production cycle in the sea will emerge.

The source of recruits may lie in the production cycle, in the oceanic divergences, in the upwelling areas, and in waters poleward of 40° latitude. However, the quantities may be modified subsequently by cannibalism or by predation. Further the magnitude of stock might be expected to affect the numbers recruited, if only mildly when stocks are over-exploited. Hence it is probably difficult to disentangle all the processes involved in the generation of recruitment.

CONCLUSION

It should not surprise us that fish stocks respond to climatic factors and to climatic change because they live their lives within the weight of the waters. Their migration circuits are tied to hydrodynamic structures, and reproductive isolation is obtained by the capacity of fishes to join or leave current systems as they will. Perturbation of a migration circuit with an unexpected advection spills fish outside their normal range where they either establish a new spawning group or die for they do not return. The West Greeland cod stock survived for thirty to forty years on the edge of its range, but the swordfish at Iceland were merely pointless migrants.

A fish stock is a multi-aged structure which conserves the variability of recruitment, which itself summarizes the variability of the environment. As the yearclass grows through the trophic levels in the ecosystem, it samples the food available much of which is affected by climatic factors, as shown in Chapter 7 on the North Sea. Thus the final magnitude of recruitment is a quantity modified by an array of factors. However, because both specific growth rate and specific death rate decline with age, larval stages and early juvenile stages are the vulnerable ones and we would expect climatic factors as they affect food to be most active at the early part of the life cycle. This is why physical factors as they modify the production cycle are effective in the establishment of recruitment.

A complex of physical and biological factors need examination. First, the effects of wind and convection upon the depth of the mixed layer upon the time of onset of the production need to be established in latitudes poleward of 40° latitude, as indicated by Woods (1980). Secondly, the development of the production cycle through the period of transient thermoclines needs analysis. Thirdly, the processes which control the magnitude of primary production as a complex of physical and biological events need further analysis with the object of predicting that magnitude to within a factor of about 0.3. Fourthly, information is needed upon the modification of recruitment by predation, although this is necessarily a minor process. It is an ambitious but not impossible programme in which physicists and biologists will have to cooperate fully, as they did in the FLEX experiment of 1976.

References

Ackefors, H. and L. Hernroth, (1972). "Estimations of the amount of zooplankton and fish in the Baltic proper." *Medd. fr. Havfiske laboratoriet Lysekil* **139**, 16.

Ahlmann, H.W. (1953). "Glaciärer och Klimat i Norden under de senaste tusentalen är." *Norske Geografisk Tidsskrift* **13**, 3–8, 56–75.

Ahlstrom, E.H. (1966). "Distribution and abundance of sardine and anchovy larvae in the California Current region off California and Baja California, 1951–64: a summary." *Spec. Sci. Rep. US Fish and Wildlife Serv. (Fisheries)* **534**, 71.

Ahlstrom, E.H. (1967). "Co-occurrences of sardine and anchovy larvae in the California current region off California and Baja California." *Calcofi Rep.* **XI**, 117–135.

Alander, H. (1948). "Swedish trawling in the Southern Baltic." *Ann. Biol. Cons. int. explor. Mer.* **3**, 111–113.

Alm, (1959). "Connection between maturity, size and age in fishes." *Rep. Inst. Freshwater Res. Drottningholm* **40**, 5–145.

Alverson, D.L. & Larkins, H.A. (1969). "Status of knowledge of the Pacific hake resources." *Rep. Calif. coop. ocean. Fish. invest.* **13**, 24–31.

Andersen, K.P. and Ursin, E. (1977). "A multi-species extension of the Beverton and Holt theory of fishing, with accounts of phosphorus circulation and primary production." *Meddr. Danm. Fisk og Havunders* **7**, 319–435.

Anon. (1909a). "Trawling Investigations, 1904–5." *Mar. Biol. Assn. UK. Intnl Fish. Invest.* 1904–1905. Part II. 114–278.

Anon. (1909b). "Trawling Investigations 1906–09." *Mar. Biol. Assn. Intnl Fish. Invest.* Fourth Report, 304–495.

Anon. (1967). "Report of the meeting of the Greenland cod working group." *ICNAF Red Book III 1966*, 3–33.

Anon. (1972). "Informe sobre la segunda reunión del panel de expertos en dinamica de poblacion de la anchoveta Peruana." *Bol. Inst. del Mar del Peru* **2**, 7, 377–457.

Anon. (1972–9). "Reports of assessment subcommittees." *ICNAF Redbook STACRES.*

Anon. (1973). "Continuous plankton records: A plankton atlas of the North Atlantic and the North Sea." *Bull. Mar. Ecol.* 7. 1–174.

Anon. (1973). "Tercera sesion del panel de expertos sobre la dinamica de la poblacion de la anchoveta Peruana." *Bol. Inst. del. Mar del Peru* 2, 9, 525–599.

Anon. (1976a). *Climatic monitor.* 5. 1. 27.

Anon. (1976b). "Report of the North-Western Working Group." ICES CM 1976/F.6 63 pp. mimeo.

Anon. (1979). "Reports of the ICES Advisory Committee on Fishery Management, 1980). *ICES Coop. Res. Rep.* **85**, 157.

Anon. (1980). "Report of the ICES Advisory Committee on Fishery Management, 1979." *ICES. Coop. Res. Rep.* **93**, 216.

Anon. (1981a). "Reports of the ICES Advisory Committee on Fishery Management, 1980." *ICES Coop. Res. Rep.* **102**, 234.

Anon. (1981b). "Yearbook of Fishery Statistics, Catches and Landings." FAO, 1979. Vol. 48, 383 pp.

Aurich, H.J. (1953). "Verbreitung und Laichverhältnisse von Sardellen und Sardinen in der südostlichen Nordsee und ihre Veränderung als Folge der Klimaänderung." *Helgolander wiss. Meeresforsch.* **4**, 175–204.

Bailey, K. (1981). "Larval transport and recruitment of Pacific hake, *Merluccius productus*." *Mar. Ecol. Progr. Ser.* **6**, 1, 1–9.

Bakken, E. (1961). "Influence of hydrographical and meteorological factors on catch and recruitment strength of the sprat stock in Western Norway." *Fisk Dir. Skr. Ser. Havunders* **14**, 61–71.

Bakun, A. (1973). "Coastal upwelling indices, West coast of North America 1946–71." *NOAA Tech. Rep.* NMFS SSRF-671 103 pp.

Bakun, A. and R. Parrish, (1981). "Environmental inputs to fishery population models for Eastern Boundary Current regions." Workshop Rep. 28 IOC, UNESCO, 1–37.

Balls, R. (1951). "Environmental changes in Herring Behaviour." *J. Cons. Int. Explor. Mer.* **17**, 3.

Bannister, R. C. A. (1978). "Changes in plaice stocks and plaice fisheries in the North Sea." *Rapp. Procès-Verb. Cons. Int. Explor. Mer.* **172**, 86–101.

Barber, R.T. and R.L. Smith (1981). "Coastal upwelling ecosystems." *In* "Analysis of Marine Ecosystems" (ed. A.R. Longhurst). Academic Press, London and New York. 31–68.

Barnett, T.P. (1977). "The principal time and space scales of the Pacific Trade Wind fields." *J. Atmos. Sci.* **34**, 221–236.

Bell, W.T. and A.E.J. Ogilvie, (1978). "Weather compilations as a source of data for the reconstruction of European climate during the mediaeval period." *Climatic Change* **1**, 331–348.

Bellamy, J.G. (1839). "The Natural History of South Devon." Simpkin Marshall, London. 455 pp.

Berg, L.S. (1935). "Rezente klimaschwankungen und ihr einfluss auf die geographische Verbreitung der Seefische." *Zoogeographica* 3, 1–15.

Berger, A. (1981). "Astronomical theories of palaeoclimate." *In* "Climatic Variations and Variability: Facts and Theories" (ed. A. Berger). Reidel, Dordrecht, Boston and London. 501–525.

Bernal, P. and J. McGowan, (1981). "Advection and upwelling in the California Current." *In* "Coastal Upwelling" (ed. F.A. Richards). Amer. Geophysical Union. 381, 399.

Betin, V.V. (1957). "Ice conditions in the Baltic and its approaches and their long term variations." *Trans. State. Oceanogr. Inst.* 41, 54–125.

Beverton, R.J.H. (1962). "Long-term dynamics of certain North Sea fish populations." p. 242–64 *In* "The Exploitation of Natural Animal populations" (ed. E.D. Le Cren and M.W. Holdgate). Blackwell, London. 242–264.

Beverton, R.J.H. and V.M. Hodder (1965). "Report of Working Group of Scientists on Fishery Assessment in relation to regulation problems." *Suppl. Ann. Proc. Int. Commn North West Atlant. Fish.* **II**, 1–81.

Beverton, R.J.H. and S.J. Holt, (1957). "On the dynamics of exploited fish populations." *Fish. Invest. Ser. 2* **19**, 533 p.

Beverton, R.J.H. and S.J. Holt (1964). "Tables of yield assessment for fishery assessment." FAO Fish. Tech. Paper 38, 49 pp.

Beverton, R.J.H. and A.J. Lee (1964). "The influence of hydrographic and other factors on the distribution of cod on the Spitzbergen Shelf." *Spec. Publ. Intern. Comm. N.W. Atlantic Fish.* **6**, 226–46.

Bidder, G.P. (1925). "The mortality of plaice." *Nature* **115** (2892), 495–96.

Bigelow, H.B. (1926). "Plankton of the Offshore waters of the Gulf of Maine." *Bull. Bur. Fish.* **40**, II. 509 pp.

Bjerknes, J. (1961). "El Niño study based on analysis of ocean surface temperatures 1935–57." *Bull. Inter-Amer. Trop. Tuna Commn* V. **3**, 219–272.

Bjerknes, J. (1966). "Survey of El Niño 1957–8 in its relation to the tropical Pacific meteorology." *Bull. Inter.-Amer. Trop. Tuna Commn* **12**, 2, 3–42.

Blacker, R.W. (1957). "Benthic animals as indicators of hydrographic conditions and climatic change in Svalbard waters." *Fish. Invest. London II* **20** (10) 1–49.

Blacker, R.W. (1965). "Recent changes in the benthos of the West Spitzbergen fishing grounds." *Spec. Publ. Int. Commn. N.W. Atlant. Fish.* **6**, 791–794.

Boaden, P.J.S., R.V. Gotto, R.G. Hartnoll and G. Williams (1964). "Northeast Ireland." *J. Anim. Ecol.* **33**, 197.

Boalch, G.T., D.S. Harbour and E.I. Butler (1978). "Seasonal phytoplankton in the western English Channel 1964–1974." *J. Mar. Biol. Assn. NS UK* **58**, 4, 943–954.

Boddeke, R. (1963). "Influence of the severe winter 1962–63 in Dutch coastal waters." ICES. CM. 1963, 119, 3 pp.

Boeck, A. (1871). "On silden og silde fiskerierne, naun lig om det norske varsildfisket." Indberetning til den Konglige norske Regierungs Department for der Indre am foretagst Praktisch-vidensk abelige Undersøgelser.

Boerema, L.K., Saetersdal, G., Tsukayama, I., Valdivia, J.E. and Alegre, B. (1967). "Informe sobre los efectos de la Pesca en el recurso peruana de anchoveta." *Bol. Inst. del Mar del Peru* **1**, 4, 136–186.

Bolster, G. and Bridger, J.P. (1957). "Nature of the spawning area of herrings." *Nature* **179**, 639.

Botsford, L.W. and D.E. Wickham (1975). "Correlation of upwelling index and Dungeness crab catch." *Fish Bull. U.S.* **73**, 901–7.

Bostrøm, O. (1955). " 'Peder Ronnestad' Ekkolodding og meldetjeneste av Skrei-forekomstene i Lofoten i tiden 1 March–2 Apr. 1955: Praktiske fiskeforsøk 1954 og 1955." *Arsberet. Vedkomm. Norges Fisk.* **9**, 66–70.

Bray, J.R. (1974). "Glacial advance relative to volcanic activity since 1500 A.D." *Nature* **248**, 42–3.

Bridger, J. (1960). "On the relationship between stock larvae and recruits in the 'Downs' herring." *Cons. Int. Explor. Mer. Herring Comm.* Paper No. 159, 9 p. mimeo.

Bridger, J.P. (1965). "The Cornish pilchard and its fishery." Lab Leaflet (NS) 9, 17 pp. Fisheries Laboratory, Lowestoft.

Buckland, F. (1875). "Report on the Fisheries of Norfolk." 84 pp. HMSO, London.

Buckland, F. (1879). Report on Sea Fisheries.

Bückmann, A. (1961). "Über die Bedeutung des Schollenlaichens in der südöstlichen Nordsee." *Kurze. Mitt. Inst. Fisch. Univ. Hamburg.* **11**, 1–40.

Budyko, J.I. (1968). "On the causes of climatic variations." *Medd. fran Sveriges meteorologiska och hydrologiska institut Serie B.* **28**, 6–13.

Burd, A.C. (1978). "Long term changes in North Sea herring stocks." *Rapp. Procès-Verb. Cons. intern. Explor. Mer* 172, 137–53.

Burd, A.C. and D.H. Cushing, (1962). "I. Growth and recruitment in the herring of the southern North Sea. II. Recruitment to the North Sea herring stocks." *Fish. Invest.* **2**, 23 (1) 71.

Burkenroad, M. (1948). "Fluctuations in the abundance of Pacific Halibut." *Bull. Bingh. Oceanogr. Coll.* **11**, 4, 81.

Carruthers, J.H. (1938). "Fluctuations in the herrings of the East Anglian Autumn fishery, the yield of the Ostend spent herring fishery and the haddock of the North Sea in the light of relevant wind conditions." *Rapp. Procès-Verb. Cons. int. Explor. Mer* **107**, 1–15.

Clark, F.N. and J.C. Marr (1956). "Population dynamics of the Pacific sardine." *Calif. Coop. Ocean. Fish. Invest. Progr. Rep.* 1 July 1953–31 March 1955. 11–48.

Cole, H.A. (1940). "Effect of the Severe Winter of 1939–40 on British Oyster Beds." *Nature* **145**, 976.

Colebrook, J.M. (1963). "Continuous plankton records: annual variations in the abundance of *Calanus finmarchicus*, 1948–1959." *Bull. Mar. Ecol.* **6**, 17–30.

Colebrook, J.M. (1965). "On the analysis of variation in the plankton, the environment and the fisheries." *Spec. Publ. Intern. Commn North West Atlant. Fish.* 6, 291–302.

Colebrook, J.M. (1978). "Changes in the zooplankton of the North Sea, 1948–1973." *Rapp. Procès-Verb. Cons. int. Explor. Mer* 172, 390–396.

Colebrook, J.M. (1979a). "Continuous plankton records: seasonal cycles of phytoplankton and copepods in the North Atlantic Ocean and the North Sea." *Mar. Biol.* 51, 23–32.

Colebrook, J.M. (1979b). "Continuous plankton records: zooplankton and environment, Northeast Atlantic and North Sea, 1948–1975." *Oceanologica Acta* 1, 1, 10–23.

Colebrook, J.M. and G.A. Robinson (1965). "Continuous plankton records: seasonal cycles of phytoplankton and copepods in the north-eastern Atlantic and the North Sea." *Bull. Mar. Ecol.* 6, 123–129.

Collins, J.W. (1884). "History of the Tile Fish." US Commission on Fish and Fisheries, Rep. Commissioner V (1882) 237–294.

Colman, J.S. (1964). "Fish mortality off the Isle of Man and in the Port Erin aquarium." *J. Anim. Ecol.* 33, 173.

Coombs, S.H. and R.K. Pike (1978). "The distribution abundance and seasonal occurrence of the eggs and larvae of blue whiting *Micromesistius poutassou* (Risso) in the eastern North Atlantic." ICES CM 1978/H45. 9 pp.

Cooper, J.C., A.T. Scholz, R.M. Horrall and A.D. Hasler (1975). "Experimental confirmation of the olfactory hypothesis with homing artificially imprinted coho salmon (*Oncorhynchus kisutch*)." *J. Fish. Res. Bd. Can.* 33, 703–10.

Cooper, L.H.N. (1948). "Phosphate and fisheries." *J. Mar. Biol. Assn. UK NS* 27, 326–336.

Crisp, D.J. (1964). "Mortalities in marine life in North Wales during the winter of 1962–3." *J. Anim. Ecol.* 33, 190–197.

Crisp, D.J. (1964). "Northwest Ireland." *J. Anim. Ecol.* 33, 197–8.

Crisp, D.J. (1965). "Observations on the effects of climate and weather on Marine communities." *In* "The Biological Significance of Climatic Changes in Britain" (ed. C.G. Johnson and L.P. Smith). Academic Press, London and New York 63–77.

Crisp, D.J. and A.J. Southward (1958). "The distribution of intertidal organisms along the coasts of the English Channel." *J. Mar. Biol. Assn. NS UK* 37, 157–208.

Crisp, D.J., and A.J. Southward (1964). "South and Southwest coast." *J. Anim. Ecol.* 33, 179–183.

Crisp, D.J., J. Moyse and A. Nelson-Smith (1964). "General conclusions." *J. Anim. Ecol.* 33, 202–210.

Csirke, J. (1980). "Recruitment in the Peruvian anchovy and its dependence on the adult population." *Rapp. Procès-Verb. Cons. int. Explor. Mer* 177, 307–313.

Cunningham, J.T. (1895). "The Natural History of the Marketable Marine fishes of the British Islands." Macmillan, London and New York, 368 pp.

Cushing, D.H. (1952). "Echo-surveys of fish." *J. Cons. int. Explor. Mer* **18**, 45–60.

Cushing, D.H. (1955). "Production and a pelagic fishery." *Fish. Invest. Lond. Ser.* 2, 18 (7) 104 pp.

Cushing, D.H. (1957). "The number of pilchards in the Channel." H.M.S.O., London, Fish. Invest., Ser. 2, Vol. 21 (5). 27 p.

Cushing, D.H. (1961). "On the failure of the Plymouth herring fishery." *J. Mar. Biol. Assn., UK NS* **41**(3), 799–816.

Cushing, D.H. (1962). "An alternative method of estimating the critical depth." *J. Cons. int. Explor. Mer* **28**, 2, 131–140.

Cushing, D.H. (1964). "The work of grazing in the sea." *In* "Grazing in Terrestrial and Marine Environments." (ed. D.J. Crisp). Blackwell, London. 207–225.

Cushing, D.H. (1966a). "Biological and hydrographic changes in British seas during the last thirty years." *Biol. Rev.* **41**, 221–58.

Cushing, D.H. (1966b). "The Arctic Cod." Pergamon, Oxford. 93 pp.

Cushing, D.H. (1967a). "The grouping of herring populations." *J. Mar. Biol. Assn. UK NS* **47**, 193–208.

Cushing, D.H. (1968). "The Downs stock of herring during the period 1955–66." *J. Cons. int. Explor. Mer.* **32**, (2), 262–69.

Cushing, D.H. (1969). "The regularity of the spawning season of some fishes." *J. Cons. int. Explor. Mer.* **33**, 1, 81–92.

Cushing, D.H. (1971a). "Upwelling and the production of fish." *Adv. Mar. Biol.* **9**, 255–335.

Cushing, D.H. (1971b). "The dependence of recruitment on parent stock in different groups of fishes." *J. Cons. int. Explor. Mer.* **33**, 340–62.

Cushing, D.H. (1972a). "The production cycle and numbers of marine fish." *Symp. Zool. Soc. London* **29**, 213–232.

Cushing, D.H. (1972b). "A history of some of the International Fisheries Commissions," *Proc. R. Soc. Edinb. Section B.* **73**, 361–390.

Cushing, D.H. (1973). "Production in the Indiàn Ocean and the transfer from the primary to the secondary level" *In* "The Biology of the Indian Ocean" (ed. B. Zeitzschel), Springer, Berlin. 475–486.

Cushing, D.H. (1974a). "The Detection of Fish." Pergamon, Oxford 200 pp.

Cushing, D.H. (1974b). "The natural regulation of fish populations." *In* "Sea Fisheries Research." (ed. F.R. Harden Jones). Elek Science, London. Pp. 399–412.

Cushing, D.H. (1975a). "The Natural mortality of the plaice." *J. Cons. int. Explor. Mer* **36**, (2), 150–57.

Cushing, D.H. (1975b). "Marine Ecology and Fisheries." Cambridge University Press, Cambridge. 278 pp.

Cushing, D.H. (1977). "The problems of stock and recruitment" *In* "Fish Population Dynamics." (ed. J.A. Gulland) John Wiley, Sydney, Toronto. 116–135.

Cushing, D.H. (1978). "The Upper Trophic levels in upwelling areas" p. 101–110 *In* "Upwelling Ecosystems." (ed. B. Boje and M. Tomczak), Springer, Berlin. 303 pp.

Cushing, D.H. (1979). "The monitoring of biological effects: the separation of natural changes from those induced by pollution." *Phil. Trans. Roy. Soc. B.* **286**, 597–607.

Cushing, D.H. (1980). "The decline of the herring stocks and the gadoid outburst." *J. Cons. int. Explor. Mer* **39**, 1, 70–81.

Cushing, D.H. (1981a). "Fisheries Biology." (2nd edit.), University of Wisconsin Press. 295 pp.

Cushing, D.H. (1981b). "The effect of El Niño upon the Peruvian anchoveta stock." *In* "Coastal Upwelling" (ed. F.A. Richards). Amer. Geophys. Union. 449–457.

Cushing, D.H. (1982a). "A simulacrum of the Iceland cod stock." *J. Cons. int. Explor. Mer.* **40**, 1, 27–36.

Cushing, D.H. (1982b). "Sources of variability in the North Sea ecosystem." (in press).

Cushing D.H. and J.P. Bridger (1966). "The stock of herring in the North Sea and changes due to fishing." *Fish. Invest.* **2**, 25 (1), 123 p.

Cushing, D.H. and A.C. Burd (1957). "On the herring of the Southern North Sea." *Fish. Invest. Lond.* **2**, 20(11) 31 p.

Cushing, D.H. and Dickson, R.R. (1976). "The biological response in the sea to climatic changes." *Adv. Mar. Biol.* **14**, 1–122.

Cushing, D.H. and J.G.K. Harris (1973). "Stock and recruitment and the problem of density dependence." *Rapp. Procès-Verb. Cons. int. Explor. Mer* **164**, 142–155.

Cushing, D.H. and T. Vucetic (1963). "Studies on a *Calanus* patch. III. The quantity of food eaten by *Calanus finmarchicus*." *J. Mar. Biol. Assn. NS UK* **43**, 349–71.

Daan, N. (1978). "Changes in cod stocks and cod fisheries in the North Sea." *Rapp. Procès. Verb. Cons. int. Explor. Mer* **172**, 39–57.

Dansgaard, W. (1981). "Palaeoclimatic studies on ice cores." *In* "Climatic Variations and Variability: Facts and Theories." (ed. A. Berger) Reidel, Dordrecht, Boston. 193–206.

Dansgaard, W., S.J. Johnson, H.B. Clausen and C.C. Langway (1971). "Climatic record revealed by Camp Century Ice Core" *In* "The Late Cenozoic Glacial Ages" (ed. K. Turekian), Yale University Press. 37–56.

Dahl, K. (1907). "The scales of the herring as a means of determining age, growth and migration." *Rep. Norweg. Fish. Mar. Invest.* **2**, (6), 1–36.

Dannevig, A. (1930). "The death of fish in very cold winters." *J. Cons. int. Explor. Mer* **5**, 194–196.

Dannevig, A. (1930). "The propagation of our Common fishes during the cold winter 1924." *Rep. Norw. Fish. Mar. Invest.* **3**, 10, 133 pp.

Day, F. (1880–1894). "The Fishes of Great Britain and Ireland." Williams and Norgate, London.

Darbyshire, M. (1967). "The surface waters off the coast of Kerala, Southwest India." *Deep Sea Res.* 14, (3), 295–320.

Defant, A. (1924). "Die Schwankungen der atmosphärischen Zirkulation." *Geografiska Annaler* **6**, 131–141.

Desbrosses, P. (1932). "Capture d'une Tortue (*Dermatochelys coriacea* L.) dans la baie d'Etel." *Bull. soc. zool. France* 57, 274–277.

Debrosses, P. (1935). "Échouage d'un Tetrodon, *Tetrodon lagocephalus* L. près de Quiberon et remarques sur la presence de cette espèce et de *Balistes capriscus* L. au nord du 44°L.N." *Bull. Soc. zool. France* **60**, 43–48.

Desbrosses, P. (1936). "Présence a l'entrée occidentale de la Manche de *Callianthias ruber* (Rafinesque 1810)." *Bull. Soc. zool. France* **61**, 406–407.

Devold, F. (1963). "The life history of the Altanto-Scandian herring." *Rapp. Procès-Verb. Cons. Intern. Explor. Mer* **154**, 98–108.

Dickie, L.M. (1955). "Fluctuations in abundance of The Giant Scallop *Placopecten magellanicus* (Gmelin) in the Digby area of the Bay of Fundy." *J. Fish. Res. Bd Can.* 12 (6), 797–856.

Dickson, R.R. (1971). "A recurrent and persistent pressure anomaly pattern as the principal cause of intermediate scale hydrographic variation in the European shelf sea." *Deutsch. Hydrogr. Zeitschr.* **24**, 3, 97–119.

Dickson, R.R. and H.H. Lamb (1972). "A review of recent hydrometeorological events in the North Atlantic sector." *ICNAF Spec. Publ.* **8**, 35–62.

Dickson, R.R., J.G. Pope and M.J. Holden (1973). "Environmental influences on the survival of North Sea cod." *In* "The Early Life History of Fish." (ed. J.H.S. Blaxter) Springer, Heidelberg. 69–80.

Dietrich, G. (1954a). "Verteilung, Ausbreitung und Vermischung der Wasserkörper in der südwestlichen Nordsee auf Grund der Ergebnisse der' "Gauss"-Fahrt' im Februar-März 1952." *Ber. deutsch. wiss. Komm. Meeresf.* **NF B13** (2), 104–29.

Deitrich, G. and K. Kalle (1957). "Allgemeine Meereskunde". Gebrüder Borntraeger, Berlin. 492 pp.

Dooley, H. (1974). "Hypotheses concerning the circulation of the northern North Sea." *J. Cons. int. Explor. Mer.* **36** (1), 54–61.

Douglas, A.V. (1980). "Geophysical estimates of sea surface temperatures off Western North America since 1671." *Calcofi. Rep.* **21**, 102–112.

Dow, R.L. (1969). "Cycle and geographic trends in sea water temperatures and abundance in American lobster catches." *Science* **164**, 1060–1062.

Easton, C. (1928a). "Periodicity of winter temperatures in western Europe since AD 760." *Proc. Roy. Acad. Amsterdam* **20**, 1092–1107.

Easton, C. (1928b) "Les hivers dans l'Europe occidentale." E.J. Brill, Leyden. 210 pp.

Eichenlaub, V.L. (1971). "Further comments on the climate of the mid-nineteenth century United States compared to current normals." *Monthly Weather Review* **99** (11), 847–850.

Eggvin, J. (1964). "Water movement in the central part of the Norwegian Sea based on recent material." *Int. Council Explor. Sea* CM 1964, 138, 12 pp. (mimeo).

Ellett, D.J. and J.H.A. Martin (1973). "The physical and chemical oceanography of the Rockall Channel." *Deep Sea Res.* **20**, 585–625.

Erlandsson, S. (1936). "Dendrochronological studies." Data 23 fr Stockholms Hogskalas Geokronol Inst. 116 pp.

Favorite, F. and D.R. MacLean (1973). "Coherence in transpacific movements of positive and negative anomalies of sea surface temperature, 1953–60." *Nature* **244**, 139–143.

Flores, L.A.P. (1967). "Informe preliminar del Crucero 6611 de la Primavera de 1966 (Cabo Blanco-Punta Coles)". *Informe Inst. del Mar del Peru* **17**, 16.

Flores, L.A.P. and Elias, L.A.P. (1967). "Informe preliminar del Crucero 6608–09 de Invierno 1966 (Mancora-Ilo)". *Informe Inst. del Mar del Peru* 24.

Foerster, R.F. (1968). "The Sockeye salmon (*Oncorhynchus nerka*)." *Bull. Fish. Res. Bd Can.* 162.

Fonselius, S.H. (1962). "Hydrography of the Baltic Deep Basins." *Fish. Bd Sweden Ser. Hydrogr. Rep.* **13**, 40.

Fonselius, S.H. (1969). "Hydrography of the Baltic Deep Basins III." *Fish. Bd Sweden Ser. Hydrogr.* **23**, 97.

Fonselius, S.H. (1970). "Observations at Swedish lightships and in the Central Baltic in 1969" *Ann. Biol. Cons. int. Explor. Mer.* **26**, 89–92.

Fonselius, S.H. (1972). "On primary production in the Baltic." *Medd. fr. Havfiskelaboratoriet Lysekil* **134**, 3.

Fonselius, S.H. and C. Tattanasen (1970). "On the water renewals in the Eastern Gotland Basin after World War II." *Int. Council Explor. Sea* CM C 8, 6 pp. Mimeo.

Fridriksson, A. (1949). "Boreo-tended changes in the marine vertebrate fauna of Iceland during the last twenty five years." *Rapp. Procès-Verb. Cons. int. Explor. Mer* **125**, 30–32.

Fritts, H.C. (1976). "Tree Rings and Climate." Academic Press, London and New York. 567 pp.

Fulton, T.W. (1897). "On the growth and maturation of the ovarian eggs of teleostean fishes." *Ann. Rep. Fish. Bd Scot.* **3**, 88–134.

Garrod, D.J. (1977). "The North Atlantic cod." *In* "Fish Population Dynamics." (ed. J.A. Gulland,) 216–242.

Garrod, D.J. and J.M. Colebrook (1978). "Biological effects of variability in the north Atlantic Ocean." *Rapp. Procès-Verb. Cons. int. Explor. Mer* **173**, 128–144.

Garstang, W. (1900–3). "The impoverishment of the sea." *J. Mer. Biol. Assn. UK NS* **6**, 1–70.

Gehrke, J. (1922). "On the after effect of ice winters upon the Deep Sea temperatures of the Kattegat." *Publ. Circ. Cons. int. Explor. Mer.* **75**, 1–58.

Gehrke, J. (1923). "Further investigations on the after effect of ice winters upon the deep sea temperatures of the Kattegat." *Publ. Circ. Cons. int. Explor. Mer.* **81**, 1–11.

Gieskes, W.W.C., G.W. Kraay and M.A. Baars (1979). "Current ^{14}C methods for measuring primary production: gross underestimates in oceanic waters." *Neth. J. Sea Res.* **13**, 1, 58–78.

Gilis, Ch. (1963). "Note préliminaire sur les pêches abondantes de soles réalisées par les pêcheurs belges dans le sud de la Mer du Nord (IVc) au cours de l'hiver 1962–1963." CM 1963 Near Northern Seas 30, 3 pp.

Glémarec, M. (1979). "Les fluctuations temporelles des peuplements benthiques liées aux fluctuations climatiques." *Oceanol. Acta* **2**, 3, 365–371.

Glover, R.S. (1957). "An ecological survey of the drift net herring survey off the northeast coast of Scotland II; the planktonic environment of the herring." *Bull. Mar. Ecol.* **5**, 1–43.

Glover, R.S. (1979). "Natural Fluctuations of Populations." *Ecotoxicology and Environmental Safety* **3**, 190–203.

Glover, R.S., G.A. Robinson and J.M. Colebrook (1972). "Plankton in the North Atlantic—an example of the problems of analyzing variability in the environment." *In* "FAO: Marine Pollution and Sea Life" (ed. M. Ruivo) pp. 439–445. Fishing News (Books). West Byfleet, Surrey and London.

Gillbricht, M. (1964). "Einwirkung des kalten Winters 1962/63 auf die Phytoplankton entwicklung bei Helgoland." *Helgol. Wiss. Meeresunters.* **10**, 263–275.

Geodecke, E. (1958). "Hydrographische Betrachtungen uber die regionale Verteilung der mittleren Eisvorbereitungs- und Eisabschmelzeit in der Deutschen Bucht und der westlichen Ostsee mit Karten." *Deutsche Hydrographisches Zeitschrift* **11**, 1–22.

Goedecke, E. (1963). "Uber die hydrographische situation in der Deutschen Bucht wahrend des Eiswinters 1962/3." *Deutsche Hydrographische Zeitschrift* **16**, 5, 249–263.

Goedecke, E. (1964). "Zur Hydrographie der Deutschen Bucht wahrend des Winters 1962/63." *Helgol. Wiss. Meeresunters.* **10**, 231–245.

Goldman, J.G., C.D. Taylor, and P.M. Gilbert (1981). "Non-linear time course uptake of carbon and ammonium by Marine Phytoplankton." *Mar. Ecol. Progress Ser.* **6**, 2, 137–148.

Goode, G. Brown (1877). "The natural and economical history of the American menhaden." *Rep. Comm. US Comm. Fish. Fisheries.* 506 pp.

Goode, G. Brown (1887). "The Fisheries and Fishery Industries of the United States. V. History and methods of the Fisheries." 808 pp.

Goode, G. Brown and J.W. Collins (1887). "The Cod, Haddock and Hake Fisheries." The Fisheries and Fishing Industries of the United States V, Vol. I.

Goode, G. Brown and T.H. Bean (1895). "Oceanic ichthyology." *Mem. Mus. Comp. Zool. Harvard* **22**, 553 pp.

Goode, G. Brown, J.W. Collins, R.E. Earll and A.M. Clark (1895). "Materials for a history of the mackerel fishery." *Rep. Comm. Fish. Fisheries* 1884, 93–531.

Gran, H.H. (1923). "Snesmeltingen som hovedaarsak til den rike produktion i vort kysthav om vaaren." *Samtiden* 34, 606–613.

Grauman, G.B. (1963). "Features of the cod spawning in the Baltic in 1961." *Ann. Biol. Cons. int. Explor. Mer* **18**, 128–9.

Grauman, G.B. (1964). "Spawning in the Southern Baltic, 1962." *Ann. Biol. Cons. int. Explor. Mer.* **19**, 98–100.

Grauman, G.B. (1966). "Some data on the spawning of cod in the Southern Baltic in 1964." *Ann. Biol. Cons. int. Explor. Mer.* **21**, 83–4.

Grauman, G.B. (1974). "Investigations on the spawning of Baltic cod in 1968–70." *Rapp. Procès Verb. Cons. int. Explor. Mer* **166**, 7–9.

Greer–Walker, M. (1970). "Growth and development of the skeletal muscle fibres of the cod (*Gadus morhua* L.)." *J. Cons. int. Explor. Mer* **33** (2), 228–44.

Grimes, C.B., K.W. Able, S.C. Turner and S.J. Katz (1980). "Tilefish: its continental shelf habitat." *Bull. Amer. Litt. Soc.* **12**, 4. 34–9.

Groot, S.J. (1964). "Diurnal activity and feeding habits of plaice." *Rapp. Procès. Verb. Cons. int. Explor. Mer* **155**, 48–51.

Gruza, G.B. and E. Ya Rankova (1979). "Data on the structure and variability of climate air temperature at sea level in the northern hemisphere." *Nauch. Issled. Inst. Gidroent. Inf. Moscow.* 204 pp.

Guillen, O. (1976). "El sistema de la corriente Peruana. Parte I. Aspectos fisicos." *FAO Fish Inform.* **185**, 243–284.

Guillen, O. and Flores, L.A.P. (1967). "Informe preliminar del Crucero 6702 del Verano de 1967 (Cabo Blanco-Arica)." *Informe Inst. del Mar del Peru* **18**, 17 pp.

Guillen, O. and Vasquez, F. (1966). "Informe preliminar del Crucero 6602 (Cabo Blanco-Arica)." *Informe Inst. del Mar del Peru* **12**, 27 pp.

Gulland, J.A. (1970). "The fish resources of the ocean." *FAO Fish. Tech. Paper* **97**, 425 pp.

Gulland, J.A. (1977). "The analysis of data and development of models." *In* "Fish Population Dynamics." (ed. J.A. Gulland). John Wiley, New York. Pp. 67–95.

Gunther, E.R. (1936). "A report on oceanographical investigations in the Peru coastal current." *Discovery Rep.* **13**, 109–276.

Hammer, C.U., H.B. Clausen and W. Dansgaard, (1980). "Greenland ice-sheet evidence of post glacial volcanism and its climatic impact." *Nature* **288**, 230–235.

Hancock, D.A. and A.E. Urquhart (1964). "Mortalities of edible cockles (*Cardium edule* L.) during the severe winter of 1962–3." *J. Anim. Ecol.* **33**, 176–178.

Hansen, P.M. (1949). "Studies on the biology of cod in Greenland waters." *Rapp. Procès. Verb. Cons. int. Explor. Mer* **129**, 5–77.

Hansen, P.M. (1954). "The stock of cod in Greenland waters during the years 1924–52." *Rapp. Procès. Verb. Cons. int. Explor. Mer* **136**, 65–71.

Hansen, P.M. (1968). "Report on cod eggs and larvae." *Spec. Publ. Int. Commn N.W. Atlant. Fish.* 7, 127–138.

Hansen, P.M., A.S. Jensen and A.V. Tåning (1935). "Cod marking experiments in the waters of Greenland, 1924–33." *Medd. fra. Komm f. Danmarks Fisk. og Havundersog.* **10**, 1, 119 pp.

Harding, D., J.H. Nichols and D. Tungate (1978). "The spawning of the plaice (*Pleuronectes platessa* L.) in the southern North Sea and English Channel." *Rapp. Procès-Verb. Cons. int. Explor. Mer* **172**, 102–113.

Hardy, A.C. and E.R. Gunther (1935). "The plankton of the South Georgia Whaling grounds and adjacent waters, 1926–1927." *Discovery Rep.* **11**, 1–456.

Heincke, F. (1913). "Untersuchungen uber die Scholle-Generalbericht I. Schollenfischerei und Schonmassregeln. Vorlaufige Kurze Ubersicht uber die wichtigsten Ergebnisse des Berichts." *Rapp. Procès-Verb. Cons. Int. Explor. Mer* **16**, 1–70.

Hermann, F., P.M. Hansen and S.A. Horsted (1965). "The effect of temperature and currents on the distribution and survival of cod larvae at West Greenland." *Spec. Publ. Intern. Commn N.W. Atlant. Fish.* **6**, 389–409.

Hickey, Barbara (1975). "The relationship between fluctuations in sea level, wind stress and sea surface temperature in the Equatorial Pacific." *J. Phys. Oceanogr.* **5**, 460–475.

Hjort, J. (1914). "Fluctuations in the great fisheries of northern Europe reviewed in the light of biological research." *Rapp. Procès-Verb. Cons. int. Explor. Mer* **20**, 1–228.

Hjort, J. and J.T. Ruud (1929). "Whaling and Fishing in the North Atlantic." *Rapp. Procès-Verb. Cons. int. Explor. Mer* **56**, 1, 123 pp.

Höglund, H. (1972). "On the Bohuslån herring during the great herring fishery period in the eighteenth century." *Rep. Inst. Mar. Res.* **20**, 1–86.

Höglund, H. (1978). "Long term variations in the Swedish herring fishery off Bohuslån and their relation to North Sea herring." *Rapp. Procès-Verb. Cons. int. Explor. Mer* **172**, 175–186.

Holden, M.J. (1978). "Long term changes in landings of fish from the North Sea." *Rapp. Procès-Verb. Cons. int. Explor. Mer* **172**, 11–26.

Holland, W.R. (1977). "Oceanic General Circulation Models." 3–46. *In* "The Sea" (ed. E.D. Goldberg, I.N. McCave, J.J. O'Brien and J.H. Steele). Vol. 6, 1048 pp.

Holme, N.A. (1961). "The bottom fauna of the English Channel." *J. Mar. Biol. Assn. NS UK* **41**, 397–461.

Horsted, S. A. (1967). "Some observations on the year class fluctuations in the West Greenland cod stock." *ICNAF Red Book 1967* **III**, 26–38.

Horsted, S. (1973). "Recent information on landings, age compositions and recruitment of Subarea I cod, and estimates of yield in 1972–5." *Int. Commn N.W. Atlant. Fish. Res. Doc.* **73/107**, 23 pp.

Horsted, S.A. and P. Kanneworff (1977). "Subarea I cod: data for 1975–6 and estimates of yield." *Int. Commn N.W. Atlant. Fish. Res. Doc.* 77, VI, 8. 31 pp.

Hurlburt, H.E. and J.D. Thompson (1973). "Coastal upwelling on a plane." *J. Phys. Oceanogr.* **3**, 16–32.

Hunter, J.R. (1972). "Swimming and feeding behaviour of larval anchovy *Engraulis mordax.*" *Fish. Bull.* **70**, (3) 821–38.

Hurlburt, H.E., J.C. Kindle and J.J. O'Brien (1976). "A numerical solution of the onset of El Niño." *J. Phys. Oceanogr.* **6**, 5, 621–31.

Iles, T.D. (1967). "Growth studies on North Sea herring. I. The second year's growth (I-group) of East Anglian herring, 1939–63." *J. Cons. int. Explor. Mer* **31**, 56–76.

Iles, T.D. (1968). "Growth studies of North Sea herring, II. 0-group growth of East Anglian herring." *J. Cons. int. Explor. Mer* **32**(1), 98–116.

Iselin, C.O'D. (1940). "Preliminary report on long period variations in the transport of the Gulf Stream system." *Pap. Phys. Oceanogr. Meteorol.* **8** (1), 40 pp.

Iselin, C.O'D. (1936). "A study of the circulation of the western North Atlantic." *Pap. Phys. Oceanogr.* 4 (4), 1–97.

Ishevskii, G.K. (1961). "The Systematic Basis of Predicting Oceanological Conditions and the Reproduction of the Fisheries." Moscow. 1–165.

Jakobsson, J. (1963). "Some remarks on the distribution and availability of the Iceland North coast herring." *Rapp. Procès-Verb. Cons. int. Explor. Mer* **154**, 73–82.

Jakobsson, J. (1969). "On herring migrations in relation to changes in sea temperature." *Jokull* **19**, 134–145.

Jamieson, A. and G. Otterlind (1971). "The use of cod blood protein polymorphism in the Belt Sea, the Sound and the Baltic Sea." *Rapp. Procès-Verb. Cons. int. Explor. Mer* **161**, 55–59.

Jamieson, A. and R.J. Turner (1981). "The extended series of Tf alleles in Atlantic cod (*Gadus morhua* L.) "*In* "Marine Organisms: Genetics, Ecology and Evolution." (ed. B. Battaglia and J. Beardmore). Plenum Press, New York.

Jefferies, D.F., A. Preston and A.K. Steele (1973). "Distribution of caesium-137 in British coastal waters." *Mar. Poll. Bull.* 4, 8. 118–122.

Jenkins, J.T. (1927). "The Herring and the Herring Fisheries." King and Son, London. 175 pp.

Jensen, A.J.C. (1952). "The influence of hydrographical factors on fish stocks and fisheries in the transition area, especially on their fluctuations from year to year." *Rapp. Procès-Verb. Cons. int. Explor. Mer* **131**, 51–60.

Jensen, Ad. S. (1925). "On the fishery of the Greenlanders." *Medd. fra. Komm. f. Danmarks Fisk. of Havundersogelser* 7 (7), 1–39.

Jensen, Ad. S. (1939). "Concerning a change of climate during recent decades in the arctic and subarctic regions, from Greenland in the West to Eurasia in the

East and contemporary biological and geophysical changes." *Det. Konglike Videnskabernes Selska Biol. Medd.* 14 (8), 75 pp.

Jensen, Ad. S. and P.M. Hansen (1931). "Investigations on the Greenland cod *Gadus callarias* (L.)." *Rapp. Procès-Verb. Cons. int. Explor. Mer* 72, 1–41.

Johansen, A.C. (1924). "On the summer and autumn spawning herrings of the North Sea." *Medd. fra. Komm. Danmarks Fisk. og Havunders.* **5**, 119 pp.

Johansen, A.C. (1925). "On the influence of currents upon the frequency of the mackerel in the Kattegat and adjacent parts of the Kattegat." *Medd. Komm. Havunders.* 7 (8), 26 pp.

Johansen, A.C. (1926). "On the remarkable quantities of haddock in the Belt Sea during the winter of 1925–26 and causes leading to the same." *J. Cons. int. Explor. Mer* **1**, 140–156.

Jones, F.R. Harden (1968). "Fish Migration." Arnold, London. 325 pp.

Jones, F.R. Harden & P. Scholes (1974). "The effect of low temperature on cod, *Gadus morhua* L." *J. Cons. int. Explor. Mer* **35**(3), 258–271.

Jones, F.R. Harden, M. Greer Walker and G.P. Arnold (1978). "Tactics of fish movement in relation to migration and water circulation." *In* "Proceedings of the Joint Oceanographic Assembly, 1976." (ed. Sir George Deacon and H. Charnock) Plenum Press, New York. pp. 185–207.

Jones, F.R. Harden, G.P. Arnold, M. Greer Walker and P. Scholes (1979). "Selective tidal stream transport and the migration of plaice (*Pleuronectes platessa* L.) in the southern North Sea." *J. Cons. int. Explor. Mer* **38** (1), 331–37.

Jones, R. (1973). "Density dependent regulation of the numbers of cod and haddock." *Rapp. Procès-Verb. Cons. int. Explor. Mer* **164**, 156–73.

Jonsson, J. (1954). "On the Icelandic stock of cod during the years 1928–33." *Rapp. Procès-Verb. Cons. int. Explor. Mer* **136**, 51–7.

Jonsson, J. (1966). "Abundance, recruitment and growth in the Icelandic stock of cod." ICES CM 1966 G.13, 3 pp. Mimeo.

Jordan, R. (1967). "The predation of guano birds on the Peruvian anchovy (*Engraulis ringens* Jenyns)." *Calcofi Rep.* **XI**, 105–109.

Jordan, R. (1976). "Biologia de la anchoveta. Parte I. Resumen del Conocimiento actual." *Actas de la Reunion de Trabajo sobre el Fenomeno Conocido como El Niño. FAO Informes de Pesca* **185**, 359–399.

Jordan, R. and Cabrera, D. (1960). "Alguanas resoltados de la anillaciones de Guanay efectudos durante 1939–41 y 1949–53." *Bol. Comp. Agg. Guana* **36**, 1, 11–26.

Jordan, R. and Fuentes, H. (1966). "La poblaciones de aves guarneras y su situacion actual." *Informe del Inst. del Mar del Peru* **10**, 30 pp.

Jordan, R. and Vildoso, A.C. (1965). "La Anchoveta (*Engraulis ringens* J.) Conocimiento actual sobre su biologia, ecologia y pesqueria." *Informe del Inst. del Mar del Peru* **6**, 52 pp.

Kalle, K. (1943). "Die grosse Wasserumschichtung im Gotland-Tief vom Jahre 1933/34." *Ann. Hydrogr. Marit. Meteorol.* Apr–June 1943.

Kasahara, H. (1962). "Fisheries Resources of the North Pacific Ocean." H. R. MacMillan Lectures on Fisheries, University of British Columbia.

Kelly, P.M. (1979). "Volcanic dust veils and North Atlantic climatic change." *Nature* **268**, 5621, 616–617.

Kesteven, G.L., Gulland, J.A., Jones, R., Barber, R., and Boerema, L. (1977). "Report of the consultative group convened by the Minister of Fisheries in Peru to advise him on the state of the stocks of anchoveta and other pelagic species and on the course of action taken for the management of the fishery." July 1977, 17 pp.

Ketchen, K.S. (1956). "Climatic trends and fluctuations in yield of Marine Fisheries in the North Pacific." *J. Fish. Res. Bd Can.* **13**, 357–374.

Killick, S.R. (1955). "The chronological order of Fraser River sockeye during migration, spawning and death." *Int. Pac. Salm. Comm.* 7, 95 pp.

King, J.E. and T.S. Hida (1957). "Zooplankton abundance in the central Pacific. II." *Fish. Bull. US Fish and Wildlife Serv.*, 57 (118), 365–95.

Kislayakov, A.G. (1961). "The relationship between hydrological conditions and variations of cod year class abundance." *Trudy soveshchani Ikhtiologich eskaya kommisiyn* **13**, 260–264.

Koch, L. (1945). "The East Greenland Ice." *Medd. om Grønland* **130**, 3, 373 pp.

Koslow, J.A. (1976). "Alaska pollack—already overfished." *Sea Frontiers* **22**, 2, 98–105.

Kothke, H. and H. Koops (1963). "Die Fischsterben in der Elbe im Winter 1962–3 eine Folge der verumreinigung des Elbwassers." *Arch. für Fischerei Wissenschaft* 14, 50–55.

Kukla, G.J. and D.A. Robinson (1981). "Temperature changes in the last hundred years." *In* "Climatic Variations and Variability: Facts and Theories." (ed. A. Berger) 193–206.

Lam, R.C. (1968). "Estudio sobre la variacion del contenido de Grasa en la anchoveta Peruana (*Engraulis ringens* J.)". *Informe del Inst. del Mar del Peru* **24**, 29 pp.

Lamb, H.H. (1959). "Our changing climate, past and present." *Weather* **14**, 299–318.

Lamb, H.H. (1963). "On the nature of certain climatic epochs which differed from the modern (1900–39) normal." *Proc. WMO/UNESCO Rome 1961 Symposium on Changes of Climate* 125–150.

Lamb, H.H. (1965). "Frequency of weather types." *Weather* **20**, 9–12.

Lamb, H.H. (1970). "Volcanic dust in the atmosphere; with a chronology and assessment of its meteorological significances." *Phil. Trans. Roy. Soc.* **266**, 425–533.

Lamb, H.H. (1972). "Climate: Past, Present and Future. I. Fundamentals and climate now." Methuen, London. 613 pp.

Lamb, H.H. (1977). "Climate: Present, Past and Future. II. Climatic history and the future. Methuen, London. 835 pp.

Larkin, P.A. (1977). "Pacific salmon." *In* "Fish Population Dynamics" (ed. J.A. Gulland) Wiley, London. 156–186.

Lasker, R. (1975). "Field criteria for survival of anchovy larvae: the relation between inshore chlorophyll maximum layers and successful first feeding." *Fish. Bull. US* 73, 453–62.

Laurs, R.M. and R.J. Lynn (1977). "Seasonal migration of albacore, *Thunnus alalunga,* into North American coastal waters, their distribution, relative abundance and association with Transition Zone waters." *Fish. Bull. US* 75, 795–822.

Le Cren, E.D. (1965). "Some factors regulating the size of populations of freshwater fish." *Mitt. Int. Verein. Limnol.* 13, 88–105.

Lee, A.J. (1952). "The influence of hydrography on the Bear Island cod fishery." *Rapp. Procès-Verb. Cons. Intern. Explor. Mer* 131, 74–102.

Lee, A.J. and A.R. Folkard (1969). "Factors affecting turbidity in the southern North Sea." *J. Cons. int. Explor. Mer* 32, 291–302.

Lett, P.F. and A.C. Kohler, (1976). "Recruitment: a problem of multi-species interaction and environmental perturbations, with special reference to Gulf of St Lawrence herring (*Clupea harengus harengus*)." *J. Fish. Res. Bd Can.* 33, 1353–71.

Li, C.C. (1955). "Population Genetics." University Chicago Press, Chicago. 366 pp.

Lindquist, A. (1960). "Salthatlen och djurvälden i Baltiska havet." *Fiskertidskr. för Finland* 3, 90–94.

Ljungman, A. (1882). "Contribution towards solving the question of the secular periodicity of the great herring fisheries." *US Comm. Fish Fisheries* 7 (7), 497–503.

Lockwood, S.J. (1972). "An ecological survey of an 0-group plaice (*Pleuronectes platessa* L.) population, Filey Bay, Yorkshire." Ph.D Thesis, University of East Anglia. 169 pp.

Lockwood, S.J. (1978). "The mackerel: a problem in stock assessment." *Min. Agr. Fish. Food Direct. Fish Res. Lowestoft Lab. Leaflet 44* 18pp.

Longhurst, A.R. (1967). "The pelagic phase of *Pleuronectes planipes* Stimpson (Crustacea, Galatheidae) in the California current." *CALCOFI Rep.* 11, 142–154.

Longhurst, A.R. (1971). "The clupeid resources of tropical seas." *Oceanogr. Mar. Biol. Ann. Rev.* 9, 349–85.

Lumby, J.R. and G.T. Atkinson (1929). "On the unusual mortality amongst fish during March and April 1929 in the North Sea." *J. Cons. int. Explor. Mer* 4, 309–332.

Lundberg, R. (1886). "The Fisheries of Sweden." US Bureau Fish. 1884.

Lyons, C. (1964). "Salmon: Our Heritage." British Columbia Packers Ltd. 768 pp.

McCreary, J. (1976). "Eastern tropical ocean response to changing wind systems with applications to El Niño." *J. Phys. Oceanogr.* 6, 632–645.

McKenzie, R.A. and R.E.S. Homans (1937). "Rare and interesting fishes and salps in the Bay of Fundy and off Nova Scotia." *Proc. Nova Scotian Inst. Sci.* 19 (3), 277.

Madsen, K. Popp (1978). "The industrial fisheries in the North Sea." *Rapp. Procès-Verb. Cons. int. Explor. Mer* 172, 27–30.

Marr, J.C. (1960). "The causes of major variations in the catch of the Pacific sardine *Sardinops caerulea* (Girard)." *Proc. World Sci. Meetings Biology of Sardines and related species* FAO III 667–79.

Marshall, P.T. (1957). "Primary production in the Arctic." *J. Cons. int. explor. Mer* 23, 173–7.

Martin, W.R. and A.C. Kohler (1965). "Variation in recruitment of cod (*Gadus morhua* L.) in southern ICNAF waters as related to environmental changes." *Spec. Publ. Intern. Commn N.W. Atlant. Fish.* 6, 833–846.

Manley, G. (1974). "Central England temperatures: monthly means 1659–1973." *Quart. J. Roy. Meteorol. Soc.* 100, 389–405.

Maslov, N.A. (1944). "Bottom fishes of the Barents Sea." *Knipovich. Inst. Murmansk, Trans.* 8, 3–186.

Mason, B.J. (1976). "Towards the understanding and prediction of climatic variations." *Quart. J. Roy. Meteorol. Soc.* 102, 473–498.

Matsumoto, W. (1966). "Distribution and abundance of tuna larvae in the Pacific Ocean." *Proc. Governor's Conf. Centr. Pac. Res. Honolulu 1966* (ed. T.A. Manier). pp. 221–30.

Meinardus, W. (1934). "Die Niederschlags-verterlung auf der Erde." *Meteor. Z. Bd* 51, 345.

Meyer, P.F. and K. Kalle (1950). "Die biologisch Umstimmung der Ostsee in den letzten Jahrzehnten-eine Folge hydrographischer Wasserumschichtungen." *Arch. für Fischerei Wissenschaft* II 1–9.

Meyer-Waarden, P.F. and K. Tiews (1965). "Die Zusammensetzung der deutschen Garnelenfischerei in den Jahren 1954–1960." *Ber. dt. wiss. Kommn Meeresforsch.* 24 (2), 124–143.

Miller, F.R. and R.M. Laurs (1975). "The El Niño of 1972–73 in the eastern tropical Pacific Ocean." *Bull. Inter-Amer. Trop. Tuna Comm.* 16, 5, 401–448.

Miller, H. (1855). "On the late severe frost." *Proc. Roy. Phys. Soc. Edinb.* 1, 10–14.

Mitchell, J.M. (1864). "The Herring: Its Natural History and National Importance." Edinburgh.

Mitchell, J.M., Jr (1963). "On the world-wide pattern of secular temperature change." *In* "Changes of Climate: Proceedings of the UNESCO and WMO Symposium, Rome", UNESCO, Paris. pp. 161–181.

Mitchell, J.M., Jr (1975). "Note on solar variability and volcanic activity as potential sources of climatic variability." *In* "The Physical Basis of Climate and Climatic Modelling." 127–131.

Meyers, Gary (1979). "Annual variation in the slope of the 14°C isotherm along the Equator in the Pacific Ocean." *J. Phys. Oceanogr.* **9**, 885–891.

Minano, J.M. (1968). "Informe preliminar del Crucero 6705–06 del Otono de 1967 Cabo Blanco-Ilo." *Informe Inst. del Mar del Peru* **21**, 13 pp.

Minano, J.B. (1968). "Estudio de la fecundidad y ciclo sexual de la anchoveta (*Engraulis ringens* J.) en la zona de Chimbote." *Bol. Inst. del Mar del Peru* **1**, 9, 505–552.

Mooers, C.N.K., C.A. Collins and R.L. Smith (1976). "The dynamic structure of the frontal zone in the coastal upwelling region off Oregon." *J. Phys. Oceanogr.* **6**, 1, 3–21.

Motoda, S. and Y. Hirano (1963). "Review of Japanese herring investigations." *Rapp. Procès-Verb. Cons. int. Explor. Mer* **154**, 249–262.

Moyse, J. and A. Nelson Smith (1964). "Effects of the severe cold of 1962–3 upon shore animals in S. Wales." *J. Anim. Ecol.* **33**, 183–190.

Murphy, G.I. (1966). "Population biology of the Pacific sardine (*Sardinops caerulea*)." *Calif. Acad. Sci., Proc., Ser. 4,* **34**(1), 1–84.

Murphy, R.C. (1926). "Oceanic and climatic phenomena along the west coast of South America during 1925." *Geogr. Rev.* **16**, 26–54.

Nakamura, H. (1969). "Tuna Distribution and Migration." Fishing News (Books) Ltd., West Byfleet, Surrey and London. 76 pp.

Namias, J. (1951). "The great Pacific anticyclone of winter 1949–50: a case study in the evolution of climatic anomalies." *J. Meteorol.* **8**, 251–261.

Namias, J. (1972). "Large-scale and long-term fluctuations in some atmospheric and oceanic variables." *In* "The Changing Chemistry of the Oceans. Nobel Symposium 20" Almqvist and Wiksell, Stockholm. 27–48.

Namias, J. (1973). "Thermal communication between the sea surface and the lower troposphere." *J. Phys. Oceanogr.* **3**, (4) 373–378.

Namias, J. (1976). "Some statistical and synoptic characteristics associated with El Niño." *J. Phys. Oceanogr.* **6**, 130–138.

Namias, J. (1981). "Case studies of exceptional climate in the United States 1975–79 and air–sea interactions." *In* "Climatic Variations and Variability: Facts and Theories." (ed. A. Berger). 369–398.

Namias, J. and R.M. Born (1970). "Temporal coherence in North Pacific Sea surface temperature patterns." *J. Geophys. Res.* **75** (30), 5952–5955.

Namias, J. and R.M. Born (1974). "Further studies of temporal coherence in North Pacific Sea surface temperatures." *J. Geophys. Res.* **79**, 6, 797–98.

Needler, G.T. (1967). "A model for thermocline circulation in an ocean of finite depth." *J. Mar. Res.* 25, 2, 329–42.

Nesis, K.N. (1960). "Variations in the bottom fauna of the Barents Sea under the influence of fluctuations in the hydrological regime." *Soviet Fishing Investigations in North European Seas, Moscow,* 129–138.

Newell, G.E. (1964). "The South coast, Whitstable area." J. Anim. Ecol. 33, 178–9.

Newman, G.G. (1977). "The living resources of the Southeast Atlantic." *FAO Fish Tech. Paper* 176, 33 pp.

Nusser, F. (1949). "Die Eisverhältnisse des Winters 1948/49 an dem deutschen Küste." *Deutsche Hydrographische Zeitschrift* 2, 6, 251–255.

Nusser, F. (1958). "Zusammenhang zwischen Grosswetterlage und Eisvorkommen an den deutschen Küsten." *Deutsche Hydrographische Zeitschrift* 11, 4, 184–194.

O'Brien, J.J. and H.E. Hurlburt (1972). "A numerical model of coastal upwelling." *J. Phys. Oceanogr.* 9, 885–891.

O'Brien, J.J., A. Busalacchi and J. Kindle (1981). "Ocean models of El Niño." *In* "Resource Management and Environmental Uncertainty Lessons from Coastal Upwelling Fisheries." (ed. Michael H. Glantz and J. Dana Thompson). Wiley-Interscience, London. 159–212.

O'Connor, J.F. (1969). "Hemispheric teleconnections of mean circulation anomalies at 700 mb." *ESSA Tech Rep* WB 10.

Olsen, O.T. (1883). "The Piscatorial Atlas." Olsen, Grimsby and London, 50 pp.

Ording, A. (1940–1). "Arringanalyser på gran og furu." *Medd. fra. det. Norske Skogforsøksvesen* VII, 105–354.

Orton, J.H. (1930/1). "On the effect of the severe winter of 1928–29 on the oyster drills of the Blackwater estuary." *J. Mar. Biol. Assn UK* NS 17, 301–313.

Orton, J.H. (1940). "Effect of the Severe Frost of the winter of 1939–40 on the Fauna of the Essex Oyster Beds." *Nature* 145, 708–9.

Otsu, T. (1960). "Albacore migration and growth in the north Pacific Ocean as estimated from tag recoveries." *Pacific Sci.* 14 (3), 257–60.

Otsu, T. and R.N. Uchida (1963). "Model of the migration of albacore in the North Pacific Ocean." *Fish. Bull. US Fish and Wildlife Serv.* 63 (1), 33–44.

Ottestad, P. (1942). "On periodical variations in the yield of the great sea fisheries and the possibility of establishing yield prognoses." *Fisk. Dir. Skrifter Havundersøg* 7, 5. 2–11.

Ottestad, P. (1969). "Forecasting the annual yield in sea fisheries." *Nature* 185 (4707), 183.

Parrish, R. (1976). "Environmental recruitment models and exploitation simulations of the California Current stock of Pacific mackerel (*Scomber japonicus*)." Ph.D. Thesis, State University, Corvallis, Oregon. 101 pp.

Patterson, A.H. (1905). "Nature in Eastern Norfolk." Methuen, London 352 pp.

Perkins, E.J. and B.R.H. Williams (1964). "The Solway Firth." *J. Anim. Ecol.* 33, 199–200.

Peterson, C.J.G., (1894). "On the biology of our flatfishes and on the decrease of our flatfisheries." *Rep. Dan. Bid. Sta.* 4, 146 pp.

Peterson, W.T. (1973). "Upwelling indices and annual catches of Dungeness crab, *Cancer magister*, along the west coast of the United States." *Fish. Bull. US* 71, 902–10.

Peterson, W.T. and C.B. Miller (1975). "Year to year variations in the planktology of the Oregon upwelling zone." *Fish. Bull. US* 73, 642–35.

Pettersson, O. (1915). "Climatic variations in historic and prehistoric time." *Svenska Hydrografisk Biologiska Komm. Skrifter* 5, 26 pp.

Pettersson, O. (1921–2). "Kosmiska orsaker till rövelserna uti hafvets och atmosfarens mellanskikt." *Svenska Hydrografisk Biologiska kommissioners fyrskepps-undersökning* 7, 23 pp.

Pettersson, O. (1926). "Hydrography, climate and fisheries in the transition area." *J. Cons. int. Explor. Mer.* 1, 4, 305–321.

Pickard, G.L. (1964). "Descriptive Physical Oceanography." Pergamon, Oxford. 199 pp.

Piechura, J. (1970). "The role of the North Sea waters in the forming of the hydrological conditions of the Baltic." *Proc. State. Inst. Hydrol. Meteorol.* 1, 3–22.

Pingree, R.D., P.R. Pugh, P.M. Holligan and G.R. Forster (1975). "Summer phytoplankton blooms and red tides along tidal fronts in the approaches to the English Channel." *Nature* 258 (5537), 672–677.

Pivovarova, Z.I. (1968). "The long term variation of intensity of solar radiation according to observations of actinometric stations." *Leningrad Trudy Glavnoi geofizicheskoi observatorie* 233, 17–37.

Postma, H. and J.W. Rommets (1979). "Dissolved and particulate organic carbon in the North Equatorial Current of the Atlantic Ocean." *Neth. J. Sea Res.* 13, 1, 85–98.

Postuma, K.H. (1978). "Immigration of southern fish into the North Sea." *Rapp. Procès-Verb. Cons. int. Explor. Mer* 172, 225–229.

Purdom, C.E., D. Thompson and P.R. Dando (1976). "Genetic analysis of enzyme polymorphisms in plaice (*Pleuronectes platessa*)." *Heredity* 37 (2), 193–206.

Quinn, W.H. (1974). "Monitoring and predicting El Niño invasions." *J. Appl. Meteorol.* 13, 825–904.

Radach, G. (1980). "Preliminary simulations of the phytoplankton and phosphate dynamics during FLEX '76 with a simple two component model." *Meteor Forschungsergebnisse A* 22, 151–164.

Radcliffe, W. (1921). "Fishing from the Earliest Times." John Murray, London. 478 pp.

Rae, K.M. (1957). "A relationship between wind, plankton distribution and haddock brood strength." *Bull. Mar. Ecol.* 4, 247–69.

Raja, B.T.A. (1969). "The Indian oil sardine." *Bull. Contr. Mar. Fish. Res. Inst.* 16, 128 pp.

Rasmussen, B. (1959). "On the migration pattern of the West Greenland stock of cod." *Ann. Biol. Cons. int. Explor. Mer* 14, 123.

Rauck, G. and J.J. Zijlstra (1978). "On the nursery aspects of the Waddensea for some commercial fish species and possible long term changes." *Rapp. Procès-Verb. Cons. int. Explor. Mer* 172, 266–275.

Rau, C. (1884). "Prehistoric fishing in Europe and North America." *Smithsonian Contributions to Knowledge* 509, 342 pp.

Rees, W.J. (1951). "The distribution of *Octopus vulgaris* Lamarck in British Waters." *J. Mar. Biol. Assn UK NS* 29, 361–378.

Rees, W.J. and J.R. Lumby (1954). "The abundance of *Octopus* in the English Channel." *J. Mar. Biol. Assn UK NS* 33, 515–536.

Reibisch, J. (1899). "Uber die Eizahl bei *Pleuronectes platessa* und die Altersbestimmung dieser Form aus den Otolithen." *Wiss. Meeres unters (Kiel)* 4, 233–48.

Reid, P.C. (1978). "Continuous Plankton Records. Large scale changes in the abundance of phytoplankton in the North Sea from 1958 to 1973." *Rapp. Procès-Verb. Cons. int. Explor. Mer* 172, 384–389.

Renk, H. (1972). "Chlorophyll a in the Southern Baltic in 1970." *Ann. Biol. Cons. int. Explor. Mer* 27, 72–3.

Renk, H. (1973). "Chlorophyll a in the Southern Baltic, 1971." *Ann. Biol. Cons. int. Explor. Mer* 28, 74–76.

Renk, H., H. Torbicki and S. Ochocki (1974). "Observations on chlorophyll a content of the Baltic waters in 1972." *Ann. Biol. Cons. int. Explor. Mer* 29, 64–67.

Renk, H., H. Torbiski and S. Ochocki (1975). "Chlorophyll contents of Baltic waters in 1973." *Ann. Biol. Cons. int. Explor. Mer* 30, 71–72.

Richardson, I.D., D.H. Cushing, F.R. Harden Jones, R.J.H. Beverton and R.W. Blacker (1959). "Echo sounding experiments in the Barents Sea." *Fish. Invest.* 2, 22 (9) 55 p.

Ricker, W.E. (1954). "Stock and recruitment." *J. Fish. Res. Bd Can.* 11 (5), 559–623.

Ricker, W.E. (1958). "Handbook of computations for biological statistics of fish populations." *Bull. Fish. Res. Bd Can.* 119, 300 pp.

Riepma, H.W. (1980). "Residual currents in the North Sea during the INOUT phase of JONSDAP '76. First results." *Meteor Forsch-Ergebnisse A* 22, 1, 19–32.

Riley, J.D. (1966). Marine Fish Culture. VII. Plaice (*Pleuronectes platessa* L.) post larval feeding on *Artemia salina* L. nauplii and effects of varying feeding levels." *J. Cons. int. Explor. Mer* 30, 204–21.

Robinson, G.A. (1970). "Continuous plankton records: variations in the seasonal cycles of phytoplankton in the North Atlantic." *Bull. Mar. Ecol.* 6, 333–345.

Rodewald, M. (1972). "Einige hydroklimatische Besonderheiten des Jahrezeits 1961–1970 im Nordatlantik und in Nordpolarmeere." *Deutsche hydrographische Zeitschrift* **25** (3), 97–117.

Roelofs, E.W. (1951). "The edible finfishes of North Carolina." Part 2. 109–39.

Rollefsen, G. (1955). "The arctic cod." *UN Sci. Conf. Conserv. Utiliz. Res., Proc., Rome,* pp. 115–17.

Rothschild, B.J. (1967). "Competition for gear in a multiple-species fishery." *J. Cons. int. Explor. Mer* **31** (1), 102–110.

Rothschild, B.J. and R.N. Uchida (1968). "The tuna resources of the oceanic regions of the Pacific Ocean." *Univ. Wash. Publ. Fish.* **4**, 19–51.

Runnstrøm, S. (1934). "The Pelagic distribution of the herring larvae in the Norwegian waters." *Rapp. Procès-Verb. Cons. int. Explor. Mer* **88** (5), 6 p.

Runnstrøm, S. (1936). "A study of the life history and migrations of the Norwegian spring herring based on the analysis of the winter rings and summer zones of the scale." *Fisk. Dir. Skr. Hav.* **5** (2), 103 pp.

Russell, F.S. (1967). "On the occurrence of the Scyphomedusan *Pelagia noctiluca* in the English Channel in 1966." *J. Mar. Biol. Assn. NS UK* **47**, 363–66.

Russell, F.S. (1973). "A summary of the observations of the occurrence of planktonic stages of fish off Plymouth 1924–72." *J. Mar. Biol. Assn. NS UK,* **53** 347–355.

Russell, F.S., A.J.S. Southward, G.T. Boalch and E.I. Butler (1971). "Changes in biological conditions in the English Channel off Plymouth during the last half century." *Nature* **234**, 468–70.

Ryland, J.S. (1966). "Observations on the development of larvae of the plaice, *Pleuronectes platessa* L., in aquaria." *J. Cons. int. Explor. Mer* **30** (2), 177–95.

Saetersdal, G. and Valdivia, J.E. (1964). "Un estudio del crecimiento, tamano y reclutamiento de la anchoveta (*Engraulis ringens* J.)". *Boll. Inst. del Mar del Peru* **1**, 4, 87–136.

Samuel, A.M. (1918). "The Herring: Its Effect on the History of Britain." John Murray, London. 199 pp.

Sars, G.O. (1869). "Indberetninger for det Indre fra Cand. G.O. Sars om de af ham i Aarene 1864–69, anstillede praktiskvidenskabdige Undersøgelser angeaande Torskefiskereriet i Lofoten." Christiania 1869. Translation by H. Jacobsen in *Rep. Comm. Fish and Fisheries* 1876–7 565–662.

Saville, A. (1959). "The planktonic stages of the haddock in Scottish waters." *Mar. Res.* **3**, 1–23.

Saville, A. and R.S. Bailey (1979). "The assessment and management of the herring stocks in the North Sea and to the West of Scotland." *Rapp. Procès-Verb. Cons. int. Explor. Mer* **177**, 112–142.

Sawyer, J.S. (1965). "Notes on the possible physical causes of long term weather anomalies." *WMO Tech. Note* 66 227–248.

Schäfer, D. (1887). "Der lübeckische Vogt auf Schonen." *Hansische Geschichtsquellen* 4, Halle.

Scherhag, R. (1950). "Die Schwankungen der atmosphärischen Zirkulation in den letzten Jahrzehnten." *Ber. des deutschen Wetter dienstes* 12, 40–44.

Scherhag, R. (1963). "Die grösste Kalteperiode seit 223 Jahren." *Naturw. Rdsch.* **16**, 169–174.

Schmidt, J. (1909). "The distribution of the pelagic fry and the spawning regions of the gadoids in the North Atlantic from Iceland to Spain." *Rapp. Procès-Verb. Cons. int. Explor. Mer* **10**, 4.

Schnoor, H. (1929). "Vernichtung von Nutzfischen der Ostsee durch den strengen Winter." Mitt. d. deutschen Seefischerei-Vereins 45. 5.

Schott, G. (1931). "Der Peru-Strom und seine nördlichen Nachbargebiete in normaler und anormaler Ausbildung." *Ann. Hydr. Mar. Met.* **59**, 161–253.

Schuurmans, C.J.E. (1981). "Solar activity and climate." *In* "Climatic Variations and Variability: Facts and Theories." (ed. A. Berger). 559–575.

Schuurman, C.J.E. (1981). "Climate of the last thousand years." *In* "Climatic Variations and Variability: Facts and Theories." (ed. A. Berger). 245–258.

Scott, A. (1910). "Note on the Flookburgh cockle beds." *Rep. 1909 Lancs Sea Fish. Lab.* 186–192.

Segestråle, S.G. (1950). "The recent increase in salinity off the coasts of Finland and its influence upon the fauna." *J. Cons. int. Explor. Mer* **17**, 1, 103–110.

Segerstråle, S.G. (1957). "Baltic Sea." *In Geol. Soc. Amer. Memoir 67* **1**, 751–800 (ed. J. Hedgpeth).

Segerstråle, S.G. (1972). "Biological fluctuations in the Baltic Sea." *Progr. Oceanogr.* **5**, 169–184.

Serobaba, I.I. (1977). "Data on the population structure of the walleye pollack, *Theragra chalcogramma*, from the Bering Sea." *J. Ichthyol.* 17 (2), 219–31.

Shelbourne, J.E. (1957). "The feeding and condition of plaice larvae in good and bad plankton patches." *J. Mar. Biol. Assn. UK NS* **36**, 539–52.

Shepherd, J.G. (1982). "A versatile new stock–recruitment relationship for fisheries, and the construction of sustainable yield curves." *J. Cons. int. Explor. Mer* **40**, 1, 67–75.

Shepherd, J.G. and D.H. Cushing (1980). "A mechanism for density dependent survival of larval fish as the basis of a stock-recruitment relationship." *J. Cons. int. Explor. Mer* **39**, 2. 160–167.

Shlumberger, E. and J.L. Reid (1981). "The Pacific shallow oxygen maximum, deep chlorophyll maximum and primary productivity reconsidered." *Deep Sea Res.* **28**, 9A, 901–920.

Simpson, A.C. (1953). "Some observations on the mortality of fish and the distribution of plankton in the Southern North Sea during the Cold Winter, 1946–1947." *J. Cons. int. Explor. Mer* **19**, 150–177.

Sindermann, C.J. (1970). "Principal Diseases of Marine Fish and Shellfish." Academic Press, New York and London. 369 pp.

Siren, G. (1961). "Skogsgränstallensom indikator för Klimatfluktuationera i norra Fennoskandien under historisk tid. Helsingfors." *Comm. Inst. Forestalis Fenniae* 54, 2.

Skud, B.E. (1975). "Revised estimates of halibut abundance and the Thompson-Burkenroad debate." *Sci. Rep. int. Pacif. Halibut Commn* 56, 36 pp.

Smith, P.E. (1971). "Distributional atlas of zooplankton volume in the California Current region, 1951 through 1966." *Calcofi Atlas* 13, v-xvi, 1–144.

Smith, P.E. (1978). "Biological effects of ocean variability: Time and space scales of biological response." *Rapp. Procès-Verb. Cons. int. Explor. Mer* 173, 117–127.

Smith, P.E. and R.W. Eppley (1982). "Primary production and the anchovy population in the Southern California Bight." *Limnol. Oceanogr.* 27, 1, 1–17.

Soetje, K.C. and K. Huber (1980). "A compilation of data on the thermal stratification at the central station in the northern North Sea during FLEX '76." *Meteor. Forsch-Ergebnisse A* 22, 69–77.

Soutar, A. and J.D. Isaacs (1974). "Abundance of pelagic fish during the 19th and 20th centuries as recorded in anaerobic sediments off California." *Fish. Bull. US Fish. Wildlife Serv.* 72, 257–75.

Southward, A.J. (1963). "The distribution of some plankton animals in the English Channel and approaches. III. Theories about long term biological changes including fish." *J. Mar. Biol. Assn UK NS* 43, 1–29.

Southward, A.J. (1967). "Recent changes in abundance of intertidal barnacles in southwest England: a possible effect of climatic deterioration." *J. Mar. Biol. Assn. UK NS* 47, 81–95.

Southward, A.J. (1974a). "Changes in the plankton community in the western English Channel." *Nature* 259 (5433), 180–1.

Southward, A.J. (1974b). "Long term changes in abundance of eggs of the Cornish pilchard (*Sardina pilchardus* Walbaum) off Plymouth." *J. Mar. Biol. Ass. NS UK* 54, 641–649.

Southward, A.J. (1980). "The Western English Channel—an inconstant ecosystem." *Nature* 285 (5784), 361–66.

Southward, A.J. and D.J. Crisp (1956). "Fluctuations in the distribution and abundance of intertidal barnacles." *J. Mar. Biol. Assn UK NS* 35, 211–29.

Southward, A.J., E.I. Butler and L. Pennycuick (1975). "Recent cyclic changes in climate and in abundance of marine life." *Nature* 253, 714–717.

Speerschneider, C.I.H. (1915). "Om Isforholdene i Danske Farvander i oeldre og nyere Til, Aarene 690–1860." Publ. fra det Danske Met. Inst. Medd. 2. Copenhagen.

Speerschneider, C.J.H. (1931). "The state of ice in Davis-Strait." Publ. fra det Danske Met. Inst. Medd. 8 Copenhagen.

Steele, J.H. (1961). "The environment of a herring fishery." *Fisheries Scotland, Mar. Res.* 1961, No. 6, 18 pp.

Steele, J.H. (1965). "Notes on some theoretical problems in production ecology." *Mem. 1st. Ital. Idrobiol. Dott. Marco de Marchi 18 Suppl.* 383–98.

Steele, J.H. (1974). "The Structure of Marine Ecosystems." Harvard University Press. 128 pp.

Stephen, A.C. (1938). "Temperature and the incidence of certain species in western European waters in 1932–1934." *J. Anim. Ecol.* 7, 125–129.

Stommel, H. (1965). "The Gulf Stream." 2nd edn. 248 pp.

Storrow, B. (1947). "Concerning fluctuations and the teaching of ecology." *Rep. Dove Marine Lab.* (3rd series) **9**, 7–58.

Strøm, H. (1782). "Sammenligning mellem de norske og svenske fiskerier." *Dan. Mus. Maanedsskr* Jan–June.

Suda, A. (1963). "Structure of the albacore stock and fluctuation in the catch in the north Pacific area." *FAO Fish. Rep.* **6** (3), 1237–77.

Suda, A., T. Koto and S. Kune (1963). "An outline of the tuna longline grounds in the Indo-Pacific." *FAO Fish. Rep.* **6** (3), 1163–1176.

Sutcliffe, W.H., Jr (1973). "Correlations between seasonal river discharge and local landings (*Homarus americanus*) and Atlantic halibut (*Hippoglossus hippoglossus*) in the Gulf of St Lawrence." *J. Fish. Res. Bd Can.* **30**, 856–859.

Sutcliffe, W.H. Jr, K. Drinkwater and B.S. Muir (1977). "Correlations of fish catch and environmental factors in the Gulf of Maine." *J. Fish. Res. Bd Can.* **33**, 98–115.

Svansson, A. (1965). "Some hydrographic problems of the Skagerrak." *Progr. Oceanogr.* **3**, 355–372.

Sverdrup, H.U. (1953). "On conditions for vernal blooming of phytoplankton." *J. Cons. int. Explor. Mer* **18**, 287–295.

Tabata, S., N.E.J. Boston and F.M. Boyce (1965). "Relation between wind speed and summer isothermal surface layer of water at Ocean Station P in the eastern Subarctic Pacific Ocean." *J. Geophys. Res.* **70**, 16, 3867–3878.

Tait, J.B. (1957). "Hydrography of the Faroe–Shetland Channel, 1927–1952." *Mar. Res. Scot.* 1957, 2.

Tait, J.B. and J.H.A. Martin (1965). "Inferential Biological effects of long term hydrographical trends deduced from investigations in the Faroe–Shetland Channel." Spec. Publ. 6 Intern. Commn N.W. Atlant. Fish (Environmental Symposium) 855–858.

Tåning, A.V. (1937). "Some features in the migration of cod." *J. Cons. int. Explor. Mer* **12**, 1–35.

Tåning, A.V. (1953). "Long term changes in Hydrography and fluctuations in Fish stocks." *Ann. Proc. Intern. Commn. N.W. Atlant. Fish.* **3**, 69–77.

Taylor, A.H. (1978). "Long term changes in the North Atlantic Current System and their biological implications." *Proc. Roy. Soc. Edin.* **76B**, 223–243.

Templeman, W. (1965). "Relation of periods of successful yearclasses of haddock on the Grand Bank to periods of success of yearclasses for cod, haddock and herring in areas to the north and east." Spec. Publ. 6 Intern. Commn N.W. Atlant. Fish (Environmental Symposium) 523–533.

Templeman, W. (1972). "Year class success in some North Atlantic stocks of cod and haddock." Spec. Publ. 8 Intern. Commn North West Atlant. Fish. 223–239.

Templeman, W. and G.L. Andrews (1956). "Jellied condition in the American plaice *Hippoglossoides platessoides* (Fabricius)." *J. Fish. Res. Bd Can.* 13(2), 147–82.

Templeman, W. and V.M. Hodder (1965). "Distribution of haddock on the Grand Bank in relation to season, depth and temperature." Spec. Publ. 6 Intern. Comm. N.W. Atlantic Fish. 171–188.

Thompson, D'A.W. (1910). "Report on the Distribution of the cod, haddock and other roundfishes." *Rapp. Procès-Verb. Cons. int. Explor. Mer* 12, 2, 42 pp.

Thompson, J.D. (1977). "Ocean deserts and ocean oases." *In* "Desertification." (ed. M.N. Glantz). Westview Press, Boulder.

Thompson, W.F. (1937). "Theory of the effect of fishing on the stock of halibut." *Rep. Intern. Fish. Commn* 12, 5–22.

Thurow, F. (1974). "Changes in population parameters of cod in the Baltic." *Rapp. Procès-Verb. Cons. int. Explor. Mer* 166, 85–93.

Tiews, K. (1971). "Weitere ergebnisse von Langzeitbeobachtungen über das Auftreten von Beifangfischen und-krebsen in den Fängen der deutschen Garnelen fishcherei (1961–1967)." *Arch. Fisch. Wiss.* 22, 214–55.

Tijssen, S.B. (1979). "Diurnal oxygen rhythm and primary production in the mixed layer of the Atlantic Ocean at 20°N." *Neth. J. Sea Res.* 13, 1, 79–84.

Tomfohrde, T. (1914). "Die Heringsfischerei periode an der Bohus-Len-Küste von 1556–89." *Arch. f. Fischereigeschichte* 3, 1–192.

Townsend, C.H. (1935). "The distribution of certain whales as shown by logbook records of American whaleships." *Zoologica* 19, 1–50.

Trenkle, H. (1956). "Naherungswerte der Zonalgeschwindigkeit im Atlantisch-Europaischen Sektor fur den Zeitraum 1881–1955." *Met. Rdsch.* 9, 153–158.

Tully, J.P. and L.V. Giovando (1963). "Seasonal temperature structure in the eastern subarctic Pacific Ocean." *In* "Marine Distributions." (ed. M.J. Dunbar.) University of Toronto Press. 10–36.

Uda, M. (1952). "On the relation between the variation of the important fisheries conditions and the oceanographical conditions in the adjacent waters of Japan 1." *J. Tokio Univ. Fish.* 38 (3), 364–89.

Uda, M. (1959). "Oceanographic seminars 2. Water mass boundaries—'Siome.' Frontal theory in Oceanography." *Fish. Res. Bd Can. MS Rep. Ser.* 51, 10–20.

Ulltang, O. (1976). "Catch per unit of effort in the Norwegian Purse seine fishery for Atlanto-Scandian (Norwegian spring spawning) herring." *FAO Fish. Tech. Paper 155*, 91–101.

Valdivia, J. (1976). "Aspectos biologicas del fenomeno 'El Niño', 1972–3. Parte II. La poblacion de la anchoveta. Actas Reunion de Trabajo sobre el fenomeno conocido como El Nino." *FAO Fish. Inform.* 185, 80–93.

Valdivia, J. and Guillen, O. (1966). "Informe preliminar del Crucero de Primavera 1965 (Cabo Blanco–Morro Sama)." *Informe Inst. del Mar de Peru* 11, 37–70.

Valdivia, J. and Poma, A. (1969). "Informe preliminar del Crucero Unanue 6711 18 Noviembre-20 Diciembre Primavera 1967." *Informe Inst. del Mar del Peru* 29, 1–14.

de Valera, M. and P. O'Ceidigh (1964). "West of Ireland." *J. Anim. Ecol.* 33, 198–9.

Vaux, D. (1953). "Hydrographical conditions in the Southern North Sea during the cold winter of 1946–1947." *J. Cons. int. Explor. Mer* 19, 127–149.

de Veen, J. (1962). "On the subpopulations of plaice in the southern North Sea." *ICES CM* 94, 1–6. Mimeo.

de Veen, J. (1978). "Changes in North Sea sole stocks." *Rapp. Procès-Verb. Cons. int. Explor. Mer* 172, 124–136.

Vildoso, A.C. and Chuman, E.D. (1968). "Validez de la lecture de otolitos para determinar la Edad de la Anchoveta (*Engraulis ringens*)." *Informe Inst. del Mar del Peru* 22, 34 pp.

Vildoso, A.C. (1976). "Aspectos biologicos del fenomeno 'El' Niño 1972–73. Part I. Distribution de la fauna." *FAO Fish. Inform.* 185, 62–79.

Vildoso, A.C. and Alegre, B.deH. (1965). "La madurez sexual de la anchoveta (*Engraulis ringens* J.) en los periodos reproductivos 1961/8." *Bol. Inst. del Mar Peru* 2, 3, 112–125.

Wagner, A. (1940). "Klimaänderungen und Klimaschwankungen." *Wissenschaften* 92. Braunschweig 221 pp.

Wahl, E.W. and T.L. Lawson (1970). "The climate of the nineteenth century United States compared to current normals." *Monthly Weather Review* 98, 259–265.

Wahl, E.W. and R.A. Bryson (1975). "Recent changes in Atlantic surface temperatures." *Nature* 254, 45–6.

Walford, L.A. (1931). "Northward occurrence of southern fish off San Pedro in 1931." *Calif. Fish and Game* 17, 401–405.

Walsh, J.J. (1975). "A spatial simulation model of the Peru upwelling ecosystem." *Deep Sea Res.* 22, 201–236.

Walsh, J.J. (1976). "Herbivory as a factor in patterns of nutrient utilization in the sea." *Limnol. Oceanogr.* 21, 1–13.

Walsh, J.J. (1978). "The biological consequences of interaction of the climatic, El Niño, and event scales of variability in the eastern tropical Pacific." *Rapp. Procès-Verb. Cons. int. Explor. Mer* 173, 182–192.

Walsh, J.J. (1981). "A carbon budget for overfishing off Peru." *Nature* 290 (5804), 300–304.

Ware, D.M. (1975). "Relation between egg size, growth and natural mortality of larval fish." *J. Fish. Res. Bd Can.* 32, 2503–12.

Watt, H. (1882). "The natural history of the herring with reference to its migrations." *Fish and Fisheries* 51–64. Int. Fish. Exhibition, Edinburgh.

Waugh, G.D. (1964). "The effect of the severe winter of 1962–1963 on oysters and the associated fauna of oyster grounds of southern England." *J. Anim. Ecol.* **33**, 173–175.

Wilcocks, J. (1882). "The History and Statistics of the Pilchard Fishery in England." Fish and Fisheries, Blackwood, 290–303.

Williamson, M.H. (1961). "An ecological survey of a Scottish herring fishery Part IV: Changes in the plankton during the period 1949–1959." *Bull. Mar. Ecol.* 5 (48), 207–29.

Wilson, D.P. (1947). "The Portuguese man-of-war, *Physalia physalis* L, in British and adjacent seas." *J. Mar. Biol. Assn UK NS* **27**, 1, 139–172.

Winstanley, D. (1973). "Recent rainfall trends in Africa, the Middle East and India." *Nature* **243**, 464–5.

Winstanley, D. (1973b). "Recent rainfall trends in Africa, the Middle East and India." *Nature* **245**, 190–4.

Woodcock, A. (1944). "A theory of surface water motion deduced from the wind induced motion of the *Physalia*." *J. Mar. Res.* **5**, 196–205.

Woodhead, P.M.J. (1964). "The death of fish and sublittoral fauna in the North Sea and the English Channel during the winter of 1962–63." *J. Anim. Ecol.* **33**, 169–173.

Woodhead, P.M.J. (1964). "Changes in the behaviour of the sole, *Solea vulgaris*, during cold winters and the relation between the winter catch and sea temperatures." *Helgol. Wiss. Meeresunters* **10**, 328–342.

Woodhead, P.M.J. (1964). "The death of North Sea fish during the winter of 1962–3, particularly with reference to the sole, *Solea vulgaris*." *Helgol. Wiss. Meeresunters* **10**, 283–300.

Woods, J.D. (1980). "Diurnal and seasonal variation of convection in the wind mixed layer of the ocean." *Quart. J. Roy. Met. Soc.* **106**, 449, 379–394.

Wooster, W.S. (1961). "Yearly changes in the Peru current." *Limnol. Oceanogr.* **2**, 222–225.

Wooster, W.S. and O. Guillen (1974). "Characteristics of El Niño in 1972." *J. Mar. Res.* **32**, 3, 387–404.

Wooster, W.S. and H.A. Sievers (1970). "Seasonal variations of temperature, drift and heat exchange in surface waters off the west coast of South America." *Limnol. Oceanogr.* **15**, 595–605.

Wyatt, T. (1972). "Some effects of food density on the growth and behaviour of plaice larvae." *Mar. Biol.* 14 (3), 210–16.

Wyrtki, K. (1954). "Der grosse Salzeinbruch in der Ostsee in November and Dezember 1951." *Kieler Meeresforschung* **X**, 1, 19–25.

Wyrtki, K. (1975). "Fluctuations of the Dynamic Topography in the Pacific Ocean." *J. Phys. Oceanogr.* **5**, 450–459.

Wyrtki, K. (1977). "Sea Level during the 1972 El Niño." *J. Phys. Oceanogr.* **7**, 779–787.

Wrytki, K. and G. Meyers (1976). "The Trade Wind field over the Pacific Ocean." *J. Appl. Met.* 15, 698–704.

Wyrtki, K., E. Stroup, W. Patzert, R. Williams and W. Quinn (1976). "Predicting and observing El Niño." *Science* 191, 4225, 343–346.

Yamanaka, I. and H. Yamanaka (1970). "On the variation of the current pattern in the equatorial western Pacific Ocean and its relationship to the yellowfin tuna stock." Proc. 2nd CSK Symposium, Tokio.

Yarrell, W. (1836). "A History of British Fishes." van Voorst, London. (2 vols).

Yoshida, K. (1967). "Circulation in eastern tropical oceans with special reference to upwelling and undercurrents." *Jap. J. Geophys.* 4, 1–75.

Yurkovskis, A. and A. Bramane (1977). "Phytoplankton pigments and some peculiarities of hydrochemical conditions in the Gulf of Riga in 1975." *Ann. Biol. Cons. int. Explor. Mer* 32, 64–7.

Zenkevitch, L.A. (1963). "Biology of the Sea of the USSR." Allen and Unwin, London. 955 pp.

Ziegelmayer, E. (1964). "Einwirkung des kalten Winters 1962/63 auf das Makrobenthos im Ostteil der Deutschen Bucht." *Helgol. Wiss. Meeresunters* 10, 276–282.

Zijlstra, J.J. (1972). "On the importance of the Wadden Sea as a nursery area in relation to the conservation of the Southern North Sea fishery resources. *Symp. Zool. Soc. Lond.* 29, 233–58.

Zupanovitch, S. (1968). "Causes of fluctuations in sardine catches along the eastern coast of the Adriatic Sea." *Anali Jadranskog Instituta* IV, 401–89.

Zuta, S. and Guillen, O. (1970). "Oceanografia de las aguas costeras del Peru." *Bol. Inst. del Mar del Peru* 2, 5, 161–323.

Zuta, S. and Meja, J. (1968). "Informe preliminar del Crucero Unanue 6708 24 Agosto-25 de Setiembre Invierno 1967." *Informe Inst. del Mar del Peru* 25, 23 pp.

Zuta, S. and Urquizo, W. (1972). "Temperature promedio del la superficie del Mar frente a la costa peruana periodo 1928–1969." *Bol. Inst. del Mar del Peru* 8, 461–519.

Index